CELLULAR NEUROPHYSIOLOGY AND INTEGRATION
An Interpretive Introduction

CELLULAR NEUROPHYSIOLOGY AND INTEGRATION
An Interpretive Introduction

WILLIAM R. UTTAL

The University of Michigan

Routledge
Taylor & Francis Group
LONDON AND NEW YORK

First published 1975 by Lawrence Erlbaum Associates, Inc.
This edition published 2014 by Psychology Press

Publisher 2022 by Routledge
4 Park Square, Milton Park, Abingdon, Oxon OX14 4RN
605 Third Avenue, New York, NY 10017

Routledge is an imprint of the Taylor & Francis Group, an informa business

Publisher's Note

The publisher has gone to great lengths to ensure the quality of this reprint but points out that some imperfections in the original copies may be apparent.

Library of Congress Cataloging in Publication Data

Uttal, William R.
Cellular neurophysiology and integration.

Bibliography: p.
Includes indexes.
1. Neurophysiology. 2. Neurons.
I. Title.
[DNLM: 1. Neurophysiology. WL102 U93c]
QP355.2.U88 612'.8 75-4673

ISBN 0-89859-429-4

ISBN 978-0-898-59429-4 (hbk)
ISBN 978-1-315-80249-7 (ebk)

For Michan

CONTENTS

PREFACE

Because of its general importance to a number of related disciplines, students of the modern science of neurophysiology have benefited from time to time from an introductory survey presented at a more elementary level than is usually found in advanced textbooks. This role was played by Brazier's (1951) pioneering book entitled *Electrical Activity of the Nervous System,* then by Stevens' (1966) *A Primer of Neurophysiology,* by Katz's (1966) *Nerve, Muscle and Synapse,* and most recently by Hodgkin's (1967) *The Conduction of the Nervous Impulse.*

The dynamism of the field is such, however, that more up-to-date statements incorporating many of the exciting new findings concerning cellular neurophysiology are required periodically. This text is aimed at filling that need. It is an outgrowth of a part of a course on the neurophysiology of sensory processes taught by the author at The University of Michigan during the last ten years. This book is an attempt to present the subject matter at a level appropriate for advanced undergraduate students and first year graduate students whose knowledge of chemistry, physics, and mathematics is limited to introductory courses.

To accomplish this, a number of the more esoteric aspects of neurophysiology have been presented in a more qualitative form than is usual. Throughout the text, a model of neural action based upon simple ionic concentration differences is emphasized, rather than the much more difficult and complete physical chemistry approach emphasizing electrochemistry found in more advanced texts. The material has also been organized in what is believed to be an improved logical sequence involving a development based on both increasing anatomic complexity and the direction of information flow in the nervous system.

Most important of all, however, an extra effort has been made to interpret and make clear the significance of the biological findings. In every instance possible the author has tried to explain why a particular finding is important and what the questions are that led to the asking of a particular experimental question. If the book is to achieve its particular goal, this is what it is hoped it does better than any of its predecessors.

The subject matter of this book—cellular neurophysiology and integration—is contemporary, relevant, and important. With regard to these issues it stands on its own feet. Though theoretical and experimental neurophysiology will probably not add to the variety of foods in our larder or solve the energy or population problems, it is a field that probably more than any other may help to provide the basis for an understanding of ourselves and of the essence of human nature. What is the origin of mental life? How is it that we are the special and unique cognitive entities that we are? How can we solve the problems that really matter—of man's relationship to himself and to his fellow man? These are the questions that will be important in the next millenium, and these are the questions whose solutions require insights from neurophysiology. In the long run (that is the target range towards which at least some of us must keep aimed, in this time of pressure for immediate "relevance") the nature of the nervous system, and with it all of the related medical, sociological, and psychological issues, will remain the major intellectual challenges.

No author of a textbook such as this can claim to have done all the work himself. The works of the many distinguished scholars that have been cited in the body of the text contributed directly to the book's organization. The indirect influence of many other scientists will be clear to any informed reader. Specifically Professor Stephen Easter of The University of Michigan's Zoology Department did me the courtesy of reading and commenting on the entire manuscript. The book is much improved because of his efforts. Professor James Ranck of The University of Michigan's Physiology Department also helped enormously by reading and commenting on a number of early draft chapters. Much of this book was written while I was a visiting researcher at the University of Hawaii's Sensory Sciences Laboratory. I am grateful to the trio of co-directors, Drs. Robert Cole, Ian Cooke, and M. E. Bitterman for their cordial hospitality in that stimulating environment. I am also grateful to Ms. Lynn Gore for her competent typing and editorial advice and to Ms. Lillian Weigum, who assisted with typing of the final manuscript. Ms. Thelma Eskin energetically helped me through the final rush of figures and bibliographies as well as compiling the subject index.

As ever, my wife May has provided the wisdom and support that makes any task of mine possible.

William R. Uttal

ACKNOWLEDGMENTS

The preparation of this book has been aided by the cooperation of a large number of authors and publishers. Many of the figures used have been made available through the courtesy and permission of their original authors and publishers, with, of course, the exception of deceased authors and defunct publishers. Chapter 6 is an abridgment of a longer version of the same material published in the author's earlier book (Uttal, 1973) and is used with the permission of Harper and Row, Publishers.

Specific acknowledgment is made for the following tables and figures. Numbers in parentheses indicate figure numbers in the present text.

Academic Press, Inc., New York

Engström, H., Ades, H. W., & Hawkins, J. E., Jr. Cellular pattern, nerve structures, and fluid spaces of the organ of corti. In W. D. Neff (Ed.), *Contributions to sensory physiology.* Vol. I. New York: Academic Press, 1965. (Copyright 1965 by Academic Press.) (3.8a, b)

Acta Oto-Laryngologica, Stockholm, Sweden

Engström, H., Bergström, B., & Ades, H. W. Macula utriculi and macula sacculi in the squirrel monkey. *Acta Oto-Larynogologica,* 1972, Supplement 301. (3.10)

Akadémiai Kiado, Budapest, Hungary

Hámori, J., & Szentágothai, J. The Purkinje cell baskets: Ultrastructure of an inhibitory synapse. *Acta-Biologica* (Academiae Scientiarum Hungaricae), 1965, **15,** 465–479. (9.1c)

Akademie-Verlag GmbH., Berlin, West Germany

Majorossy, K., Réthelyi, M., & Szentágothai, J. The large glomerular synapse of the pulvinar. *Journal für Hirn Forschung,* 1965, **7,** 415–432. (Copyright 1965 by Akademie-Verlag GmbH.) (9.1b)

American Association for the Advancement of Science, Washington, D.C.

Barlow, H. B., Narasimhan, R., & Rosenfeld, A. Visual pattern recognition in machines and animals. *Science,* 1972, **177,** 567–575. (Copyright 1972 by the American Association for the Advancement of Science.) (Table 10.1)

Bodian, D. The generalized vertebrate neuron. *Science,* 1962, **137,** 323–326. (Copyright 1962 by the American Association for the Advancement of Science.) (3.17)

Bullock, T. H. Neuron doctrine and electrophysiology. *Science,* 1959, **129,** 997–1002. (Copyright 1959 by the American Association for the Advancement of Science.) (5.3)

De Robertis, E. Molecular biology of synaptic receptors. *Science,* 1971, **171,** 963–971. (Copyright 1971 by the American Association for the Advancement of Science.) (9.6)

Fox, S. S., & Norman, R. J. Functional congruence: An index of neural homogeneity and a new measure of brain activity. *Science,* 1968, **159,** 1257–1259. (Copyright 1968 by the American Association for the Advancement of Science.) (8.7)

Fox, S. S., & O'Brien, J. H. Duplication of evoked potential waveform by curve of probability of firing of a single cell. *Science,* 1965, **147,** 888–890. (Copyright 1965 by the American Association for the Advancement of Science.) (8.6)

Hillman, D. E., & Lewis, E. R. Morphological basis for a mechanical linkage in otolithic receptor transduction in the frog. *Science,* 1971, **174,** 416–419. (Copyright 1971 by the American Association for the Advancement of Science.) (3.9)

Maffei, L., Fiorentini, A., & Bisti, S. Neural correlate of perceptual adaptation to gratings. *Science,* 1973, **182,** 1038–1083. (Copyright 1973 by the American Association for the Advancement of Science.) (10.10a, b, c)

Singer, S. J., & Nicholson, G. L. The fluid mosaic model of the structure of cell membranes. *Science,* 1972, **175,** 720–731. (Copyright 1972 by the American Association for the Advancement of Science.) (3.24)

Wald, G. Molecular basis of visual excitation. *Science,* 1968, **162,** 230–239. (Copyright 1968 by the American Association for the Advancement of Science.) (6.1, 6.2, 6.3, 6.4)

American Elsevier Publishing Co., Inc., New York

Pinching, A. J. Myelinated dendritic segments in the monkey olfactory bulb. *Brain Research,* 1971, **29,** 133–138. (Copyright 1971 by the American Elsevier Publishing Co., Inc.) (3.16)

Scheibel, M. E., & Scheibel, A. B. Some structural and functional substrates of development in young cats. *Progress in Brain Research,* 1964, **9,** 6–25. (Copyright 1964 by the American Elsevier Publishing Co., Inc.) (3.22c)

American Physiological Society, Bethesda, Maryland

Brown, J. E. Rat retinal ganglion cells: Receptive field organization and maintained activity. *Journal of Neurophysiology,* 1965, **28,** 1091–1100. Figure 3. (3.22b)

Casby, J. U., Siminoff, R., & Houseknecht, T. R. An analogue cross-correlator to study naturally induced activity in tact nerve trunks. *Journal of Neurophysiology,* 1963, **26,** 432–448. Figures 2, 3, and 10. (8.9, 8.10, 8.11)

Frazier, W. T., Kandel, E. R., Kupfermann, I., Waziri, R., & Coggeshall, R. E. Morphological and functional properties of identified neurons in the abdominal ganglion of *Aplysia californica. Journal of Neurophysiology,* 1967, **30,** 1288–1351. Figure 2. (11.4)

Kandel, E. R., Frazier, W. T., Waziri, R., & Coggeshall, R. E. Direct and common connections among the identified neurons in the abdominal ganglion of *Aplysia. Journal of Neurophysiology,* 1967, **30,** 1352–1376. Table 1. (Table 11.2)

American Psychological Association, Washington, D.C.

Uttal, W. R. A comparison of neural and psychophysical responses in the somesthetic system. *Journal of Comparative and Physiological Psychology,* 1959, **52,** 485–490. (Copyright 1959 by the American Psychological Association.) (10.1)

Scheibel, M. E., & Scheibel, A. B. Some structuro-functional correlates of develop-

ASP Biological and Medical Press, Elsevier Division, Amsterdam, Netherlands

ment in young cats. *Electroencephalography and Clinical Neurophysiology,* 1963, Suppl. 24, 235–246. (3.22e)

Baillière Tindall, London, England

Bernard, M. C. *Lessons on the physiology and pathology of the nervous system.* London: Baillière Tindall, 1858. (2.3)

Biometrika, London, England

Cox, D. R., & Smith, W. L. On the superposition of renewal processes. *Biometrika,* 1954, **41,** 91–99. (8.3)

Cambridge University Press, New York

Cauna, N. The free penicillate endings of the human hairy skin. *Journal of Anatomy,* 1973, **115,** 277–288. (3.13)

The Ciba Foundation, London, and Churchill Livingstone, Edinburgh, England

Andres, K. H. Discussion. In G. E. W. Wolstenholme and J. Knight (Eds.), *Taste and smell in vertebrates, A Ciba foundation symposium.* London: J. & A. Churchill, 1970, Pp. 25–30. (Copyright 1970 by J. & A. Churchill.) (3.15)

The Clarendon Press, Oxford, England

Eccles, J. C. *The neurophysiological basis of mind.* Oxford: The Clarendon Press, 1956. (Figure used is after Liley and North.) (10.7)

Clark University Press, Worcester, Mass.

Bard, P. Emotion. In C. Murchison (Ed.), *Foundations of experimental psychology.* Worcester, Mass.: Clark University Press, 1929. Figure 2 after W. B. Cannon, *Bodily changes in pain, hunger, fear and rage,* New York: D. Appleton and Company, 1915, p. 24. (3.3)

Cold Spring Harbor Laboratory, Cold Spring Harbor, New York

Bodian, D. Introductory survey of neurons. *Cold Spring Harbor Symposium on Quantitative Biology,* 1952, **17,** 1–13. (Copyright 1952 by Cold Spring Harbor Laboratory.) (3.22d)

Cosmetics, Toiletry and Fragrance Association, Washington, D.C.

Amoore, J. E. Elucidation of the stereochemical properties of the olfactory receptor sites. *Proceedings of the Scientific Section of the Toilet Goods Association,* 1962 (October), Special Supplement to #37, 13–23. (6.12)

W. H. Freeman & Co., San Francisco, Calif.

Amoore, J. E., Johnston, J. W., Jr., & Rubin, M. The stereochemical theory of odor. *Scientific American,* 1964, **210,** 42–49. (Copyright 1964 by Scientific American, Inc. All rights reserved.) (6.11)

Horridge, G. A. *Interneurons: Their origin, action, specificity, growth, and plasticity.* San Francisco: W. H. Freeman & Co., 1968. (Copyright 1968 by W. H. Freeman & Co.) (3.22a)

Young, R. W. Visual cells. *Scientific American,* 1970, **223,** 80–91. (Copyright 1970 by Scientific American, Inc. All rights reserved.) (3.4a, b)

Harper and Row, Publishers, New York

Thompson, R. F. *Foundations of physiological psychology.* New York: Harper & Row, 1967. (10.11)

Uttal, W. R. *Real-time computers: Technique and applications in the psychological sciences.* New York: Harper & Row, 1968. (4.5, 4.10, 8.8)

Uttal, W. R. *The psychobiology of sensory coding.* New York: Harper & Row, 1973. (4.11, Tables 3.1, 3.2, 6.1, 6.2)

Medical Division, Harper and Row, Publishers, Hagerstown, Maryland

Pappas, G. D. Electron microscopy of neuronal junctions involved in transmission in the central nervous system. In K. Rodahl & B. Issekutz (Eds.), *Nerve as a tissue.* Hagerstown, Md.: Medical Division, Harper & Row, 1966. Pp. 49–87. (9.2)

Holt, Rinehart and Winston, Inc., New York

Davis, H., & Silverman, S. R. *Hearing and deafness.* (3rd ed.) New York: Holt, Rinehart & Winston, 1970. (Copyright 1947, © 1960, 1970 by Holt, Rinehart and Winston, Inc.) (6.6)

Journal of Physiology, Cambridge, England

Coombs, J. S., Eccles, J. C., & Fatt, P. The specific ionic conductances and the ionic movements across the motoneuronal membrane that produce the inhibitory post-synaptic potential. *Journal of Physiology,* 1955, **130,** 326–373. (9.5)

Iggo, A., & Muir, A. R. The structure and function of a slowly adapting touch corpuscle in hairy skin. *Journal of Physiology,* 1969, **200,** 763–796. (3.12)

Kandel, E. R., & Tauc, L. Heterosynaptic facilitation in neurones of the abdominal ganglion of *Aplysia depilans. Journal of Physiology,* 1965, **181,** 1–27. (a) (11.13)

Kandel, E. R., & Tauc, L. Mechanism of heterosynaptic facilitation in the giant cell of the abdominal ganglion of *Aplysia depilans. Journal of Physiology,* 1965, **181,** 28–47. (b) (11.14)

Kehoe, J. Three acetylcholine receptors in *Aplysia* neurones. *Journal of Physiology,* 1972, **225,** 115–146. (b) (11.5)

Lea and Febiger, Philadelphia

Porter, K. R., & Bonneville, M. A. *Fine structure of cells and tissues.* (4th ed.) Philadelphia: Lea & Febiger, 1968. (3.21, 3.23)

Elizabeth Licht, Publisher, New Haven, Conn.

Galvani, L. *Commentary on the effect of electricity on muscular motion* (as translated by R. M. Green). New Haven, Conn.: Elizabeth Licht, 1953. (2.1)

Macmillan (Journals) Limited, London

Ozeki, M. Hetero-electrogenesis of the gustatory cell membrane in rat. *Nature,* 1970, **228,** 868–869. (6.15)

Tauc, L., & Epstein, R. Heterosynaptic facilitation as a distinct mechanism in *Aplysia. Nature,* 1967, **214,** 724–725. (11.3)

McGraw-Hill Book Company, New York

Hyman, L. H. *The invertebrates: Mullusca I.* Vol. I. New York: McGraw-Hill, 1940. (Copyright 1940 by McGraw-Hill Book Company.) (11.2)

Katz, B. *Nerve, muscle, and synapse.* New York: McGraw-Hill, 1966. (Copyright 1966 by McGraw-Hill Book Company.) (5.2)

Plonsey, R. *Bioelectric phenomena.* New York: McGraw-Hill, 1969. (Copyright 1969 by McGraw-Hill Book Company.) (7.1)

von Békésy, G. *Experiments in hearing.* New York: McGraw-Hill, 1960. (Copyright 1960 by McGraw-Hill Book Company.) (6.5)

M.I.T. Press, Cambridge, Massachusetts

Kandel, E. R. An invertebrate system for the cellular analysis of simple behaviors and their modifications. In F. O. Schmitt & F. G. Worden (Eds.), *The neurosciences: Third study program.* Cambridge, Mass.: M.I.T. Press, 1973. Chapter 31. (Copyright 1973 by the Massachusetts Institute of Technology Press.) (11.1, 11.8)

Kiang, N. Y.-S. *Discharge patterns of single fibers in cat's auditory nerve.* (M.I.T. Research Monograph #35) Cambridge, Mass.: M.I.T. Press, 1965. (Copyright 1965 by the Massachusetts Institute of Technology Press.) (4.8b, 10.2)

The New York Academy of Sciences, New York

Uttal, W. R., & Cook, L. Systematics of the evoked somatosensory cortical potential: A psychophysical-electrophysiological comparison. In R. Katzman (Ed.), *Sensory evoked response in man.* New York: New York Academy of Sciences, 1964. Pp. 60–80. (8.2)

Norton, W. T. Myelin. In R. W. Albers, G. J. Siegel, R. Katzman, & B. W. Agranoff (Eds.), *Basic neurochemistry.* Boston: Little, Brown & Co., 1972. Pp. 365–386. (3.19)

Pergamon Press Ltd., Oxford, England

Bradbury, S. *The evolution of the microscope.* Oxford: Pergamon Press, 1967. (Copyright 1967 by Pergamon Press Ltd.) (2.4)

de Lorenzo, A. J. Studies on the ultrastructure and histophysiology of cell membranes, nerve fibers and synaptic junctions in chemoreceptors. In Y. Zotterman (Ed.), *Olfaction and taste.* Oxford: Pergamon Press, 1963. Pp. 5–17. (Copyright 1963 by Pergamon Press Ltd.) (3.14)

Takagi, S. F. Are EOG's generator potentials? In T. Hayashi (Ed.), *Olfaction and taste II.* Oxford: Pergamon Press, 1967. Pp. 167–179. (Copyright 1967 by Pergamon Press Ltd.) (6.13)

Peters, A., Palay, S. L., & Webster, H. de F. *The fine structure of the nervous system, the cells and their processes.* New York: Harper & Row, 1970. (3.20)

Prentice-Hall, Inc., Englewood Cliffs, New Jersey

Kandel, E. R., & Wachtel, H. The functional organization of neural aggregates in *Aplysia.* In F. D. Carlson (Ed.), *Physiological and biochemical aspects of nervous integration.* Englewood Cliffs, N.J.: Prentice-Hall, 1968. Pp. 17–o5. (Copyright 1968 by Prentice-Hall, Inc.) (11.6, 11.7)

Willows, A. O. D. Behavioral acts elicited by stimulation of single identifiable nerve cells. In F. D. Carlson (Ed.), *Physiological and biochemical aspects of nervous integration.* Englewood Cliffs, N.J.: Prentice-Hall, 1968. (Copyright 1968 by Prentice-Hall, Inc.) (11.9a, b, c, 11.10a, b, 11.11)

Psychonomics Press, Austin, Texas

Dzendolet, E. A structure common to sweet-evoking compounds. *Perception & Psychophysics,* 1968, **3,** 65–68. (6.14)

Rockefeller University Press, New York

Davis, L., Jr., & Lorente de No, R. Contribution to the mathematical theory of the electrotonus. In R. Lorente de No, *A study of nerve physiology,* Studies from the Rockefeller Institute for Medical Research, Vol. 131. New York: Rockefeller University Press, 1947. (7.2)

de Lorenzo, A. J. The fine structure of synapses in the ciliary ganglion of the chick. *Journal of Biophysics, Biochemistry, and Cytology,* 1960, **7,** 31–36. (b) (9.1a)

De Robertis, E. Some observations on the ultrastructure and morphogenesis of photoreceptors. *Journal of General Physiology,* 1960, **43** (suppl.), 1. (3.5)

Hartline, H. K., Wagner, H. G., & Ratliff, F. Inhibition in the eye of *Limulus. Journal of General Physiology,* 1956, **39,** 651–673. (10.8)

Loewenstein, W. R., & Rathkamp, R. The sites for mechano-electric conversion in a Pacinian corpuscle. *Journal of General Physiology,* 1958, **41,** 1245–1265. (6.8, 6.9, 6.10)

Purpura, D. P. Comparative physiology of dendrites. In G. C. Quarton, T. Melnechuk, & F. O. Schmitt (Eds.), *The neurosciences: A study program*. New York: Rockefeller University Press, 1967. Pp. 372–393. (3.22f)

Smith, D. S., Järlfors, U., & Beránek, R. The organization of synaptic axoplasm in the lamprey (Petromyzon marinus) central nervous system. *Journal of Cell* Biology, 1970, **46,** 199–219. (9.4)

Tilney, L. G., & Porter, K. R. Studies on the microtubules in heliozoa. II: The effect of low temperature on these structures in the formation and maintenance of the axopodia. *Journal of Cell Biology,* 1967, **34,** 327–343. (3.18)

W. B. Saunders Co., Philadelphia

Patton, H. D. Special properties of nerve trunks and tracts. In T. C. Ruch & H. D. Patton (Eds.), *Physiology and biophysics.* (19th ed.) Philadelphia: W. B. Saunders Co., 1965. Pp. 73–94. (7.10, 8.4, Table 3.3)

Ranson, S. W. *The anatomy of the nervous system: Its development and function,* as revised by S. L. Clark. (10th ed.) Philadelphia: W. B. Saunders Co., 1959. (3.1, 3.2)

Woodbury, J. W. The cell membrane: Ionic and potential gradients and active transport. In T. C. Ruch & H. D. Patton (Eds.), *Physiology and biophysics.* (19th ed.) Philadelphia: W. B. Saunders Co., 1965. Chapter 1. (5.1)

Springer-Verlag, Berlin

Eccles, J. C. *The physiology of synapses.* Berlin: Springer-Verlag, 1964. (9.8)

Springer-Verlag, Inc., New York

Munger, B. L. Patterns of organization of peripheral sensory receptors. In W. R. Loewenstein (Ed.), *Principles of receptor physiology.* New York: Springer-Verlag, 1971. (3.11)

Uttal, W. R. Oscillations in the amplitude of human peripheral nerve action potentials during repetitive stimulation. *Kybernetik,* 1966, **3,** 24–27. (10.3)

Charles C Thomas, Publisher, Springfield, Illinois

Hodgkin, A. L. *The conduction of the nervous impulse.* Springfield, Ill.: Charles C Thomas, 1967. (7.7, 7.9)

University of California Press, Berkeley, California

Cole, K. S. *Membranes, ions, and impulses.* Berkeley: University of California Press, 1968. (Copyright 1968 by the University of California Press.) (7.8)

Segundo, J. P., & Perkel, D. H. The nerve cell as an analyzer of spike trains. In M. A. B. Brazier (Ed.), *The interneuron.* Berkeley: University of California Press, 1969. Pp. 349–390. (Copyright 1969 by the University of California Press.) (11.15, 11.16, 11.17)

The Williams and Wilkins Company, Baltimore

Truex, R. C., & Carpenter, M. B. *Strong and Elwyn's human neuroanatomy.* (5th ed.) Baltimore: The Williams & Wilkins Co., 1964. (Copyright 1964 by The Williams and Wilkins Company.) (8.5)

1
INTRODUCTION

A. THE CONTEMPORARY SIGNIFICANCE OF NEUROPHYSIOLOGY

An extraordinary scientific development has been underway for the last 30 years. After centuries of speculation and frail theories devoid of empirical foundation, it is now possible to examine directly the cellular basis of the action of the nervous system, and thus to begin considering the true relation between the neural substrate and behavior. Perhaps no science holds such intrinsic excitement or hope for the long-range solution of both theoretical and practical human problems as does modern cellular neurophysiology—the inquiry into the basic nature and function of neurons. It is this class of specialized cells that ultimately defines the very basis of human nature, and it is at this level that many of the solutions to the social and personal problems afflicting people may ultimately be found.

Each neuron, the basic building block of the nervous system, is a member of a class of cells that includes a remarkable variety of cellular units found throughout the peripheral and central portions of all multicellular organisms. The essential feature of the neuron is its greatly heightened ability to react to events external to itself, and thus to interact with events occurring in neighboring neurons, as well as with the external environment. The reactive capacity of this class of highly specialized single cells has evolved from the primitive excitability, elicited by relatively gross amounts of physical stimuli, characteristic of all cells, to a sensitivity that allows neurons to respond to stimuli whose energy content approaches, in many cases, the absolute limits of physical energy. A photoreceptor in the eye, for example, is sensitive to a *single quantum* of light when conditions are optimized, and the ear is sensitive to a mechanical oscillation an *order of magnitude smaller* than the diameter of a hydrogen atom.

The highly economical power levels required by the individual neuron allow an information-processing system to be built into the human brain that is probably the most complicated single structure to have evolved in the history of the universe. Only at levels of cosmological interactions do we find comparable complexities, but even at that level there is no evidence for the emergence of any sentient awareness.

This economy of energy, however desirable from an evolutionary point of view, is nevertheless not the essential characteristic of nervous action. The same events that are represented by patterns of neural action could conceivably be constructed from completely different kinds of components, which are far less efficient, without any change in the nature of thought. The logical operations could remain invariant in an information-processing system, no matter what sort of parts were used in its construction. The essential aspect of neural processes is that they are able to synthesize complex events, including thoughts and self-awareness, by concatenating a huge number of quite simple processes. While the individual neuron can perform only a few simple logical processes, logicians have established that a multiply connected net of neurons can represent virtually any concept, no matter how complicated.

The presumption that a net of neurons ultimately gives rise to cognitive behavior is based on a great logical leap, a statement of belief, or more formally, an axiomatic premise, rather than a rigorously established "proof" of the relationship between brain structure and mind. Nevertheless, this premise is based on a compelling set of arguments. Many developments in a variety of sciences have coalesced on this issue in the last 30 years in a way that cannot be ignored any longer. Indeed, we may now be in the midst of a psychobiological revolution, comparable to the great religious and industrial revolutions of the past, that will have profound effects on society in the centuries to come. The notion of brain-mind identity may have as powerful an influence on social development as the notions of personal freedom and democracy have had in the last few centuries.

The notion that complex interactions between simple neural units may underlie much more complex cognitive processes can be understood only if the problem of neural activity is approached from a point of view that emphasizes networks of neurons. Yet for many technological and historical reasons, our knowledge is still limited to the actions and interactions of a relatively few neural units. We know quite a bit about the nature of the membrane of the neuron and the processes that allow the membrane to perform its specialized functions. We know quite a bit about how neurons interact with each other at their points of synaptic contact. However, the limits of our knowledge are rapidly reached when attempts are made to understand how even as few as three neurons interact collectively. Of course a far greater number must be involved in even the simplest information-processing act in the mammalian nervous system.

It is for these reasons that researchers often study relatively simple structures to elucidate some aspect of neural interaction. Invertebrates are commonly used, not so much because of any intrinsic interest in their function (how few, indeed, worry about the problem of the innervation of the muscles that masticate food in the stomach of the lobster for its own sake?) but because these simpler creatures model some aspect of nervous function that may be generally applicable to the solution of the mind-brain problem in man. Even when working with higher animals, including man, we tend to study the function of relatively simple, well-ordered sensory receptor structures to find out about some special aspect of a more general nervous function. Much of what is known about inhibition in nervous nets, for example, has arisen from studies of crustacean neurons and vertebrate sensory systems. But obviously inhibition is a process of great significance throughout the phylogenetic tree, including all parts of the vertebrate nervous system, and we can gain important understanding by generalizing from the primitive to the more complex.

As more is learned about the basic mechanisms that underlie complex processes, the focus of laboratory attention has shifted perceptibly in emphasis. During the 1950's and early 1960's an observer at a neurophysiological conference would have seen an almost complete absorption with the study of action potentials produced by single neurons. In the last 10 years, however, increasing attention has been directed at interactions among neurons.

In light of this modern emphasis on information processing in nets of neurons, it is important not to overlook another vital function of neurons, which is clearly interwined with the processing aspect, but which in many regards is quite different. This, of course, is the function of communication. The necessity for neuronal communication links is obvious. Receptors and effectors are located at the periphery of the body. Yet the complicated central information processing that links inputs from receptors to the biologically appropriate, adaptive effector responses generally occurs in the most centralized, interior portion of the organism. It is to and from the brain, in particular, that information must be communicated. In man this distance may be five or more feet, and in other animals even greater distances must be traversed. To depend upon the slow spread of excitable responses along an unspecialized cellular net would be behaviorally catastrophic; thus there is strong evolutionary pressure toward the development of rapid communication links in multicellular animals.

To make the distinction between the processes of communication and integration more specific, let us define these terms more explicitly at this point:

Communication—The process by which information is transmitted from one point to another in the nervous system. The prime criterion for excellence of communication is that the transmitted information be the same when it arrives at its destination as when it left the source.

Integration—The process by which information from several different sources interacts at some processing point to produce an output that is improved in its utility, economy of representation, or biological effectiveness. The prime criterion for excellence of integration is that multiple inputs result in outputs of increased biological value.

It must be realized that there is a basic incompatability between the requirements for good communication and the requirements for good information processing in the nervous system. The *communication* aspects of the nervous system should be ideally designed to have the highest possible fidelity. The communicated information pattern should be accurately reproduced. On the other hand, *integrative* information-processing mechanisms function to alter, to convert, or to transform the incoming signal from one pattern to another, and thus they run counter to the demands for transmission fidelity. The ability to adapt to the great variety of environmental stimuli encountered by organisms is critically dependent upon this integrative processing of information. From the evolutionary point of view it is virtually useless for an organism to spew out the same information pattern that was taken in. Without integration, interaction, and synthesis, without the deletion of irrelevant information and the enhancement of incoming information patterns by assimilation with old memories, behavioral evolution would probably come to a standstill as animals failed to modify their responses to environmental stresses.

The analogy between neural integration and the genetic combination process in sexual reproduction is clear. Both processes achieve variety, flexibility, and adaptability because they combine a large number of components into a virtually infinite number of combinations, in a way that leads to continuous improvement in the species. Just as genetic combination is the essence of evolutionary development, so neural integration is certainly the really unique aspect of an animal's mental life. Though it may seem far-fetched to some, it is reasonable to argue that the essence of human consciousness is largely this ability to combine and permute input signals at the neuronal level.

Older models that view the nervous system as simply a switching network (such as classic conditioning models or more recent computer models) miss this essential point. In the view of most psychobiological philosophers today, man cannot be analogized by a telephone system or a computer. The manner in which man is able to program his behavior and invent new responses generates surprising results in a way that still does not have the remotest artificial analog.

B. PLAN AND ORGANIZATION OF THE BOOK

This book is directed toward those students at the university level who have already had some introduction to the biological and physical sciences. It is assumed that the reader has a basic, introductory knowledge of electricity and cellular biology. The foundations of cellular neurophysiology will be presented in a way that will allow students of specialities other than physiology to acquire an understanding of the basic concepts of this elegant and intricate body of knowledge. The main effort in this book is to interpret extensively the meaning of the various aspects of neurophysiology to be considered. The historical, philosophical, and conceptual aspects of cellular neurophysiology will be emphasized.

Modern neurophysiology is unquestionably a difficult body of material to master. Much of it is not only highly technical but also involves a number of interacting concepts that make some aspects of the subject quite complex. Unfortunately there is no way to avoid this detail without losing a solid understanding of the subject. But it is a body of knowledge of increasing importance that grows almost daily in new findings, and the time has come for a statement that attempts to extract the important concepts and general principles to a greater degree than usual. Methods and findings will necessarily play an important role in our discussion, but the author's principal intent is to use these materials more to help in achieving conceptual understanding than as goals in themselves.

The only hope for grasping the fundamentals of cellular neurophysiology is to keep rigid control over the sequence of ideas as they are presented. The problem is in deciding along which of the alternative dimensions this sequence of control should be exerted. There are two obvious candidates. One possible thread along which the discussion might be organized is that of increasing levels of cellular anatomy and complexity. This thread passes from the basic biochemical properties of the neuronal membranes to the interactive mechanisms that allow cells to communicate with each other. An alternative sequence of discussion is one imposed by the intrinsic directionality of information flow within the nervous system. Receptors "transduce" physical energy. The long axons of certain specialized neurons then transmit that information through synaptic junctions to the central nervous system, where it is integrated into output patterns that are again transmitted through a similar channel—but in a reverse direction, to the periphery.

From this point of view, the study of cellular neurophysiology is dominated by its directionality, ratherd than by levels of anatomical complexity.

In this book the author has tried to maintain both threads of the discussion simultaneously. In a few instances, the thread must deviate from a strictly parallel route, in order to avoid premature discussion of material for which the prerequisites have not yet been introduced. But in general we shall be progressing from the microscopic to the more macroscopic while simultaneously following the stream of information flow from the periphery to the central nervous system and then to the periphery once again.

Chapters 2 and 3 contain the first digressions. In order to set the stage for the cellular neurophysiological details to follow, two bodies of background information must first be considered. The first, presented in Chapter 2, is a brief historical survey, to place the discussion in the context of its intellectual origins. Though some of the basic ideas of modern neurophysiology are surprisingly antique (for example, the notion of the cell membrane was originally proposed around the turn of the century), much supporting evidence for some of the most widely accepted theories did not appear until the 1940's. Chapter 3 briefly presents some details of the gross anatomy of the nervous system, in only sufficient detail to introduce the reader to the locus of various tissues that are discussed in later chapters. A brief anatomy of the microstructure of neurons and receptor cells is then presented. Chapter 3 then concludes, at an even finer level of microanatomy, with a discussion of the anatomy of the cell or plasma membrane, a tissue of the utmost importance to our later discussion.

Chapter 4 presents a discussion of modern neurophysiological research techniques, information required for understanding the experimental data to be discussed in later chapters. Chapter 5 first specifically introduces what is meant by the *neuron theory,* the now universally accepted notion that the nervous system is made up of cytoplasmically separate cellular units rather than an interconnected reticulum or network. The discussion then turns to a consideration of how the distributions of ions across the membrane are produced by the various passive and active forces that operate on chemical ions and how those distributions of ions give rise to potential differences, particularly the so-called resting potential. Finally, a taxonomy of membrane potentials is presented.

Once it has been established how resting potentials can be developed by metabolic engines within the membrane, several subsequent chapters are dedicated to explorations of how the resting potential can be perturbed to carry information to other parts of the nervous system. Chapter 6 discusses the initiation of membrane action potential by external physical energies. This, of course, is the function generally known as transduction. Transductive mechanisms are considered separately for each of the senses, after a brief introduction to the general nature of the process. This is necessary because each of the sensory receptors has evolved a somewhat different solution to its own biological transduction problem.

The transduction processes are the main means through which neural information patterns begin to course through the nervous system. Once the process has started, the main question is: How is this neural information transmitted from place to place? Over short distances, passive and decremental signal spread appears to be used, but over longer distances some other mechanism is required. The need is filled by the propagating spike action potential, a response class evolved specifically for communication over the long axons of the body. Chapter 7 is entirely devoted to the topic of transmission, including both general qualitative and quite specific explanations of how decremental and propagating action potentials occur. These explanations are phrased in terms of ionic and membrane properties.

In Chapter 8 we consider how single cell responses can be mixed to produce compound or pooled action potentials. These classes of electrical signals reflect the action of many cells, rather than one. A particularly important issue is whether or not the group statistics of mixed neurophysiological responses truly represent the characteristics of individual cells. A consideration of some appropriate research shows that the two are highly correlated, and thus some important links can be made between the two kinds of response.

Having completed the discussion of transmission properties along neurons and mixed, but noninteracting, neuronal responses, we then are prepared to consider, in Chapter 9, the problem of transmission between neurons. The synapse is a highly specialized region between neurons that allows the electrochemical activity in one cell to initiate a comparable pattern of activity in a nearby cell. Both chemically and electrically mediated synapses are discussed, as well as some important topics concerning the ionic basis of synaptic action.

Chapter 10 also emphasizes interactions among neurons, but from a very different point of view. Rather than dealing with the ultramicroscopic action of the synapse, we are concerned with the interactions of highly idealized neurons in networks. This topic is largely the domain of the theoretician at present but is rapidly coming under experimental attack. The concept of interneurons—neurons that are specialized not for communication over long distances, but for the integration of multiple inputs—is more fully developed. We also discuss in detail some functions of interactive networks that are relatively well understood, such as rhythmicity, plasticity, and inhibition. Finally, we consider some factors that affect the fidelity of information processing in the nervous system. These are the processes that limit the accuracy of information pattern representation in the various communication and information-processing stages of the central nervous system. Chapter 10 closes with a discussion of the neuroeffector junction, the final stage in the system, through which neurophysiological information patterns are communicated to the muscles or glands, and thence to the outside world.

Chapter 11 attempts to make many of the concepts introduced in Chapter 10 more concrete by describing some experimental data specific to neural integration, which has been obtained from a single genus of mollusk, the *Aplysia*. By discussing a body of data obtained solely from this animal, a high degree of comparability, consistency, and order is achieved.

Finally, the author would like to make it clear that the purpose of the book is tutorial. No new global perspective or theory is presented; rather this text strives to be interpretive, in the sense that the meaning and significance of the various physiological facts are stressed. The discussion is intended to extend beyond the bare data of the neurophysiological laboratory and offer a more detailed discussion of the significance and meaning of these data than is found in the conventional textbook. To enhance this interpretative approach, Chapter 12 closes with a summary of those general principles of cellular neurophysiology that seem to be emerging from the wealth of information accumulated in the last three decades.

2

A BRIEF HISTORY OF CELLULAR NEUROPHYSIOLOGY

In all the sciences there is a critical dependence between the data obtained and the theories developed on the one hand, and the available technological instrumentation on the other. In neurophysiology this is particularly true, because the difference between the scale of common observation and the scale of events at which neurons operate is so disparate. The human observer must deal with inches and centimeters unless he has some sort of magnifier, and with seconds unless he has some device to convert brief occurrences into longer-lasting records. Neurons, however, are microscopic cells, and many of the critical components are submicroscopic. Micrometers and nanometers are the appropriate units for measurement at this end of the scale. Similarly, the salient events occur not over periods of seconds, but in milliseconds; yet even these brief events have temporal subcomponents that are essential to their understanding. The voltages and the amounts of chemical involved are likewise vanishingly small. A few millivolts or a few molecules may represent the critical scales of electrical or chemical events at the neural level.

It is easy to understand, therefore, how the history of neurophysiology was influenced, and powerful constraints lifted, by the technological developments that allowed this disparity in scale to be surmounted. The nature of the individual microscopic neuron could not be appreciated until the microscope was invented. The electrochemical nature of the neuron could not be understood until the nature of electricity and the relationship between chemical and electrical events were understood. The rapid electrical events themselves could not be analyzed in any detail until electronic amplifying and display instrumentation became available in the twentieth century. Even a cursory examination of scientific history shows that the major conceptual and philosophical questions were being asked about the nervous system long before the answers could be acquired. But, as in other sciences, satisfactory answers to neurophysiological puzzles could not be obtained until data based on direct observation were forthcoming. To achieve a direct answer with direct observation required a technology of magnification in both time and space. This brief neurophysiological history, therefore, necessarily emphasizes the development of those instruments that were crucial to the development of direct, more refined observational techniques.

A full review of the enormous scope of neurophysiological history, even one limited mainly to technological influences, is not possible here. The interested reader is directed especially to the first chapter in the monumental *Handbook of Physiology* (Brazier, 1959), or to Fulton and Wilson's (1966) translations of some of the classic biological source documents, for a more complete treatment.

In the brief history that follows, a single main theme is stressed: the origin of the concept that neural function has an electrical basis. However, we should note that, from one point of view, this theme is a misstatement of the physical actuality. As we learn more about the nervous system, it becomes increasingly clear that the electrical signals that are so easily recorded with even the simplest of currently available instruments are only reflections of processes that should more properly be considered to be fundamentally chemical. While our technology allows, in fact, forces, us to stress electrical terminology, many of the electrical signs are, to a considerable degree, only concomitant signs of ionic migrations. Unfortunately, there is not yet an equivalent technology for direct measurement of ionic ebb and flow without reference to the associated electrical signals. Perhaps some day there will be chemically sensitive electrodes that have the precision and delicacy of modern electrically sensitive ones; until then, we are bound by this technological constraint to observe secondary electrical signals in situations where ion behavior may be more fundamental. But this is a modern caveat; let us see how the idea that neurons are electrochemical in nature developed over the years.

A. CLASSIC VIEWS OF NERVOUS ACTION

The peripheral nerves, the spinal cord, and the brain were known to exist as anatomic entities by the early Greek scientist-philosophers. That is about all that needs to be said of their view of the nervous system, for ancient ideas of mental activity dealt with it as a separate, noncorporate entity and usually associated thought processes with other parts of the body. Our language and culture still reflect the predominance of the view that the heart, for example, is more important to intellectual and emotional function than the brain.

Although there was some earlier philosophical speculation that the nerves and the brain might be specifically associated with mental activity, it was not until Galen's time (A.D. 130–200) that what we now consider to be scientific ideas about the nervous system's function began to emerge. Galen, like most theoreticians past and present, was influenced by the contemporary technology familiar to him, and his models and ideas reflected the technology of his times. In the second century A.D. the predominant technologies were hydraulics and mechanics, and it was on these foundations that Galen built his theory. Though he accepted the crucial notion that the nerves were necessary concomitants of human thought and action, Galen assumed that the nerves actually were part of a hydraulic system through which fluids, produced by various organs of the body, were conveyed to produce muscular movements or perceptions. His notions of a hydraulic nervous system that was directly related to behavior had predecessors in antiquity, but the details of his theory were based largely on his own dissections and observations of the variety of fluids that seemed to permeate the body.

The sheer persistence of hydraulic theories of nervous action was incredible. These ideas were dominant throughout the dark ages following the decline of Rome. When these problems were considered at all, Galen's views went unchallenged, even well into the European Renaissance. Harvey's description in 1628 of the circulatory system, and his emphatic confirmation of the notion of closed fluid cycles in the body, reinforced these views even more. The notion of a nervous system that operated on the basis of fluid transport was strongly supported by both Descartes, that universal mathematical, physiological, and philosophical genius, and by the mathematician Borelli (who, incidentally, eliminated the notion that nerves carry gases rather than fluids, by dissecting nerves under water and showing that no bubbles were released). And so this wet view of the nervous system persisted until the end of the eighteenth century.

B. THE ORIGIN OF THE IDEA OF THE ELECTRICAL BASIS OF NERVOUS ACTION

The histories of physiology and physics are extraordinarily intertwined from the late eighteenth century on. Discoveries in physiology stimulated findings in physics, which in turn provided impetus for new areas of physiological research. The highlight of these exciting times occurred in 1791, when Luigi Galvani published his observations on a series of experiments carried out on a neuromuscular preparation from the frog. Galvani showed that preparations, consisting only of amputated frog legs and the attendant stumps of the severed nerves, could be made to respond to electrical stimuli that were produced in any number of ways. Electricity was manipulated (if not completely understood) well enough by this time that it could be produced from electrostatic generators, stored in Leyden jars (the early capacitor), or acquired from lightning rods.

Figure 2.1 shows one of Galvani's experiments. Clearly the electrical stimulus produced by the lightning was the direct antecedent of the observed action in the frog legs; and although this did not necessarily establish electrical action *in* the nerves (that discovery would have to wait until the tiny neuronal responses could be electronically amplified and displayed), it was a minuscule logical step to the assumption that the physical energetics of the neuron were the same as the stimulus. In fact we now know that this logical leap is not necessarily valid. The modern view of transduction is based upon the premise that the physical energy of the stimulus is converted to other forms of physical energy in the nervous system. Indeed, the variety of stimulus energies to which the receptors are normally sensitive should have warned against such a deduction.

Nevertheless, the twitch of Galvani's frog legs signaled an end to hydraulic theories of nervous action and a beginning to the electrical theories, versions of which have persisted until our time. It is now understood that the electricity of the nervous system does not function in the same way

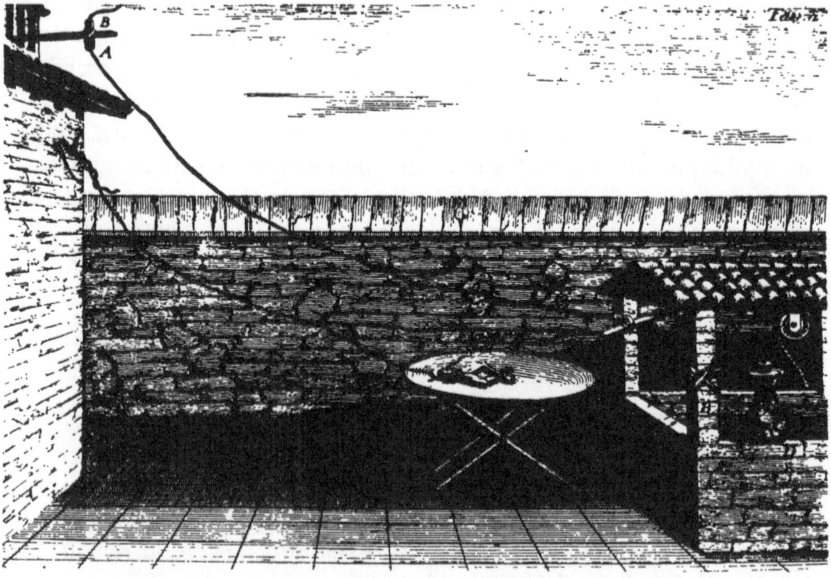

FIG. 2.1. Etching of one of Galvani's experiments in which the nerves of a frog's leg were stimulated by atmospheric lightning. (From Green's 1953 translation of Galvani's original work.)

as the electricity in a metallic conductor. Though both are electrical and can be described by the same mathematical functions, in each case the carriers are different. Electrons and "holes" convey charge in metals, while ions convey charge in the neuron. In addition, the properties of the conductors are different. Metals have low resistances, while neurons have mixed resistive and capacitive elements of appreciable magnitude. The capacitive and resistive elements in a neuron and the nature of the carrier specify that conduction velocities will be far slower in biological tissues than in a wire. Furthermore, the neuron is a metabolic system that in many instances produces its own power from energy sources located within the cell itself, unlike the metallic conductor, which must be powered by an external power source.

At the end of the eighteenth century, this sort of physicochemical sophistication was not highly enough developed to allow contemporary scientists to appreciate this subtlety, yet the magnitude of the intellectual leap stimulated by Galvani's discoveries must not be underestimated. The notion of the electrical concomitant of neural action is, quite simply, the key idea in the development of all that is yet to be discussed in this book.

As noted earlier, Galvani's observations were also extremely influential in the development of the physical science of electricity. One experiment carried out by Galvani involved the stimulation of a frog's nerve-leg preparation by hanging it on a metal railing. The source of the electricity in this case was initially obscure, but a contemporary of Galvani, Alessandro Volta, correctly suggested that the source of the electrical stimulus was the junction between the two different metals of the hook and the railing. Stimulated by Galvani's biological data, Volta went on to invent the bimetallic battery, a device that for the first time in history provided a steady electrochemical source of electricity. This invention led to the many discoveries and technological advances in electricity of the next two centuries, eventually culminating, among other innovations, in the modern electronic instruments that now provide answers to some of the major problems in neurophysiology. Thus the circle of interaction between electrophysiology and electrotechnology was completed.

By the middle of the nineteenth century there was little remaining controversy regarding the electrical basis of nervous action. Most major researchers had accepted the notion without argument and went on to special problems of their own interest. Some workers suggested roles for electricity in nerve action that sound surprisingly modern. Among the most notable of these was Johannes Müller, the formulator of the *law of specific energies of nerves* (the notion that the sensation produced by a response in an afferent nerve is a function of which nerve is involved, not which stimulus triggered the response). According to Brazier (1959), Müller was also one of the first to suggest specifically that electricity "was an artificial excitant that had no part in natural excitation." Nevertheless, the notion of electricity as the prime correlate of nervous action was firmly established in the science of neurophysiology by this time. As it turned out, this view was to be reinforced by a continuing series of developments in neurophysiological instrumentation in the twentieth century. Many of these developments involved electrical stimulators and recorders that were sensitive to the electrical signals from the nervous system.

C. STIMULATORS AND RECORDERS

Once the electrical nature of the neuron was established, investigators began a search for the perfect nerve stimulator, which continues to this day. Electricity is so easily controlled, compared to other forms of physical energy, and it is so effective in giving the experimenter control over the stimulus pattern, whether or not it is an "artificial" stimulant, that the technology of electrical stimulation is still under active development today.

Indeed, the history of electrical stimulators mirrors the history of neurophysiology even when it was not so intended. It is amusing to note that one of the earliest recorded nerve stimulators was thought by its inventor in the seventeenth century to be purely mechanical, but in fact, it was probably electrical. This early work was carried out by a young student at the University of Leyden by the name of Swammerdam. Though the results of his experiments were not published for over 100 years, it seems fairly certain that in the year 1660, Swammerdam was unwittingly stimulating nerves with what most probably was an electrical stimulus. His "mechanical" stimulator is best described by his own words:

> If, instead of the heart, we should chuse [sic] to make use of some other muscle, we may proceed in the manner represented in the eighth figure [see Fig. 2.2], where the glass siphon, *a*, contains within its hollow the muscle, *b*, and the nerve hanging from the muscle is fastened, without being cut or bruised to a slender twisted silver wire, *c*, that runs at the other end, an eye made in a piece of brass wire, soldered to the embolus or piston of the siphon, *d*. Things being thus made ready, a drop of water, *e*, must be let into the slender tube of the siphon by a very fine funnel. Now, if after this, the silver wire be cautiously drawn with a leisurely hand *f* through the ring or eye of the brass wire, till the nerve is irritated by the compression, it must by this means undergo, the muscle will contract itself in the same manner with the inflated heart, whose alterations, upon a similar occasion, I have already described, even the drop of water will in some measure sink, though afterwards it never rises again [cited in Fulton & Wilson, 1966, p. 212].

However, there is a catch in this description. The astute reader will have noticed that the muscle hung on a silver wire, which was in turn passed through a brass eye. The necessary conditions for electrical stimulation with current from a bimetallic battery were therefore present, and it certainly seems that Swammerdam may have inadvertently stumbled upon an

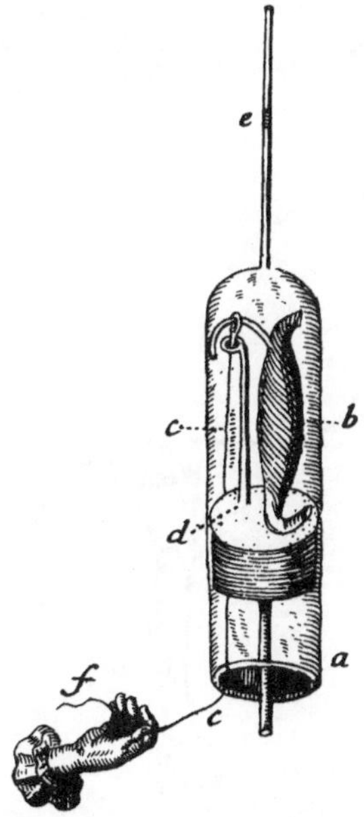

FIG. 2.2. Etching of one of Swammerdam's experiments in which he may have inadvertently stimulated the nerve with an electrical voltage produced by a bimetallic contact. c = silver wire; d = brass support. (From Swammerdam's book *Biblia Naturae* published in 1737, as reproduced by Fulton & Wilson, 1966.)

electrical stimulus 100 years before Galvani. However, since the idea of electricity had not yet been established in the seventeenth century, there was no way in which this could have made sense to him.

Another early electrical stimulator was simply another animal. Galvani himself used the spinal cord or the crushed end of a frog's leg nerve to produce muscular contractions in a second frog's leg. These experiments were especially important because they showed that the signals involved in the electrical action of nerves had nothing intrinsically to do with the metals of some bimetallic battery or other parts of the stimulating mechanism. Rather, they suggested that the process of the stimulus was also intrinsic to the nerve and occurred even though no wire, bimetal junction, or lightning was present.

The conceptual link between electricity and nerve action having been made, the design of electrical stimulators became more specifically directed. After Volta's description of the bimetallic battery effect, Claude Bernard developed some small units especially suitable for stimulating nerves. A photograph of one of his miniature battery stimulators is shown in Fig. 2.3.

The next important development in stimulator technology occurred as a result of the rapidly evolving physical science of electromagnetism. The transformer, a voltage magnifier consisting simply of two coils of wire wrapped around the same core, came into general neurophysiological use in the form of the induction coil, which was introduced by Emil du Bois-Reymond in the nineteenth century. The idea of the induction coil is simple. A circuit containing a relatively low voltage source, typically a battery

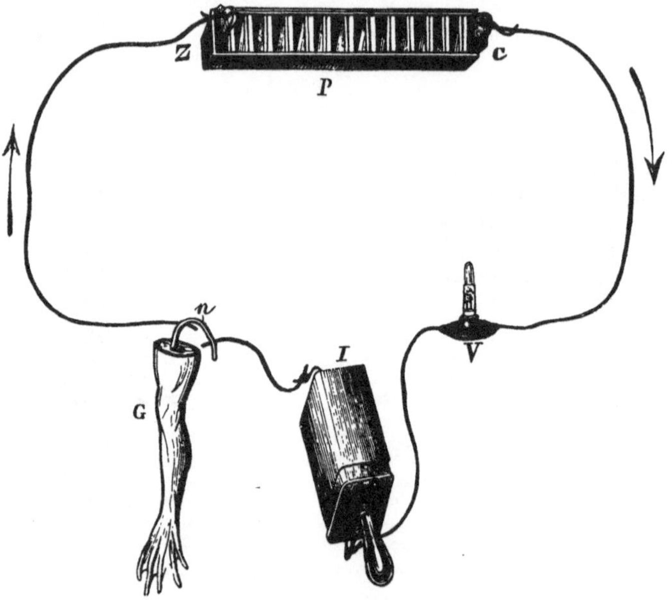

FIG. 2.3. Etching of one of Claude Bernard's voltaic pile electric stimulators, showing the switch (I = interrupter), the voltaic pile (P), and the frog's leg preparation (G). (From Bernard, 1858.)

and a coil of wire wound with relatively few turns, could be interrupted with a simple switch. During the turn-off and turn-on times (the periods in which the flow of current through the first coil is varying), a fluctuating magnetic field is produced. That magnetic field is capable of inducing a much larger voltage (though a smaller current) in a second coil with a larger number of turns wound concentrically with the first. Since the switched interruptions could be controlled by mechanical rotators, the time sequence of a pattern of electrical stimuli could also be varied.

The induction coil was used as the main electrophysiological stimulator well into the twentieth century, when it was finally replaced by all-electronic stimulators that allowed much better control over the amplitude and timing of the stimuli. The earliest units were constructed from vacuum tubes, but this technology ultimately gave way to one using transistors, which became available in the 1960's. In recent years such luxuries as constant-current and voltage-isolated features have come to be routinely included in even the least expensive stimulators (see Chapter 4).

Although stimulating instrumentation has changed greatly in appearance, and its convenience has increased enormously, there is little difference in concept between the earliest induction-coil stimulators and the most modern integrated circuit designs. Both are means of exciting nervous action. On the output end of the experiment—the amplification, display, and recording of faint signals from neurons—the situation is entirely different. Recording and display equipment has progressed to a level of such complexity and utility that a truly qualitative and conceptual change has occurred in the neurophysiologist's ability to see what is happening.

The earliest recording devices were, once again, other organic units. The muscle twitch induced in the frog's leg was the very first display indicative of neural activity. Credit for the initial application of this organic "scope" must be assigned either to Swammerdam or to his contemporary, Thomas Willis.

The next major step in recording technology was the subsequent invention of the electrometer, a simple device composed of two leaves of thin metal foil suspended in a glass jar from a metallic bulb. Electrometers act as detectors of electricity, because of the repulsive forces generated by like electrostatic charges. When the bulb at the top of the jar is charged, the two foil leaves become identically charged to the same potential, since they are electrically interconnected to the bulb. The electrostatic repulsion between equally charged objects, even if slight in amplitude, is sufficient to force the delicate leaves apart. Similarly, a charged electrometer (and thus one in which the leaves were spread apart) would be discharged in the presence of any appropriate conductive pathway that allowed the electrometer's stored charge to leak away. Unfortunately, neither the frog's leg nor the electrometer could respond rapidly enough to keep up with the speed of a neural response, nor could they give a quantitative measure of the magnitude of the signal. Neurophysiologists thus eagerly capitalized on the development of a new instrument, the moving coil galvanometer.

The moving coil galvanometer is essentially a small motor consisting of a smaller coil of wire mounted between the poles of a permanent magnet. When the current from a voltage source, such as a neurophysiological preparation, passes through the smaller coil, it turns it into a small electromagnet. The interaction of the two electromagnetic fields is sufficient to produce a mechanical force, and if the small magnet is free to rotate (restrained only by a light spring) and has a pointer attached to it, the rotation of the coil produced by the tiny neuropotential can be measured. Moving coil galvanometers were used until the electronic instrumentation revolution began in the 1930's.

Other forms of galvanometers and electrometers have also been developed. The moving coil galvanometer can be modified so that it operates in a "ballistic" fashion. In the ballistic galvanometer, the maximum deflection of the coil is proportional to the total amount of current that has flowed through the coil. The peak of the deflection is, therefore, a time integral of all the current produced by the driving voltage.

Another innovation that came into vogue around the turn of the twentieth century was the capillary electrometer. This device is essentially a modern analog of the old leaf electrometers. In the case of the capillary electrometer, however, the moving component is a tiny bubble in a column of fluid. The position of the bubble in the capillary tube is also proportional to the applied voltage. Capillary electrometers rivaled in sensitivity some of the moving coil galvanometers and were particularly useful in neurophysiological research because of the small mass of the bubble.

Indeed, a major problem inherent in any electromechanical device is the considerable mass of the moving element, no matter how lightly it is constructed. The fact that there is inertia associated with bodies of any significant mass means that a substantial amount of energy is required to accelerate these bodies, even if the mass is only as slight as that of a small metal pointer or a bubble. There was a great need, therefore, to develop instruments that had little mass and low inertia, which therefore could move fast enough to keep up with the electrochemical processes under observation.

This problem was not solved until the twentieth century, with the development of electronic measuring and display instruments. Two separate innovations, the high gain amplifier and the cathode ray oscilloscope, resulted in the development of recording instruments of virtually zero mass, and thus nearly instantaneous response time. The first application of cathode ray oscilloscopes to neurophysiological recording is usually attributed to Erlanger and Gasser (1924), who studied compound (multifiber) responses in the phrenic and sciatic nerves of the cat. Erlanger and Gasser took advantage of a newly-developed electronic technology to make some extraordinary neurophysiological discoveries. In the years that have followed, the general recording and display techniques they pioneered have become universally used in this kind of research. Today there is a substantial industry providing not only these recording instruments but also the electrodes and other elements needed to run a complete laboratory. Where once the recording of the simplest neural potential was a technical *tour de force,* now even undergraduate students make use of equipment and procedures that would have astounded senior investigators only a decade or two ago.

Certainly there are new developments on the horizon, and not all that is to be done has yet been done. Perhaps no single technological development is needed so badly or is so imminent in the neurophysiological laboratory as the use of the small digital computer to process information acquired by amplifiers, so that more convenient and meaningful displays of data summaries can be produced.

The histories of the computer and of neurophysiological research in the past 20 or 30 years are also curiously intermingled, continuing the trend of parallel development in electrical technology and neurobiology. In the 1940's and 1950's there was a tendency for computer scientists and neuro-

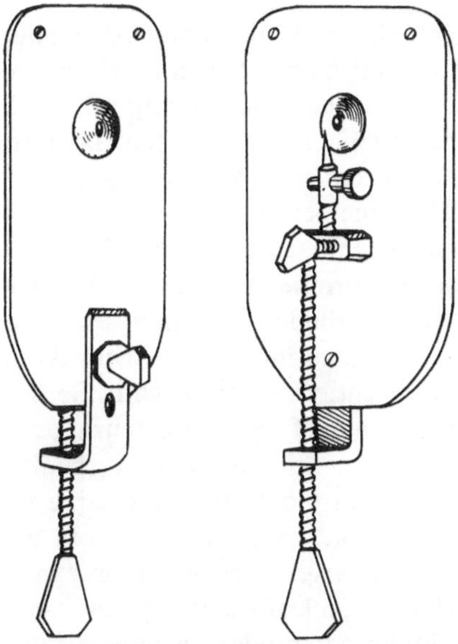

FIG. 2.4. Drawing of one of Leeuwenhoek's 2-inch long, simple (single spherical lens) microscopes showing the screw arrangement that was the antecedent of the modern mechanical stage. This type of microscope had a maximum magnification of about 275×. (From Bradbury, 1967.)

physiologists to compare what appeared to be many common features of computers and nervous tissue. Today, however, it is clear that the digital computer, as it currently exists, is a poor model of even the simplest brain. Nevertheless, there is no question that the computer will have a profound impact on neurophysiology as it becomes more widely used as a data processing and display device, and as a generator of stimulus patterns. The many powerful ways in which the digital computer can analyze and display summaries of large amounts of data promise to be particularly important in future studies of the interactions of large numbers of neurons.

D. INTRACELLULAR TECHNIQUES

Leeuwenhoek's development of the first high-power simple microscope in 1682 opened a window into the microscopic world that was to revolutionize all of biology. Though his best models consisted of only a single lens (see Fig. 2.4) and had magnification of only about 275×, the world they opened to scientists was amazing in both its beauty and complexity. By the early 1800's, higher-power compound microscopes consisting of multiple lens elements* had evolved to the point where they provided a level of magnification sufficient for observation of the microstructure of animal tissues. It became obvious to nineteenth-century investigators that nerves, which previously were believed to be relatively homogenous tissues,

* Surprisingly, *low power* compound microscopes were originally developed by Janssen in 1590, and were in common use by such scientists as Schemer in 1628 and Hooke in 1660 long before the Leeuwenhoek developments.

were actually multiple cables of smaller fibers. In fact, the microstructure of the axonal filament of a single neuron was beginning to be understood by that time. Schwann, among others, had observed the outer fatty covering, the myelin sheath, of the axon formed from those specialized nonneural cells that now bear his name.

As the nineteenth century passed, the notion that the really important processes occur at the cellular level grew stronger, along with what must have been enormous frustration. Though researchers knew that they should be looking at single cell responses, these single cells were simply too small to be investigated individually with the techniques then available.

In the twentieth century, with the development of oscilloscopes and amplifiers, the task of recording neural activity from within single cells became a major, but continually elusive, goal of many neurophysiologists. A few investigators in the 1930's were able to dissect a small bundle, containing only a single active fiber, free from a compound nerve, and by laying it across some large electrode (for example, a saline-soaked cotton wick), were able to pick up extracellular action potentials from single cells. But extracellular potentials are only weak reflections of what is really going on across the cell membrane. Most workers in the field realized that neurochemical theory would have to remain speculative until it became possible to record potentials across cell membranes, and that meant placing an electrode inside the neuron.

The task was finally accomplished because of an anatomical freak. Young (1936) described a most unusual neuron, a giant cell in the squid nervous system which had an axon as large as 1 mm in diameter. This axon was sufficiently macroscopic that even a relatively large glass tube could be inserted into the intracellular space. If this tube contained a salt solution, it could act as the necessary internal electrode. Tubular electrodes were used to record transmembrane, intracellular potentials by two groups almost simultaneoulsy (Curtis & Cole, 1942; Hodgkin & Huxley, 1939). The experimental procedure is sufficiently important to warrant quoting their description in its entirety:

> . . . A 500 μ axon was partially dissected from the first stellar nerve and cut half through with sharp scissors. A fine cannula was pushed through the cut and tied into the axon with a thread of silk. The cannula was mounted with the axon hanging from it in sea water. The upper part of the axon was illuminated from behind and could be observed from the front and side by means of a system of mirrors and a microscope; the lower part was insulated by oil and could be stimulated electrically. Action potentials were recorded by connecting one amplifier lead to the sea water outside the axon and the other to a microelectrode which was lowered through the cannula into the intact nerve beneath it. The microelectrode consisted of a glass tube about 100 μm. in diameter and 10–20 mm. in length; the end of the tube was filled with sea water, and electrical contact with this was made by a 20

> μm. silver wire which was coated with silver chloride at the tip. . . . A small action potential was recorded from the upper end of the axon and this gradually increased as the electrode was lowered, until it reached a constant amplitude of 80–95 mv. at a distance of about 10 mm. from the cannula. In this region the axon appeared to be in a completely normal condition, for it survived and transmitted impulses for several hours [Hodgkin & Huxley, 1939, p. 710].

This technique was astonishingly influential in its impact on theory, for virtually the first observation—the fact that the electrical potential across the membrane went positive rather than simply retreating to zero during the production of a spike action potential—necessitated a drastic reformulation of the contemporary theory of membrane action. Modern membrane theory and the full impact of this discovery are discussed in greater detail in Chapters 5 and 7.

As productive as the procedure was, this sort of intracellular recording required a freak neuron, and the study of neurophysiology could advance only if a procedure for the intracellular recording of neuronal action potentials of broader applicability could be developed. The important step of nondestructive penetration of a cell by an electrode was accomplished by Gerard and his colleagues (Graham & Gerard, 1946; Ling & Gerard, 1949) on muscle fibers, but the technique is identical to that currently used on neurons. Their procedure involved the use of a tiny fluid-filled glass tube. These tiny tubes were made by a process in which tubes of soft glass were heated and then pulled evenly from both ends until they thinned and broke in the middle. The fluid characteristics of molten glass act as a "demagnifier" to produce a replica of the original tube, but of a very small size. These microelectrodes, or *micropipettes* as they are now called, are so tiny that they may be literally pushed through the membrane of even a very small neuron without destroying it. The membrane forms a seal around the micropipette, like a self-sealing tire; this seal is sufficiently robust to allow the neuron to continue operating for hours or even days after being impaled. Modern micropipettes may be smaller than 1/10 of a μm (one ten-millionth of a meter).

Again, the technique is of sufficient importance that excerpts from an early Ling and Gerard (1949) publication are still extremely interesting:

> . . . we have pushed in the direction of drawing and filling microelectrodes of well under 1 μ tip diameter and properly tapered and were rewarded by obtaining highly constant membrane potentials.
>
> . . . The making of the microelectrodes was critical. Extensive trial led to a tip tapering over a terminal millimeter from a diameter of 25 μ to a few tenths at the opened end. This is stiff enough to handle and to obtain easy penetration and slender enough so that deep insertion does no further damage and that the electrode is still usable after the extreme tip is lost.
>
> Electrodes are drawn from capillary tubing of .5 to .7 mm. o.d. (Schaar

and Co.). Instead of a microburner, the flame of an air-gas blow torch is used. The capillary is gradually moved in from the side toward the flame and pulled abruptly when sufficiently soft. With some practice, two good needles, separated and with open ends, are obtained fairly regularly. The needle is bent at 90°, with the aid of forceps, and about 30 at a time are mounted for filling on a glass plate with cotton thread wound on it; other materials produce sediment which clogs the needles.

Filling is easily achieved by boiling vigourously in KCl solution under slight intermittent back pressure. The holder is placed in half isotonic KCl in a porcelain evaporating dish of 12″ diameter to which another, of 10″ diameter, serves as cover. The dish is heated with the blow torch for about half an hour, by which time the volume is reduced to half and the air in the needles replaced by vapor. Cool isotonic KCl is then added and the needles promptly fill. This treatment, associated with building up and sudden release (as the lid lifts at intervals) of some pressure, leaves all the needles completely filled with approximately isotonic KCl.

Each tip is then examined for shape and for absence of bubbles under the microscope and is tested for resistance (normally 100 megohms, usually falling with use but still satisfactory at 20 megohms) on the amplifier in a convenient holder [pp. 383–385].

This description is sufficiently detailed to be used as a manual for contemporary microelectrode production. A more detailed discussion of microelectrodes of several modern kinds is given in Chapter 4.

E. THE MEMBRANE THEORY OF NEURONAL ACTION

From a purely historical perspective, it is astonishing how fast new knowledge accumulated and new theories developed in the twentieth century concerning the cellular and biochemical basis of nervous action. The general framework of the contemporary theory was spelled out by Bernstein in 1902 when he suggested that the basis of neural action was the selective passage of various ions through the semipermeable plasma membrane of the neuron. However insightful this notion was, it was based on what were then some very flimsy ideas. Even the existence of an intracellular fluid filled with mobile ions was not firmly established until Höber's work in 1910 and 1912 on the red blood cell. Although there was some suggestive evidence for the existence of the membrane itself, its physical properties were not measured until the mid-1920's by Fricke. He was the first to suggest that the membrane is only a molecule or two thick and to estimate (correctly, as it turned out) its electrical capacitance. Direct viewing of the cell membrane with an electron microscope has been possible only since the mid-1940's.

Once routine entry of the cell with micropipettes was possible, many speculations of the past were opened to direct experimental inquiry. Hodgkin and his co-workers, as we have noted, quickly demonstrated the voltage overshoot, and, on the basis of impedance measurements, were able to show that the whole course of the spike action potential could be explained solely on the basis of variations in membrane permeability and the resulting flow of particular ions (see Chapter 7). This series of ideas still forms the basis for most contemporary theory, although a number of details are still topics of active investigation (for example, the nature of the channels through which the ions flow and the nature of the active pumping processes that produce the electrochemical driving forces are still not entirely understood).

These, then, are the major technological developments and the associated conceptual breakthroughs that have occurred up to now. The routine intracellular microelectrode penetration of even very small mammalian cells opened a floodgate of new knowledge about the basic electrochemical dynamics of the membrane that accounts for the action of individual neurons. This technique ultimately made possible the study of interactions between numbers of cells. It is this latter topic that is the main theme of contemporary neurophysiological research.

3
NEUROANATOMY

Each of the various types of cells that make up the body is specialized to perform one of the general life functions in a particularly efficient manner. Neurons, the cells of greatest interest in this text, have been specialized under the influence of evolutionary forces so that the stimulation of one portion of a cell may result in action at some distant locus. Thus, the generalized excitability of the prototypical cell has evolved into a capacity for specialized information transmission and integration between and among networks of neurons. We may therefore consider the primitive excitability of a single generalized cell as the antecedent of all psychological and behavioral responses.

What are the fundamental premises of the current theory of neuronal anatomy? Basically, it is agreed that each neuron is anatomically separate and distinct from all others. Specifically, this means that each neuron is completely surrounded by its own cell membrane. This is not to say that cells are also physiologically independent because a basic axiom of modern neuron theory is that neurons do interact. However, it has now been incontrovertibly established that they do so primarily at specialized intercellular regions called *synapses*. The interaction between the cells may be quite intimate; small molecules and charge carriers such as ions do move between the gap between two cells, but most of the intracellular material, as well as cell metabolism, is sharply delimited at the membrane. Even when the membrane of one neuron is fused to the next one (a tight junction), the intracellular contents of the two cells are mainly isolated from each other.

Interaction at a synapse is an intricate process in which the information being communicated or processed by the neuron is encoded, transmitted, and decoded several times. The information may be represented by spike, graded potentials, or even the amount of certain chemical transmitter substances. Although we discuss all of these matters later, the important point to note here is that intercellular transmission is a communication process that does not take place through protoplasmic continuity. While most synapses have only a very small gap between the cell membranes of the transmitting and the receiving neurons, there is almost always a protoplasmic discontinuity present. This cleft in a chemical synapse may be less than $1/10$ of the wavelength of visible light—only a few hundred angstrom units. Nevertheless, in spite of these small dimensions, there are measurable delays, observable electrochemical events, and direct electronmicroscopic evidence that demonstrate that neurons in the nervous system are separate and distinct cells.

The high level of neuronal specialization and unity of function has not led to a single standard nerve cell type. There are many different neuron shapes to be found in various portions of the nervous system. Some are highly specialized for communicating information over very long distances, while others are specialized for the integration of several sources of nervous activity within a local region of some central nucleus. The former, the communication type, is usually characterized by very long fibers that propagate information by nondecremental, i.e., constant amplitude, conduction. The latter, the integrative type, is usually characterized by the extreme arborization of some of its terminal branches. Integration is, however, a property of most neurons; even receptors and the last neuron in a motor chain can perform integrative functions. In addition, receptor cells have unique structural adaptations for carrying out the special function of converting one or another kind of physical energy into patterns of neurochemical energy.

Despite great diversity of anatomical form, all neurons contain homologous parts. Some of the common features are shared by all living cells. Others, however, are specific to neurons and reflect the nature of the specialization they have undergone during their evolution.

In this chapter we first briefly introduce some of the gross anatomical features of the nervous system. We then consider various examples of the receptors, and neurons specialized for transmission and integration, respectively, with the intent of alerting the reader to the great variety that can be observed in neuronal anatomy. This chapter then concludes with a discussion of the ultramicroscopic nature of the neural membrane—that region of demarcation, only a few molecules thick, that plays such an important role in nervous action.

A. GROSS ANATOMY OF THE HUMAN NERVOUS SYSTEM

It has been said that the human nervous system is perhaps the highest development of organic evolution and the most complex structure that has ever existed. Although the human brain is not the largest in either an absolute or in a relative sense (the whale's brain is considerably larger than man's; the tree shrew has a higher brain-to-body-weight ratio than man), the unique attributes of the human brain have culminated in a thinking, history-keeping, and creative animal that has no known analog.

To understand fully and appreciate the nervous system we should first note that it is widely dispersed throughout the body. The nervous system is seen to be composed of a central core consisting of the brain, the spinal cord, and (because of their embryological origins), the retinae of the eyes, as well as a peripheral network of sensory and motor nerves. The central nervous system, although it does serve communication functions (many of the tracts that lie within it are simply links from one part to another), may be characterized as being more responsible for the integration of incoming and outgoing information.

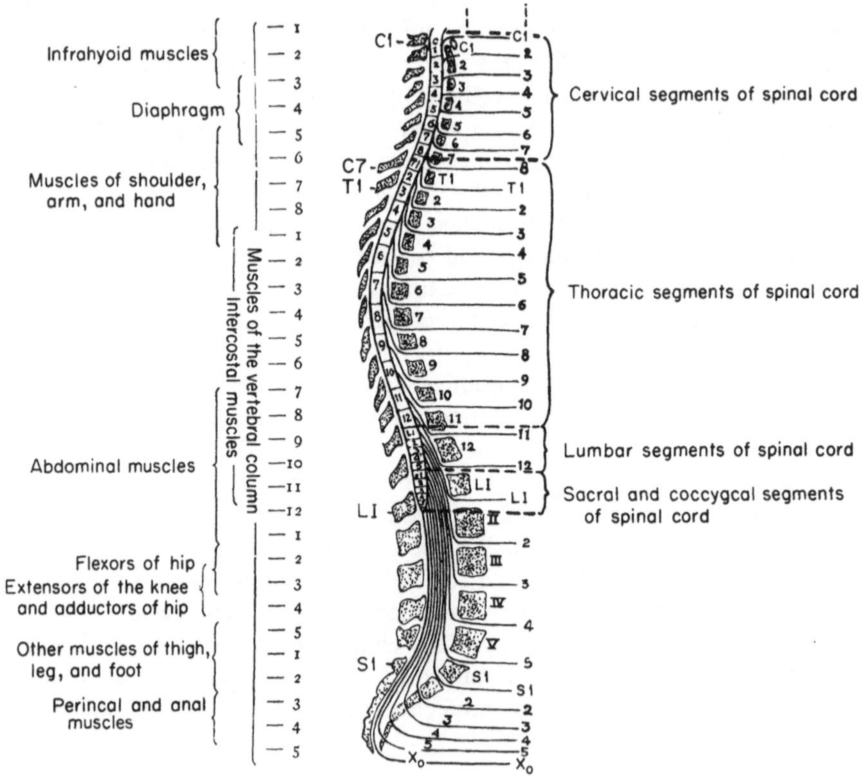

FIG. 3.1. Drawing shows the ventral spinal motor roots and indicates the various muscles driven by each root and its level of origin. Letters labeling each root represent the general level from which it emanates: C = cervical; T = thoracic; L = lumbar; S = sacral; and X = coccygeal. Numbers indicate the particular root. A similar system of dorsal roots is the pathway for incoming sensory messages. (From Ranson, 1959.)

The peripheral nervous system, the network of nerves that extends to all portions of the body, consists of two parts. One part is composed of the somatic nerves sending motor or *efferent* (outward-going) signals to the voluntary muscles and receiving sensory or *afferent* (inward-going) signals from the sensory receptors. The other part of the peripheral nervous system is also both efferent and afferent and both sends motor signals to and receives sensory messages from the smooth muscles and the internal organs of the body. This second part is known as the *autonomic system.* Its motor component, in turn, is composed of two parts, the *sympathetic* and *parasympathetic* portions.

Communication between the central and peripheral nervous system is accomplished through two sets of nerves. One set, the spinal nerves, enter or leave the central nervous system at the level of the spinal cord. Figure 3.1 shows the many spinal nerves and defines the various regions of the cord from which they emanate, as well as the motor groups activated by each. The other set of nerves communicating with the central nervous system are those that enter at the level of the brain stem. These are referred

TABLE 3.1
The Cranial Nerves[a]

Number	Name	Functions
I	Olfactory	Smell
II	Optic	Vision
III	Oculomotor	Eye movement
IV	Trochlear	Eye movement
V	Trigeminal	Masticatory movements Sensitivity of face and tongue
VI	Abducens	Eye movement
VII	Facial	Facial movement
VIII	Auditory vestibular	Hearing Balance
IX	Glossopharyngeal	Tongue and pharynx
X	Vagus	Heart, blood vessels, viscera
XI	Spinal accessory	Neck muscles and viscera
XII	Hypoglossal	Tongue muscles

[a] Adapted from Thompson (1967).

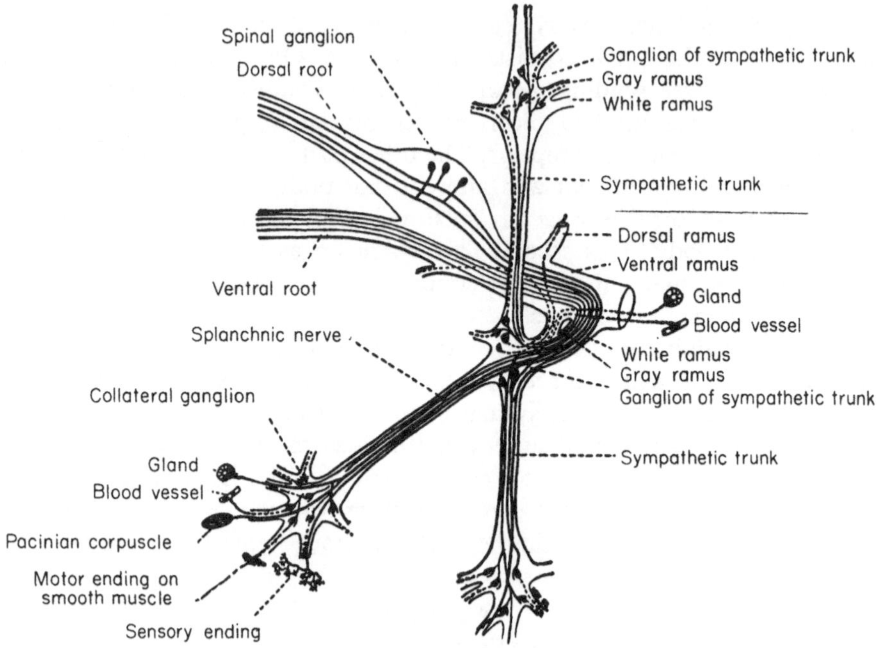

FIG. 3.2. Arrangement of the autonomic nervous system showing one of the chain of ganglia running parallel to the spinal cord. The autonomic system contains both sensory and motor fibers. (Drawing from Ranson, 1959.)

to as the cranial nerves. Table 3.1 lists the cranial nerves and the functions of each.

The spinal nerves enter and leave the spinal cord through an efficient duplex system shown in Fig. 3.2. The fibers are segregated on the basis of their function according to a law first enunciated by Magendie, but (because of one of the nastiest priority disputes in scientific history) now formally known as the Bell-Magendie law. According to this functional-anatomical law, sensory fibers enter through the dorsal roots and motor fibers emanate from the ventral roots.

The two portions of the autonomic system generally function antagonistically to each other, but nowadays it is somewhat more usual to emphasize the anatomic rather than the functional distinction between the two. As shown in Fig. 3.3, sympathetic nerves originate in the thoracic and lumbar regions of the spinal cord, and parasympathetic nerves have their origins in the cranial and sacral regions of the spinal cord. This figure also shows one of the dual chains of sympathetic ganglia that run parallel to the spinal cord.

The central nervous system is a complex net of transmission pathways, usually referred to as *tracts* (as opposed to *nerves,* which are pathways

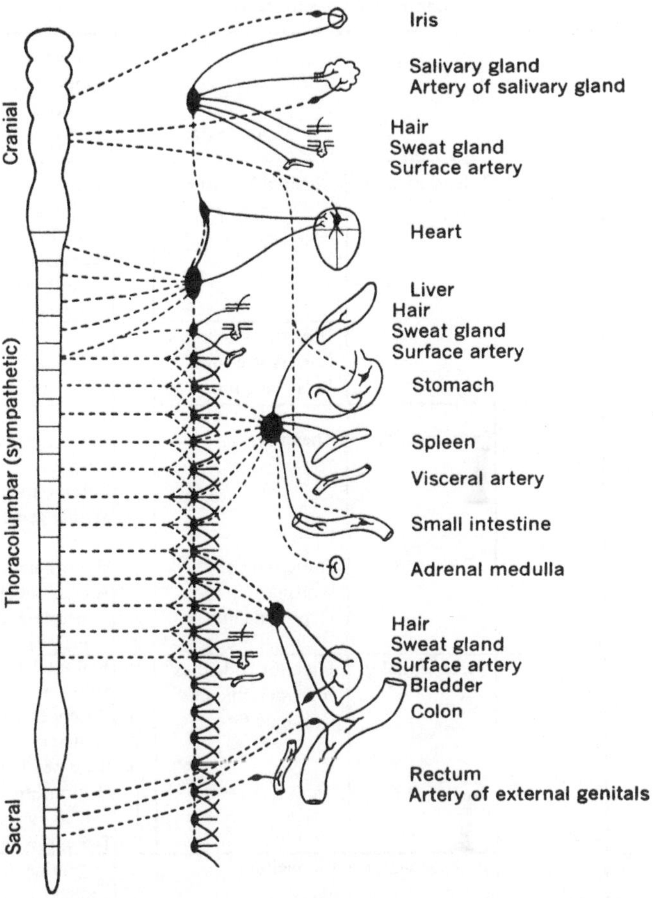

FIG. 3.3. Arrangement of the efferent or motor portions of the autonomic nervous system. This drawing shows the anatomic division of the sympathetic (thoracolumbar) and parasympathetic (sacral and cranial) components and the internal organs supplied by each division. (From Mountcastle, 1968; after Cannon from Bard.)

outside the central nervous system), and of *nuclei* consisting of clusters of cell bodies connected by these tracts. Neuronal cell body clusters outside the central nervous system are referred to as *ganglia*. The term *ramus* is used to refer to a smaller branch of a nerve or tract, while *trunk* is the term used to describe a short, major, and undivided portion of a nerve.

Surprisingly little can be seen of the complex of nuclei and tracts in the central nervous system when it is initially dissected. Special procedures must be used to clarify the extent of the various parts. These procedures include neural degeneration induced by cutting a portion of the nervous system, electrophysiological stimulation and recording, and special staining

TABLE 3.2
The Various Nuclei and Centers of the Central Nervous System[a]

			Major centers		Further subdivisions and nuclei
The brain	Forebrain	Telencephalon	Cerebral hemisphere		Sensory and motor regions of cerebral cortex Association regions of cerebral cortex
			Olfactory bulb Corpus straitum Basal ganglion		
		Diencephalon	Thalamus	Reticular formation	Lateral geniculate body Lateral nucleus Medial geniculate body
			Optic tract and retina Hypothalamus		Hypothalamus Arcuate nucleus Ventrolateral nucleus
	Midbrain	Mesencephalon	Superior colliculus[b] Inferior colliculus[b] Crus (the floor)[b]		Red nucleus Substantia nigra Mesencephalic nucleus Basis pedunculi Oculomotor nucleus Trochlear nucleus
	Hindbrain	Metencephalon	Cerebellum Pons[b]		Trigeminal nerve nucleus Parabrachial nucleus
		Myelencephalon	Medulla[b]		Gracile nucleus Cuneate nucleus Dorsal, ventral, and cochlear nuclei Solitary nucleus Superior, medial, and lateral vestibular nuclei Superior olive Inferior olive Trapezoid body
The spinal cord					

[a] From Uttal (1973).

[b] Components of the brain.

procedures that discriminate one region from another by selective coloration. Fluorescence and autoradiographic techniques are also in current vogue.

The central nervous system may be considered to be a tube along which various portions have swelled into major centers. The most extraordinary growth, of course, is in the most anterior portion—the great centers known as the cerebral hemispheres, which under ordinary conditions overhang and hide the underlying brain stem. The brain stem itself consists of a number of important nuclei that perform various communication and integration functions. These have been summarized in Table 3.2.

Ascending through the core of these direct pathways is the reticular formation, an anatomically unstructured system (the cellular network is very diffuse, and no clear lines of demarcation between its various regions have ever been found) that seems to acquire input signals from almost all the senses and then to disperse these messages to most other portions of the brain. The reticular formation has been assumed to serve a general activating function and to be associated particularly with the maintenance of consciousness.

Armed with this brief gross anatomy, our attention can now turn to the main theme of this book: the structure and function of the individual building blocks, the microscopic neurons that make up these macroscopic structures.

B. ANATOMY OF RECEPTORS

The receptor cell is the natural starting point for any discussion of the microanatomy of the nervous system. Receptor cells are located in the sense organs, where the initial points of contact between the physical stimuli of the outside world and the internal, electrochemical environment occurs. Thus, it is at this point that information patterns first may be considered to be neurally encoded.

An important general point that must be considered at the beginning of this discussion is that the receptor cells of the sense organs are themselves neurons. They are specialized, of course, and differ from transmission neurons and interneurons in a number of ways. Rather than having a synaptic input, the most peripheral portion of each has been specialized to convert a physical stimulus of one form or another into neural signals.

Almost all receptor cells terminate in elongated structures, or they occur at the end of an elongated neural fiber. This is a useful adaptation since the receptor's function is to probe the external environment. Many receptor cells are, in addition, ciliated and may have a cluster of small extensions at their end. Ciliation serves to increase the surface area of the receptor without requiring a major increase in the volume of the cell. Each kind of receptor has other anatomic and biochemical features specialized for its particular function.

Much of the information concerning receptor cell structure has become available only in the last few years, with the availability of the electron microscope. The recent development of the scanning electron microscope, with its ability to display structures in almost three-dimensional perspective and great depth of field, has been very influential in clarifying subcellular details.

In the following discussion we shall concentrate mainly on the receptors of the vertebrate. As far as possible, we shall also limit our examples to samples drawn from the primate sense organs; however, in a number of cases the best photographs obtained have come from other vertebrates. In general, the points made by any sample figure may be considered to be typical of all vertebrates, and in many cases, they also describe invertebrate cellular properties.

1. Visual Receptors

The vertebrate retina often contains two kinds of visual receptor cells, *rods* and *cones*. Though we now know that there are important functional differences between the two cell types, the original dichotomy was based simply on observations of their different structures.

Typical shapes of a rod and a cone are illustrated in the diagrammatic displays of Figs. 3.4a and 3.4b. These figures illustrate a number of important points. Both the rod and cone are seen to be composed of two main parts: an inner segment containing the cell nucleus and the synaptic contacts, and an outer segment containing discs in which the light-sensitive photochemicals are enclosed. The structure of the rod, however, can be seen to be not only architecturally different from the cone, but a close examination also suggests a developmental difference of considerable importance. Apparently, the outer segments of rods are in a constant state of disintegration at their apical end and reconstitution from their basal end. Figure 3.4b shows that the outer segment of the rod is composed of a series of discs or plates, which are mostly unconnected to the outer membrane of the cell. These intracellular discs are apparently produced by invaginations of the cell membrane at the base of the outer segment. Radioisotope studies indicate that the newly formed discs migrate up through the outer segment, subsequently to be cast loose into the intercellular environment. Discs in the rods, therefore, are both unconnected and transient structures in a constant state of replacement.

Discs in the outer segment of the cone, on the other hand, are always partially attached to the cell membrane. These discs do not migrate but

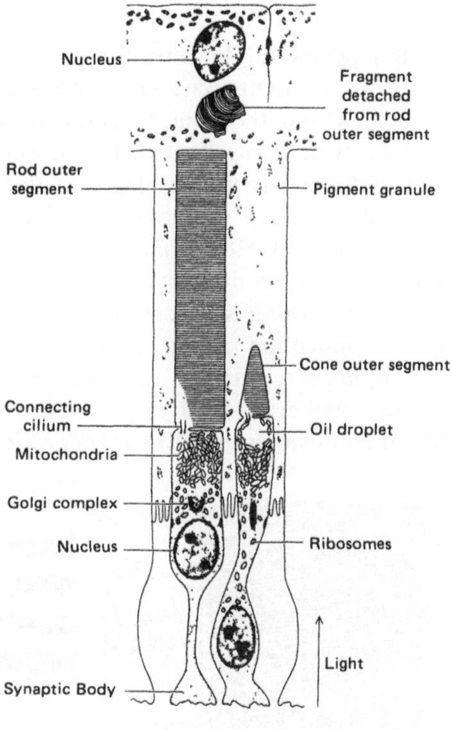

(a)

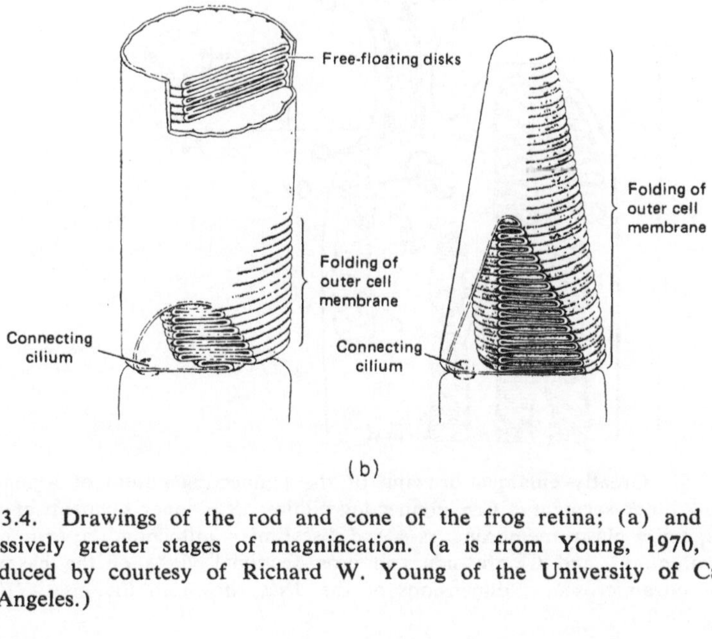

(b)

FIG. 3.4. Drawings of the rod and cone of the frog retina; (a) and (b) are at successively greater stages of magnification. (a is from Young, 1970, while b is reproduced by courtesy of Richard W. Young of the University of California at Los Angeles.)

are fixed in their initial position. The intradisc photochemicals are replenished by a circulation of nutrient materials throughout the cone's entire outer segment. It may be that the conical shape itself is due to the fact that the outermost discs were formed early in the embryological development of the animal, and the larger ones later in the animal's history.

Although the general structure of the rod and cone has been known for some time, it was not until 1960 that de Robertis, using high-power electron microscopy, was able to demonstrate the presence of the narrow neck between the inner and outer segments of both the rod and the cone. This narrow neck, or connecting cilium, shown greatly enlarged in Fig. 3.5, has a rather complex structure of its own; it is threaded with fine filaments. In spite of its small cross-section, all the nutrient materials necessary for normal cell metabolism and for the regeneration of photochemicals

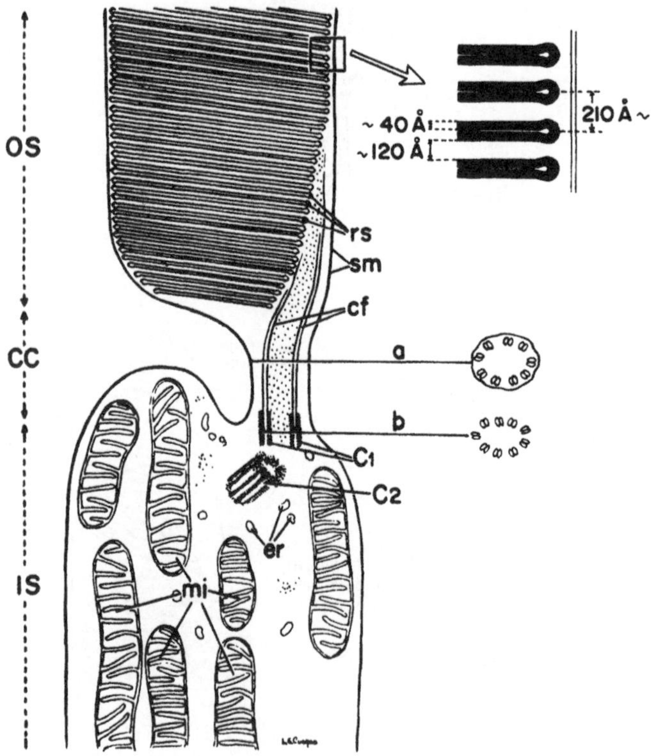

FIG. 3.5. Greatly enlarged drawing of the connecting cilium of a mammalian rod: OS = outer segment; CC = connecting cilium; IS = inner segment; cf = ciliary filaments; sm = plasm membrane; rs = rod discs; mi = mitochondria; er = endoplasmic reticulum; C_1 and C_2 are side and cross-sectional views of the basal bodies. Note the ultramicroscopic dimensions of the discs shown in the inset. (From De Robertis, 1960.)

following visual stimulation must flow through this narrow cilium since the metabolic controller, the cell nucleus, is located in the inner segment.

Figure 3.5 also illustrates the fineness of the microscopic scale at this level of cellular ultramicroscopy. The smallest discriminable structures are the membranes of the discs within the outer segment. These are only about 40 angstrom units in thickness. This is only one one-hundredth of the wavelength of any light that might be an effective stimulus for a given visual receptor.

2. Auditory Receptors

Figure 3.6 is a scanning electronmicrograph of the receptors of a monkey's basilar membrane, showing the regular arrangement of the auditory hair cells. This view was made possible by removal of the tectorial membrane, a flap of cells that usually covers the hair end of these auditory transducer cells. Here, as in the eye, the receptor cells may be divided into two categories: the *inner* and *outer* hair cells. This dichotomy was also originally based on an observed anatomical difference, but it has now been suggested that there might also be a functional difference between the two types of hair cell. Dallos, Billone, Durrant, Wang, and Raynor (1972) have observed that the outer hair cells appear to be sensitive to the absolute displacement of the cilia. However, they report that the inner hair cells seem to respond to the velocity of the sound wave rather than to its displacement. Such a dual system would add considerably to the dimensions of the incoming information that would be available to the auditory system in representing the auditory stimulus.

A closer view of the "organ pipe" arrangement of hair-like cilia at the end of an *outer* hair cell is shown in Fig. 3.7. The arrangement of these cilia into a W-shaped tuft may also help to convey additional information about the stimulus to the central nervous system. For example, it may add a directionality to the receptor action that may help in specifying the phase of the acoustic stimulus.

More diagrammatic representations of inner and outer hair cells are shown in Fig. 3.8a and Fig. 3.8b, respectively. Figure 3.8a shows a number of important features of the inner hair cell, not all simultaneously visible in a single scanning electronmicrograph. The cilia can be seen to terminate on the *basal body,* a structure that may play a central role in the actual transduction process (see Chapter 6). Furthermore, there appear to be two different kinds of synaptic connection at the base of the hair cell. One contains granules (which may possibly be packets of synaptic transmitter substance; see Chapter 9) in the hair cell on one side of the synapse, but not in the neuron labeled *NE1* on the other. The second kind of synapse seems to contain granules in the other neuron (*NE2*), but not in the hair

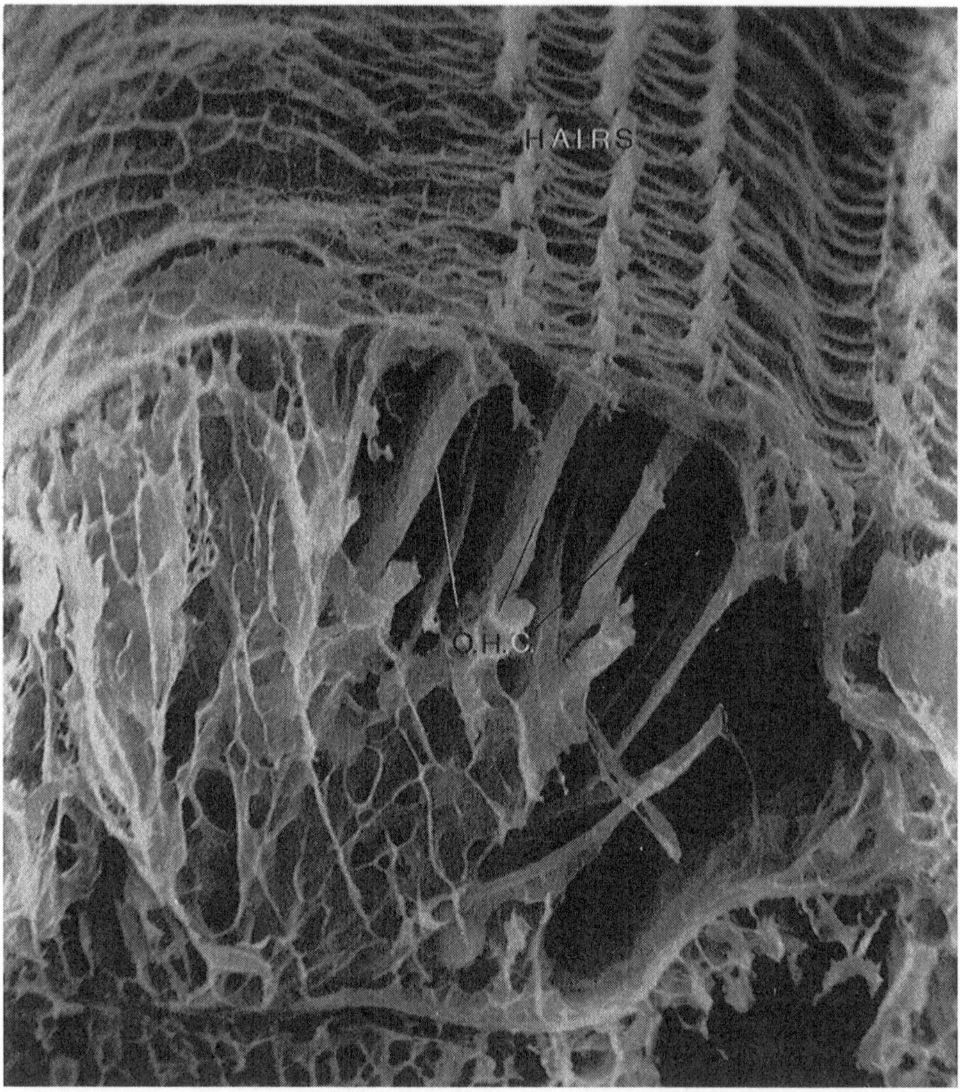

FIG. 3.6. Scanning electronmicrograph of a view down the tunnel of corti of the basilar membrane of the cochlea. Note the three columns of outer hair cells (OHC) and the rows of hairs protruding above the membrane. Preparation from a Rhesus monkey; display magnified approximately 1100×. (Courtesy of Robert E. Preston of The University of Michigan.)

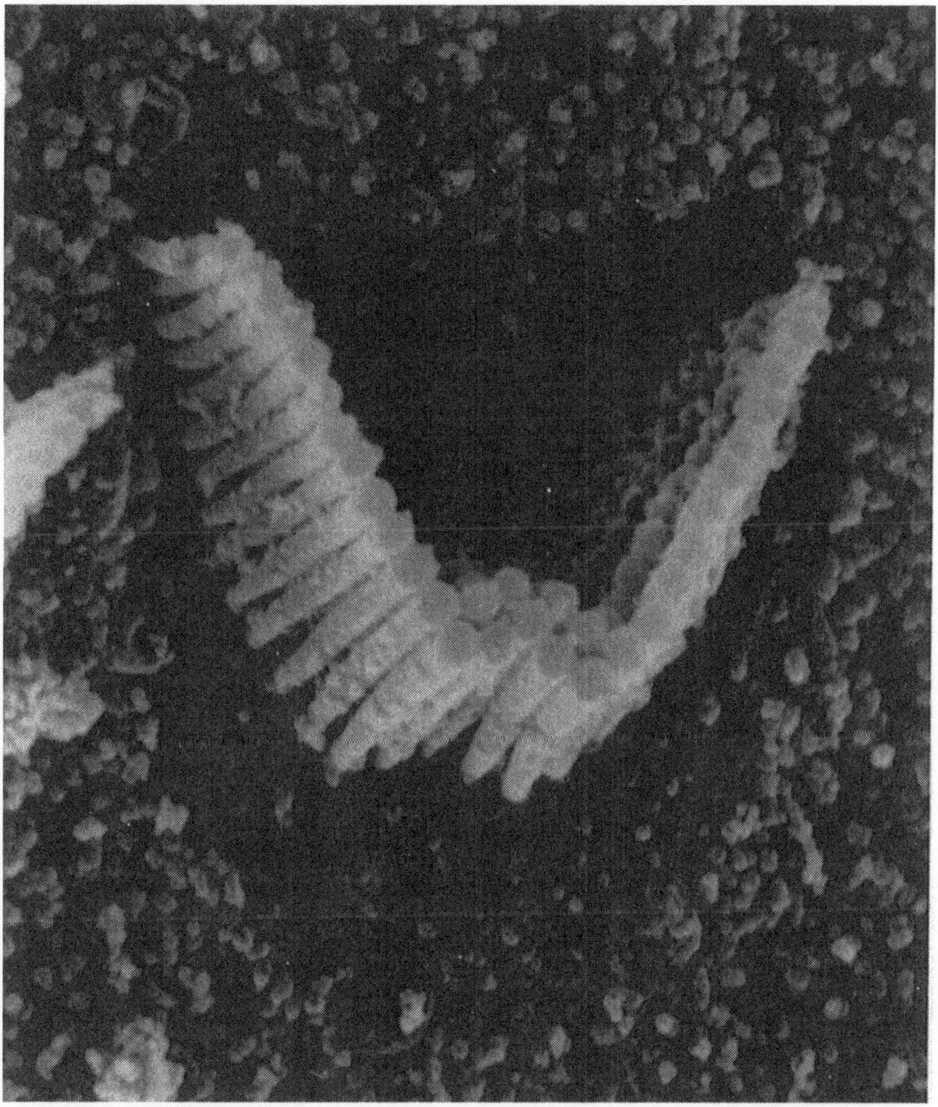

FIG. 3.7. Highly magnified scanning electronmicrograph showing the W-shaped tuft of hairs of a single outer hair cell. Preparation from a Rhesus monkey; display magnified approximately 22,000×. (Courtesy of Robert E. Preston of The University of Michigan.)

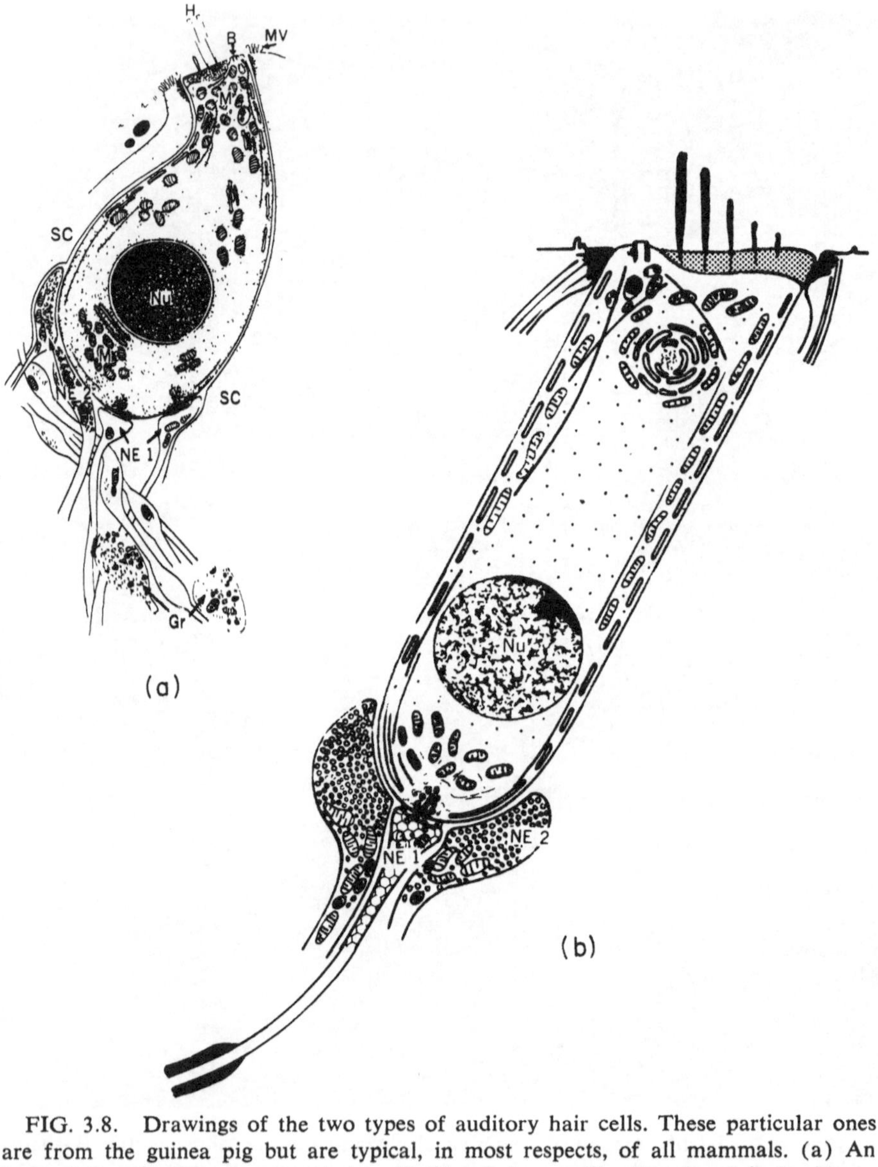

FIG. 3.8. Drawings of the two types of auditory hair cells. These particular ones are from the guinea pig but are typical, in most respects, of all mammals. (a) An inner hair cell. (b) an outer hair cell. H = hair or cilia; B = basal body; Nu = nucleus; SC = supporting cells; NE1 and NE2 = afferent and efferent synapses, respectively; MV = microvilli; Gr = a granulated structure of unknown function. (From Engström, Ades, & Hawkins, 1965.)

cell. Some authorities have suggested that these two forms of synaptic contact may represent afferent and efferent connections, respectively. Thus, the hair cell might be capable not only of sending signals to the second-order transmission neurons in the auditory nerve, but also of receiving signals from some efferent axons in that nerve. These efferent signals could conceivably modulate or control the sensitivity of the hair cell upon command from the central nervous system.

The outer hair cell, shown in Fig. 3.8b, has a somewhat more cylindrical form than the flask shape that is typical of the inner hair cell, but its features are otherwise much the same. Differences in function, if they exist between the two types of hair cell, are based on molecular organization too small to be seen even at this fine level of magnification, or on the topographical placement of the hair cell on the basilar membrane of the cochlea.

3. Vestibular Receptors

Not too surprisingly, considering its embryological history and its anatomic propinquity, receptor cells of the vestibular system are very similar to those of the auditory system. Figure 3.9 is an electromicrograph of the field of cilia of a group of vestibular hair cells embedded in the epithelial tissue of the frog's vestibular receptor. Unlike auditory hair cells, the cilia of vestibular hair cells are not all of equal size. There is one exceptionally large *kinocilium* with a bulbous end,. which appears to lie atop the ends of the other, slightly smaller *stereocilia.* Hillman and Lewis (1971), who took this extraordinary picture, believe that the kinocilium is physically connected to the stereocilia in a way that performs an important biological function. Because of the levering action of the connection between the bulb and the stereocilia, stimulus motions that are primarily directed tangentially to the top of the hair cell are converted into an up-and-down, plunger-like motion of the stereocilia. This redirection of the stimulus force greatly increases the physical energy available at the base plate, magnifying the overall sensitivity of the vestibular system to small accelerations.

There are two types of vestibular hair cells, defined by shape rather than differences in location or observed function. Figure 3.10a (HC I) shows a typical Type I vestibular hair cell displaying an unusual enveloping of the hair cell by a flask-like calyx formed from the end of the second-order neuron. As with the auditory hair cells, both afferent and efferent synapses appear to be present, but in this case the efferent synapse probably exerts its effects on the calyx rather than directly on the hair cell. Type II vestibular hair cells, an example of which is also shown in Fig. 3.10b (HC II), are similar to the typical auditory hair cell, with both afferent and efferent synapses terminating at the base of the cell itself.

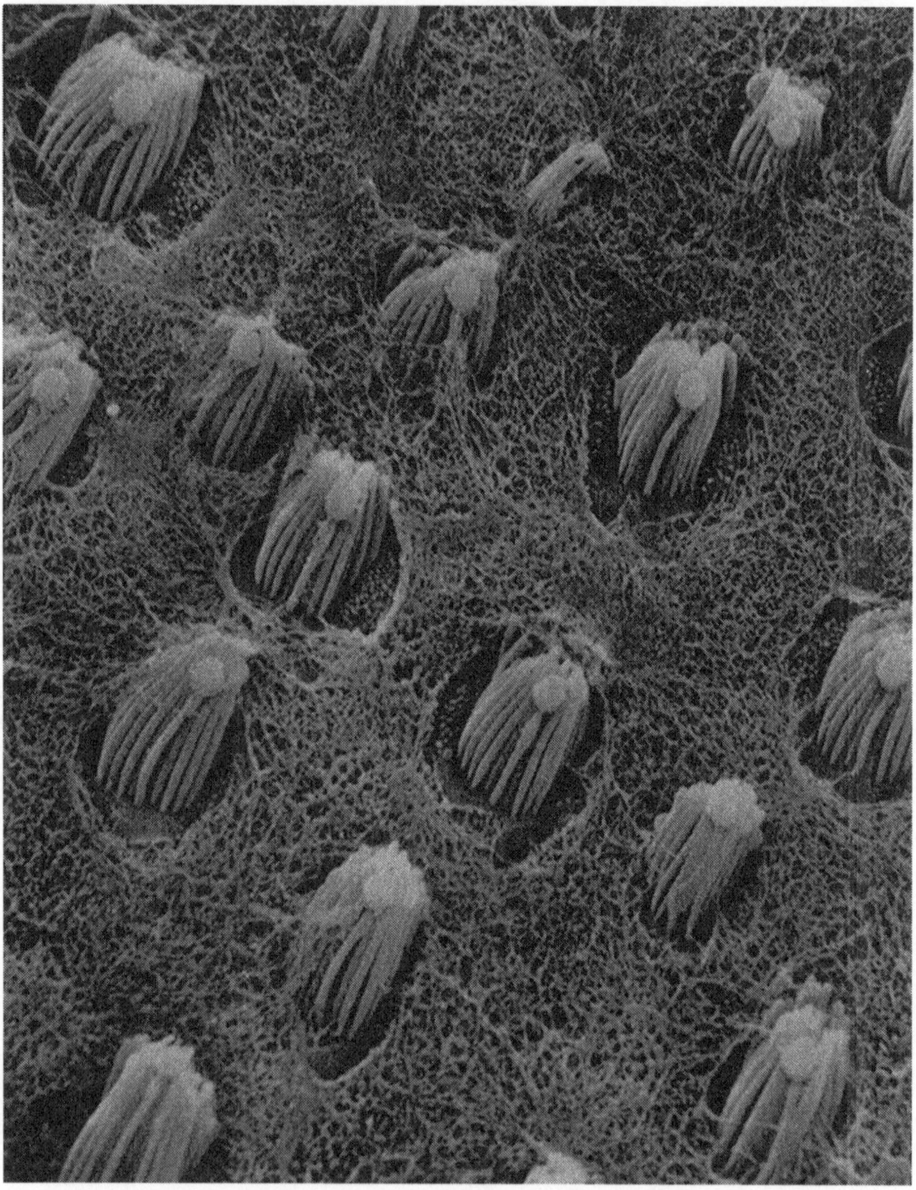

FIG. 3.9. Scanning electronmicrograph of a field of vestibular hair cells, showing the bulbous endings of the kinocilia. (From Hillman & Lewis, 1971.)

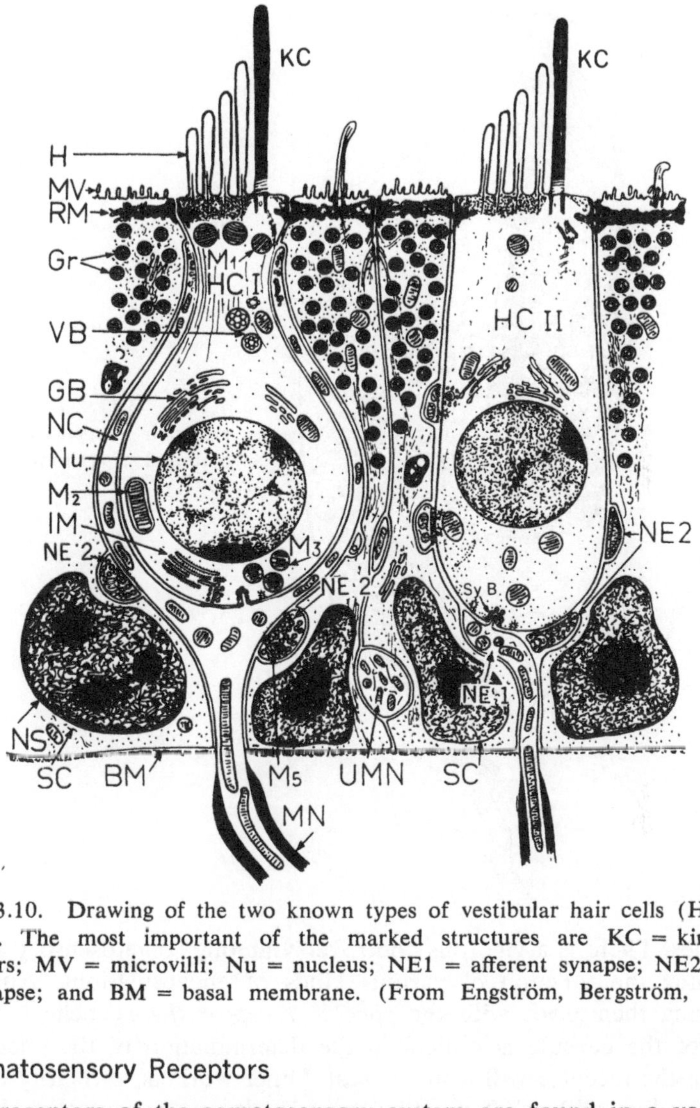

FIG. 3.10. Drawing of the two known types of vestibular hair cells (HC I and HC II). The most important of the marked structures are KC = kinocilium; H = hairs; MV = microvilli; Nu = nucleus; NE1 = afferent synapse; NE2 = efferent synapse; and BM = basal membrane. (From Engström, Bergström, & Ades, 1972.)

4. Somatosensory Receptors

The receptors of the somatosensory system are found in a variety of shapes and forms and are generally sensitive to both mechanical and thermal stimuli. Figure 3.11 is a drawing of a typical Pacinian corpuscle. This is one example of an *encapsulated* receptor structure, since the nerve ending is completely enclosed in an accessory structure—an onion-like, multilayered capsule. In the Pacinian corpuscle, this capsule does not seem to be involved in the electrochemical aspects of neural transduction, but rather to act as a preneural mechanical modifier of the stimulus (Loewenstein & Rathkamp, 1958), on the basis of its elastic and other mechanical

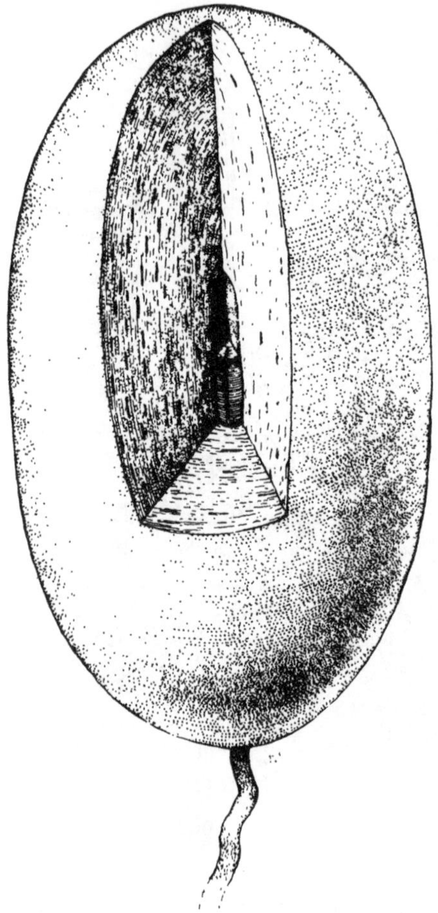

FIG. 3.11. Drawing of a Pacinian corpuscle from a mammal. (From Munger, 1971.)

properties. There is a vigorous controversy among somatosensory theorists concerning the function of various kinds of encapsulations. Authorities associating themselves with the *specific receptor theory* believe that the nature of the capsule is critical in the determination of the energy type to which the receptor will respond best. Other workers, primarily associating themselves with what is called *pattern theory,* believe that these encapsulations are of no significance. In fact, it has been suggested that "capsules" are nothing more than simple benign tumors (neuromas) created by the abnormal conditions existing at the end of the neuron.

The structure of some somatosensory receptors can be very complex. Figure 3.12 is a drawing of the Iggo corpuscle, a complex of several free neurons and several receptor cells that possesses an exceedingly fine sensitivity to mechanical stimuli. On the other hand, somatosensory receptor structures may be very simple; some are referred to simply as free nerve

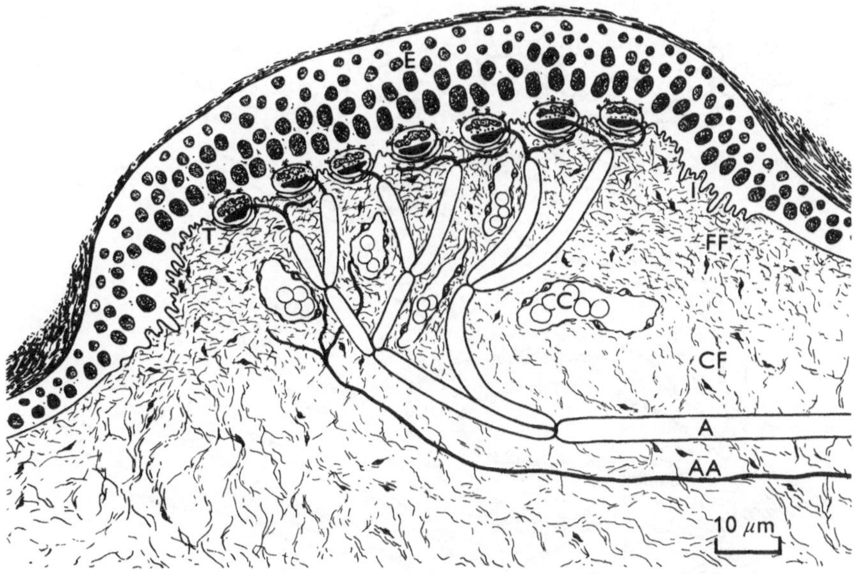

FIG. 3.12. Drawing an Iggo corpuscle from a mammal. Parts of the structure indicated are: E = epidermis; T = tactile cells; A = myelinated fibers; AA = unmyelinated fibers; C = capillary; CF and FF = coarse and fine bundles of collagen, respectively. (From Iggo & Muir, 1969.)

endings. In some parts of the body (for example, the cornea), only free endings are present. Figure 3.13 is a drawing of a group of free nerve endings that shows, however, that even these so-called "free" structures actually have accessory Schwann cell envelopes.

5. Chemoreceptors

The olfactory and gustatory organs have receptor neurons that are well-adapted for their role as sensors of chemical molecules. Figure 3.14 is a schematic drawing of some olfactory receptor cells in the mucosa of the nasal cavity. One of the most important characteristics of this cell is that, like the cutaneous receptors, there is no synapse at its base. It is the axon of the receptor cell itself that constitutes the primary transmission link back to the central nervous system. The gustatory receptor, although it too is a chemoreceptor, differs anatomically in this regard. Gustatory receptor cells do have a synapse at their base, as shown in Fig. 3.15, a diagrammatic cross-section of what is believed to be a typical *taste bud*. The taste bud, it should be noted, is not the receptor, but rather is a cluster of gustatory receptor cells and their precursors—undifferentiated epithelial cells.

Gustatory receptors are extraordinary in a very important regard. They seem to be constantly in a state of decay and regeneration: not regeneration

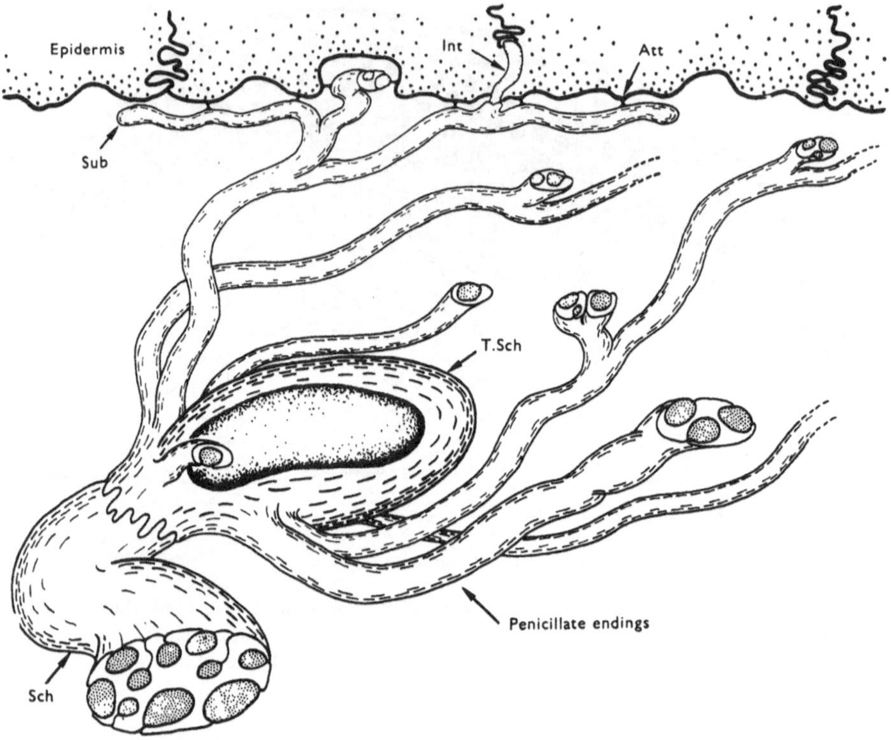

FIG. 3.13. Drawing of a group of "free penicillate" or tuft-like nerve endings in the human hair skin, showing the Schwann cell envelope around even these fine fibers. Sch = Schwann cell enclosing ten fine fibers; T.Sch = terminal Schwann cell. The free endings in this sample terminate either in the subepidermal zone of the corium (Sub), in the epidermis (Int), or may be loosely attached to the epidermis (Att). (From Cauna, 1973.)

of parts of the cell, like discs in the visual rod, but of the entire receptor cell. Radioautographic techniques (Conger & Wells, 1969; de Lorenzo, 1963) have shown that these cells die, decompose, and are replaced by new receptors that metamorphose from the supporting epithelial cells originally surrounding the receptors. The adaptive utility of the synapse at the base of the gustatory receptor cell becomes evident in this dynamic context. If the cell body of the receptor died and there were no synapse, the axon would degenerate too. However, with a synapse, the second-order fiber may simply migrate to the new cell, which replaced the dying receptor, and connect to it, thus maintaining the functional continuity of the receptor unit.

This, then, is a brief sampling of some of the most important anatomical features of some of the most important receptor cells. We now turn to a consideration of the anatomy of the transmission neurons, the class of

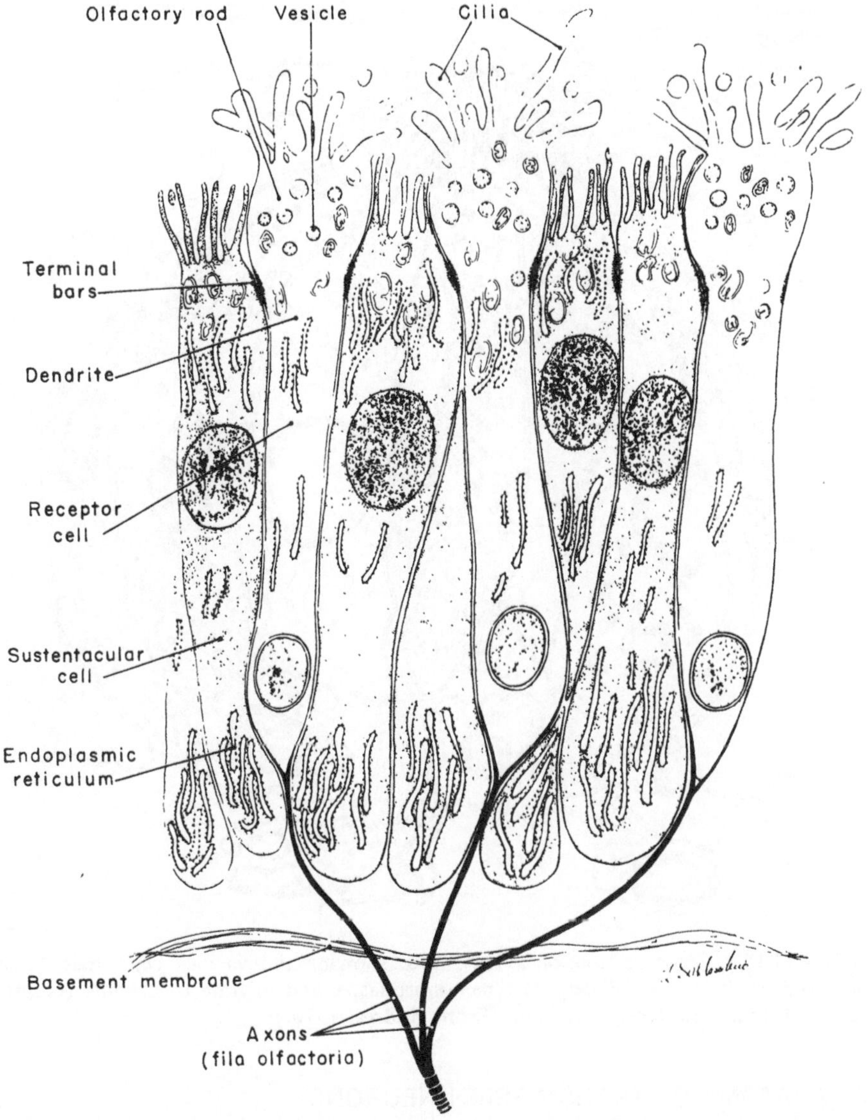

FIG. 3.14. Drawing of an array of receptors and supporting cells from the olfactory epithelium. (From de Lorenzo, 1963.)

cells that conveys information from these receptors toward the central nervous system, or from the central nervous system to the effectors. We shall also use this opportunity to introduce a number of the other cellular anatomic terms that are important in understanding this level of microanatomy.

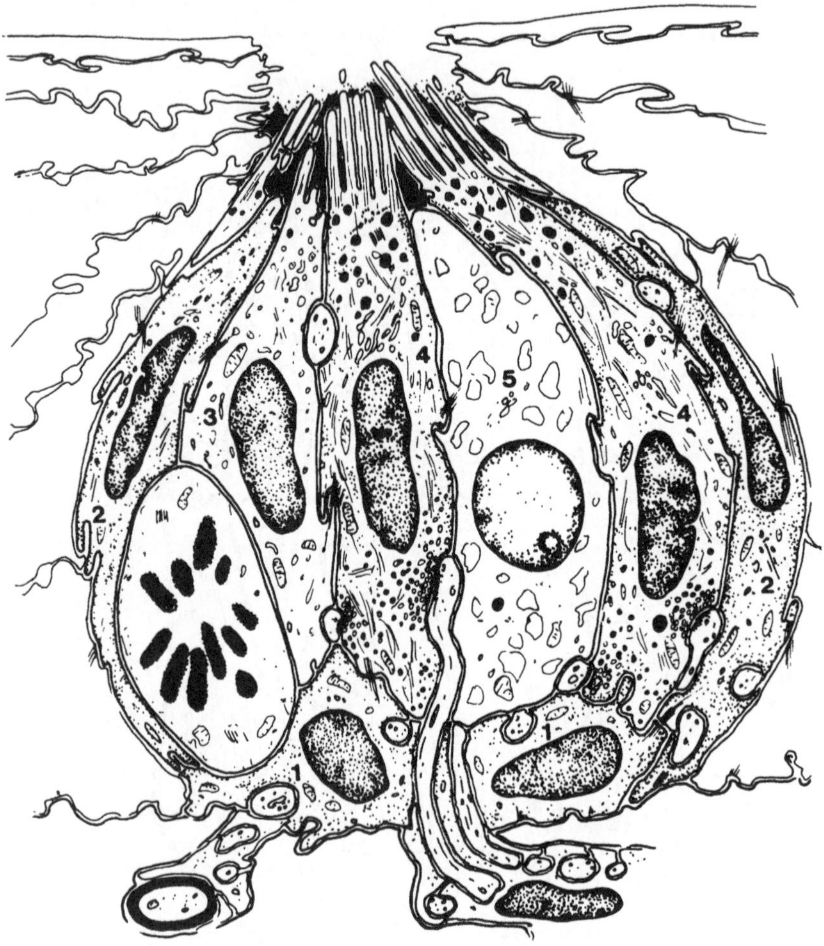

FIG. 3.15. Cross section of a taste bud, showing the various cell types (1–5) observed as the epithelial cells (2) metamorphasize and migrate to become receptor cells (4), and then degenerate (5). (From Andres, 1970.)

C. ANATOMY OF TRANSMISSION NEURONS

Transmission neurons have been specialized for the purpose of conveying information from one place to another in the body. To achieve this they have evolved into the longest cells in the body. A typical somatosensory neuron, for example, may convey information from a receptor in the skin of the foot to a synaptic contact in the spinal cord. Such a cell may be as long as three or four feet. It need not necessarily be of any great volume, however, since the diameter of that part of the cell that contributes most to its length, the axon, is only a few microns.

The first point to consider when defining the parts of transmission neurons is a functional one. Activation of such a neuron must be initiated at a certain point, and information about this activation must be conveyed to some other point if the neuron is to perform its information-transmission function adequately. This set of functional concepts—initiation at an origin, transmission to a distant destination, and utilization of the transmitted information—defines a directionality for the neuron's main action.

The concept of the functional directionality of a neuron is important to the following discussion because it is used in the definition of a number of key terms in neuronal anatomy. Transmission neurons are connected in neural nets so that they will conduct normally in only one direction.* Transmission in the usual direction is called *orthodromic* conduction, while activity moving in the opposite direction, which (though abnormal) can be produced by experimental procedures, is referred to as *antidromic* conduction.

It has generally been assumed that antidromic and orthodromic spike action potentials are equal in all regards, but recent evidence (Mulloney & Selverston, 1972) indicates that this may not be the case, at least in the stomatogastric ganglion of the spiny lobster. Not only did these workers discover that the shape of the spikes in a neuron differed, depending on whether they were spontaneously triggered orthodromic responses or the result of antidromic invasion of the cell, but they also observed that some of the normal synaptic effects of the orthodromic spikes did not occur when they were produced by antidromic stimulation of other neurons. The particular cell they were studying had previously been shown to inhibit the activity of four other cells when it was triggered by an orthodromic spike. However, antidromic elicitation of the response of this cell resulted in a spike that did not produce this inhibition.

This study is of considerable importance because, as Mulloney and Selverston note, it destroys the premise that antidromic and orthodromic responses are equivalent in all respects other than direction of travel. It thus raises the possibility that the absence of observed interaction between two neurons when one is conducting antidromically is equivocal. The lack of interaction may now be interpreted in either of two ways: in terms of the ineffectiveness of antidromic spikes, or completely to the contrary, as reflecting a true absence of interaction between the two cells. A classic and often-used laboratory test of interneuronal interaction is thus opened to question. Perhaps even more important, the nonequivalence of spikes traveling in different directions suggests the presence of some differential

* It should, however, be understood that this directionality is not so much a function of the cell as it is of the synapse. In many cases, neurons may be bidirectional, possessing both presynaptic and postsynaptic terminals within micra of each other.

membrane property that is not yet part of the generally accepted theoretical mode of neural action.

The notion of directionality has also been historically used to define the two major types of transmission fibers found on a single neuron. The region that carries information from a synapse toward the cell body (also called the soma or perikaryon) is referred to as the *dendrite.* The *axon,* on the other hand, was conventionally considered the output element of the neuron, carrying information away from the dendritic region. Axons are usually single fibers of greater diameter than even the largest of the dendrites. In some cells the axon may extend outward on both sides of the cell body; such cells are said to be *bipolar.* In addition, axons are often, unlike dendrites, covered with a fatty sheath of myelin. Although dendritic structures are not usually myelinated, in some cases even this counter-generalization fails to hold. For example, Pinching (1971) reports electronmicrographic evidence of the myelination of dendrites of the mitral cells of the olfactory bulb. Figure 3.16 shows the very fine dendritic myelination he observed at a magnification of 45,000×.

Bodian (1962, 1967) has added considerable precision to these rather vague definitions, while maintaining the importance of the criterion of directionality. For example, he defines the dendrite as that part of the neuron that is uniquely capable of receiving information from presynaptic regions. This definition, of course, is also general enough to include receptor

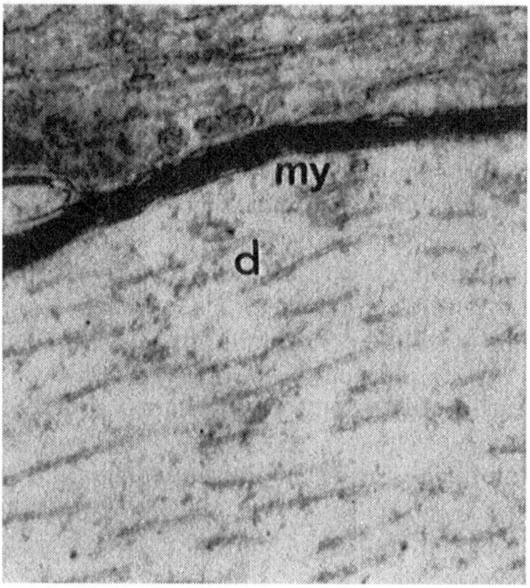

FIG. 3.16. Electronmicrograph of a dendrite (d) of a neuron from a monkey's olfactory bulb, showing the delicate myelination (my) at a magnification of 45,000×. (From Pinching, 1971.)

cells themselves, but the context in this case is that of a synaptic junction in contact with a preceding cell rather than an energy transducer in contact with the physical environment.

The best way to summarize these modern and precise definitions of the neuronal parts is to quote Bodian's definitions in full:

Dendritic zone: the receptor membrane of a neuron, either consisting of a set of tapering cytoplasmic extensions (dendrites) which receive synaptic endings of other neurons or differentiated to convert environmental stimuli into local-response-generating activity.* Mitochondrial concentrations may be present.

Axon: a single, often branched and usually elongated, cytoplasmic extension morphologically and perhaps uniquely differentiated to conduct nervous impulses away from the dendritic zone. It is characteristically uniform in caliber and ensheathed by neuroglial or neurilemma cells. Sheath differentiation and axon caliber are related to speed of conduction.

Perikaryon: the "cell body" or nucleated cytoplasmic mass, which is usually characterized by the presence of chromidial substance [i.e., chromosomes]. This portion of the neuron is the focal point of embryonic outgrowth of dendrites and axon, of axon regeneration, and perhaps of enzymatic synthesis in the differentiated neuron. It may be located in the dendritic zone or within the axon, or it may be attached to the axon.

Axon telodendria: the usually branched and variously differentiated terminals of axons which show membrane and cytoplasmic differentiation related to synaptic transmission or neurosecretory activity. Mitochondrial concentrations, "synaptic vesicles," or secretory granules are commonly present in bulb-like terminals. Like the dendritic zone, the telodendria do not transmit all-or-none conducted potentials. They transmit electrical or chemical signals capable of producing generator potentials in the dendritic zone of other neurons and in muscle, and stimulatory effects in innervated glandular cells or in distant cells via the humoral route (neurohormones) [Bodian, 1962, p. 326].

Figure 3.17 further clarifies these definitions by presenting a number of examples of the anatomy of various transmission neurons.

The perikaryon contains the nucleus that produces the enzymes that are responsible for the metabolic processes that nourish and cleanse the neuron of wastes. The nucleus also stores genetic information in the darkly-staining chromosomes. Because the nucleus is the sole source of the metabolic enzymes, if an axon or a dendrite is separated from the cell body accidentally

* Note: Some exceptions to this rule have been observed. Spikes have been detected in certain dendrites in the brain of the alligator, for example.

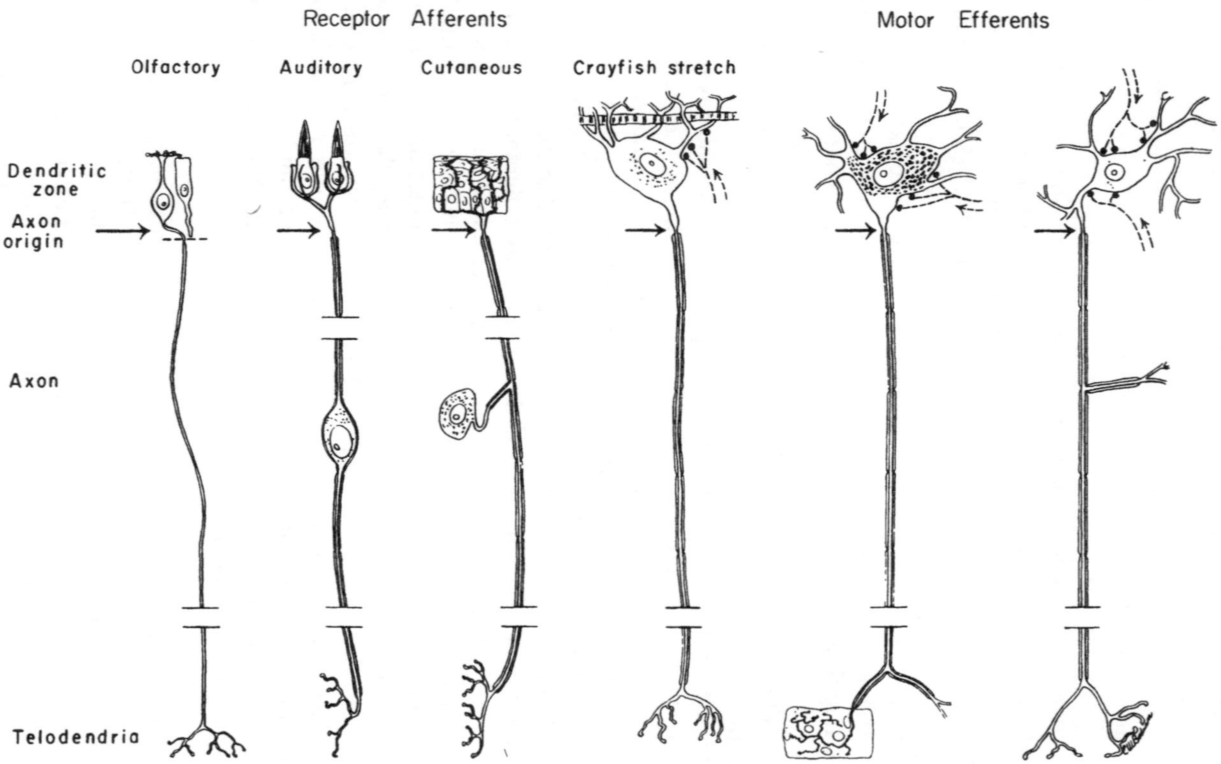

FIG. 3.17. Drawing depicting various forms of transmission neurons. (From Bodian, 1962.)

or by surgery, it will deteriorate. On the other hand, great damage can be done to peripheral parts of a neuron, and yet those segments that remain connected to the cell body and nucleus will still have a good chance of survival, even though some form of "retrograde" degeneration may occur. The cell body also is responsible for the generation of new intracellular fluid, the cytoplasm. As this material is created in the cell body, it streams into the axons and dendrites, maintaining internal pressure. The velocity of cytoplasmic streaming or *axoplasmic transport,* as it is often called, has been estimated to vary from 1 to 100 mm per day.

Closely associated with axoplasmic streaming are some newly discovered structures called *microtubules*. Microtubules are observable only under enormous electron microscope magnification, for although they may be as long as the axon itself, they are only about 250 angstrom units in diameter. Yet they do seem to be true tubes with an inner diameter of about 130 angstrom units. Some authorities have speculated that metabolites are transferred from the perikaryon to the terminals of the axon through these tubules. The microtubules are composed of a heavy protein with a molecular weight of about 55,000, called *tubulin.* The way in which tubulin molecules polymerize to form the tubules can be manipulated by varying the temperature and chemical composition of sample preparations. For example, if a tubule is forced to depolymerize by a rise in temperature to about 40° centigrade, when it repolymerizes it will form an abnormally large microtubule. A useful summary of recent work on microtubules may be found in Schmitt and Samson (1969).

Microtubules appear to play no direct role in the electrochemical action potentials of a neuron. It takes several hours after they have been destroyed before any change in the action of the neuron can be detected. Structurally, tubulin molecules appear to be relatively simple helical formations, but the best electronmicrographs show only their general shape. In neurons, microtubules appear to be randomly distributed. In other cells, however, they may be arranged in very orderly patterns as shown in Fig. 3.18.

The region of interaction between the dendrites of a cell and the axons of a cell preceding it is called a *synapse,* a term that has crept into the preceding discussion several times already. Synaptic action and anatomy are discussed in detail in Chapter 9. However, it is important to stress at this point that the synapse is the cellular region mainly responsible for the conduction directionality of the neuron. As noted above, axons can be stimulated so that they conduct either orthodromically or antidromically. The most common form of chemical synapse, however, can conduct in only one direction. Signals can go only from the presynaptic to the postsynaptic terminals. If the cell is a final motor transmission neuron, it does not terminate with the usual interneural synaptic junction, but rather with a special type of terminal referred to as a *neuromuscular junction.* This

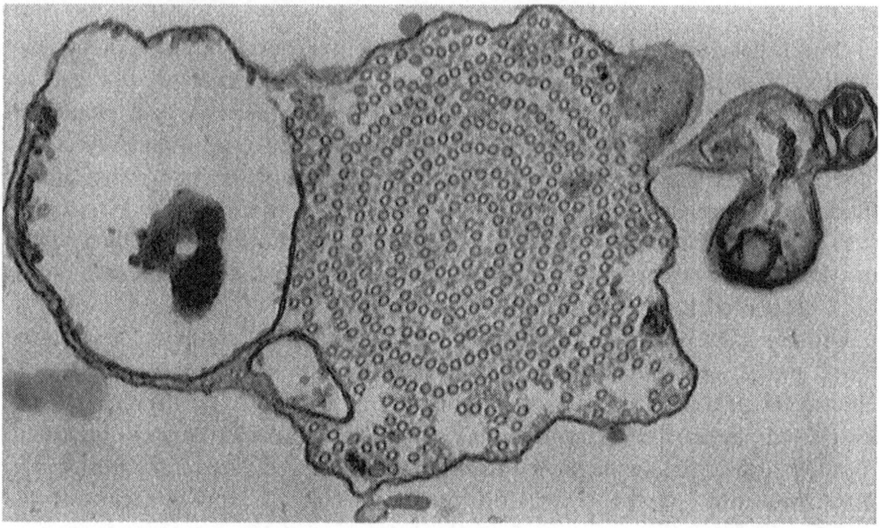

FIG. 3.18. Electronmicrograph of the highly regular arrangement of microtubules in a protozoan at a magnification of about 95,000×. Similar microtubules can be found in the neurons of more complex animals. (From Tilney & Porter, 1967.)

structure is similar to the usual "interneural" synapse (see Chapter 9), but it does have some specialized features that enable signals to be transmitted from the neuron to the motor cells of muscles.

In transmission neurons the large axon plays the central role, for it is the link that allows the transmitting cell to perform its main function—communication. A number of anatomic peculiarities of the axon deserve specific mention. When an appropriately prepared vertebrate neuron is observed microscopically, the most obvious feature is the coat that surrounds the axonal fiber. This coat, the *myelin sheath,* is a multilayered structure of fused lipid and proteinaceous layers, which serves a number of purposes. It insulates the cell from the electrical and chemical activity of surrounding cells, thus reducing "cross-talk" between cells. Since it is not continuous but rather is structured more like a chain of linked sausages, it allows a very rapid form of transmission termed *saltatory conduction.* The interruptions between the links of the myelin "sausages" are called the *nodes of Ranvier*. Internode distance may be about 1.5 mm in a typical vertebrate neuron. The nodes act as local regions of low resistance to which excitatory ionic currents can skip, rather than being propagated along the entire length of the axon. This sort of conduction by skipping greatly reduces the conduction time over a given length of axon.*

* Saltatory conduction was first described by Lillie (1925), and more fully developed as an important concept by Huxley and Stämpfli (1949) and by Tasaki and his colleagues (Tasaki, 1953).

The myelin surrounding peripheral axons is an extraordinary substance. It is formed by a sequential build-up of multiple and overlapping layers of the wall of specialized accessory cells, called *Schwann cells*. Schwann cells appear in close association to the neuron early in its embryological development. The Schwann cell has the dynamic capacity to encircle the axon by means of both physical growth and actual migration. Figure 3.19 shows this enveloping process. The Schwann cell is only loosely attached to the neuron at the beginning of development, but as it matures it rotates and grows in such a way that successive layers of its own cell membrane are deposited around the axon. Finally, the Schwann cell itself becomes little more than a nucleus hidden away among the folds of its own hypertrophied cell wall.

It is thought that Schwann cells may also perform a metabolic service for the axon. This notion, while not yet well established, is based on the idea that some parts of the axon are so far from their own cell body that they cannot adequately be serviced by the usual intracellular flow of metabolites. Intercellular perfusion of metabolites from the Schwann cell nucleus might be utilized by the axon for its own maintenance.

The layered appearance of the myelin sheath is shown especially clearly in the electronmicrograph of Fig. 3.20. Darker lines are due to contacts

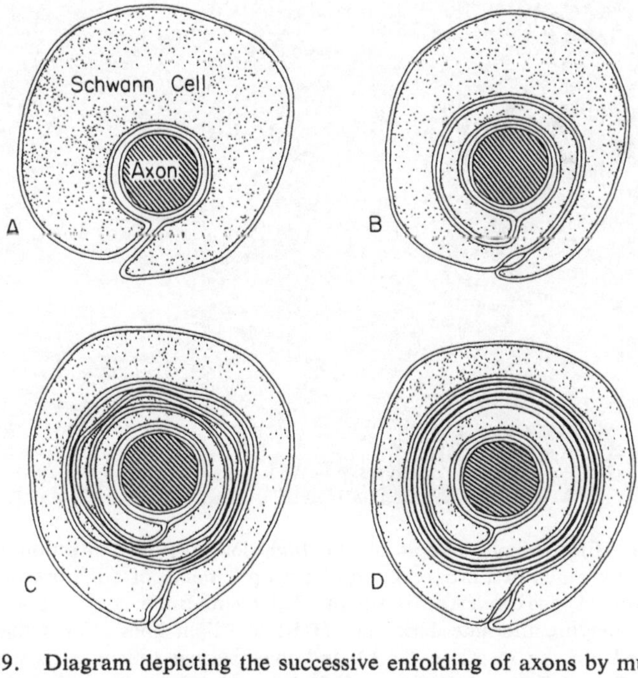

FIG. 3.19. Diagram depicting the successive enfolding of axons by multiple layers of the Schwann cell to build up a layer of myelination. (From Norton, 1972.)

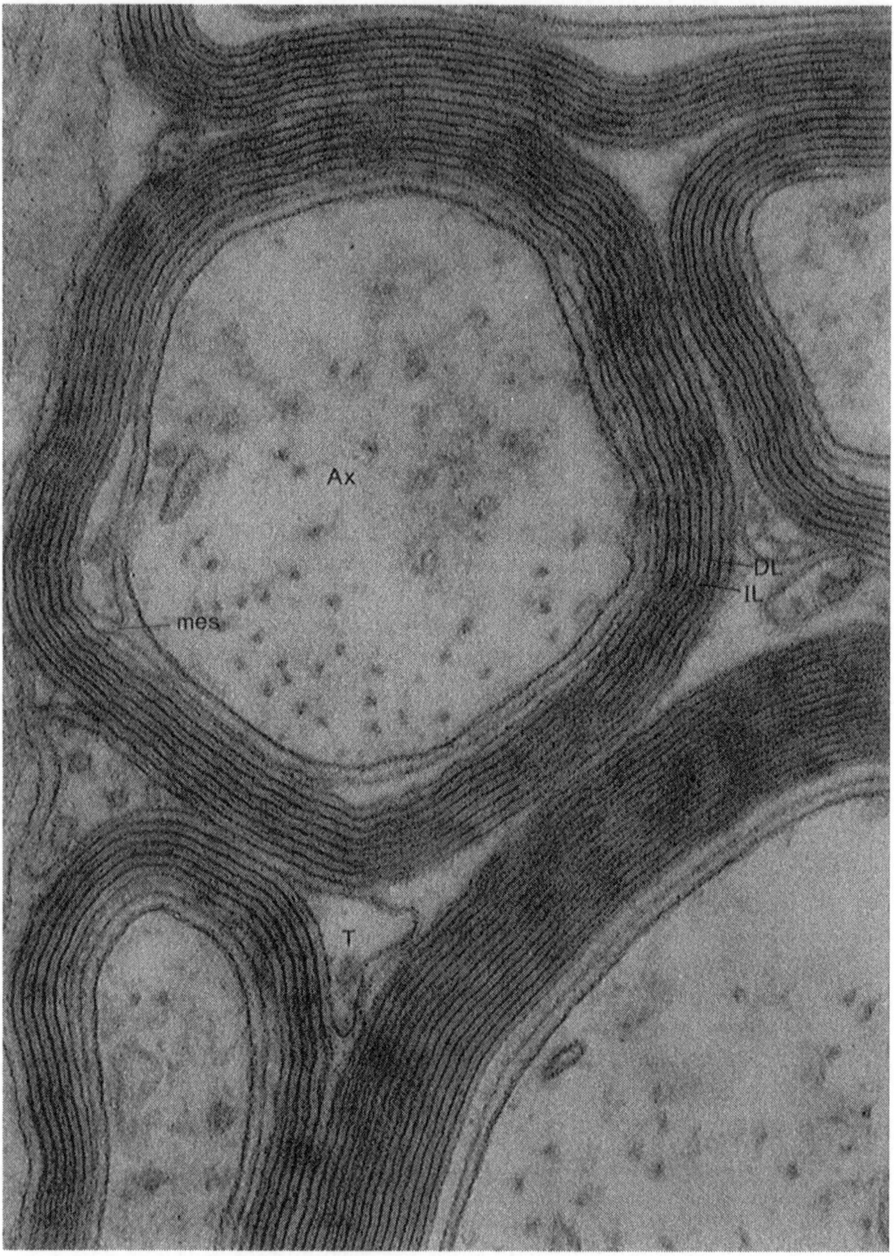

FIG. 3.20. Electronmicrograph of the high level of myelinization that can be observed, for example, in a cross section of the optic nerve of a mature rat. Magnified about 12,000×. The axon (Ax) is surrounded by multiple layers of a protein-lipid membrane displaying alternate dark lines (DL) and light lines (IL). Other characteristics indicated are the mesaxon (mes) and the external tongue of a myelin sheath (T). (From Peters, Palay, & Webster, 1970.)

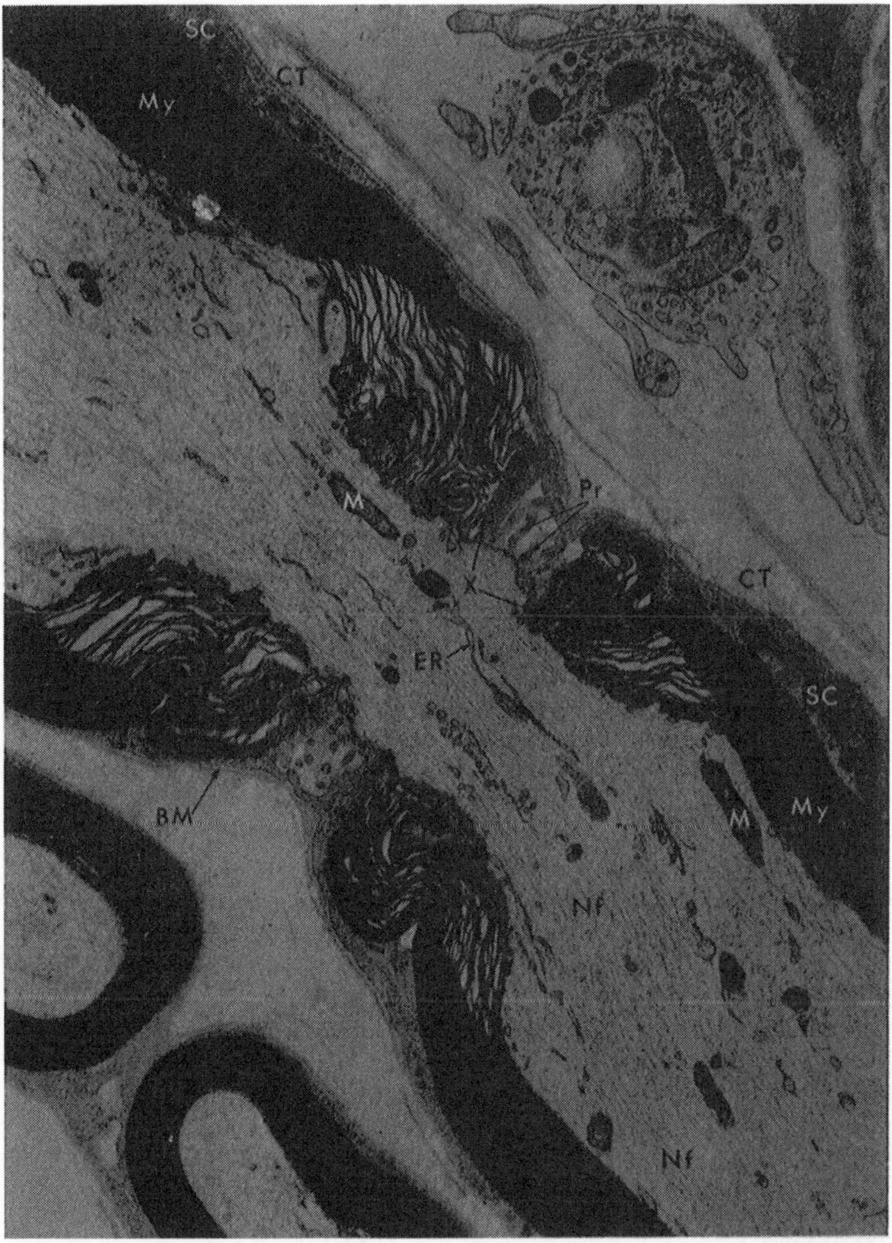

FIG. 3.21. Electronmicrograph of a longitudinal section of a node of Ranvier of a mouse neuron. Structures indicated include CT = connective tissue; Pr = parts of other Schwann cells; X = folds of myelin; ER = endoplasmic reticulum; Nf = neuronal filaments; and BM = basement membrane. (From Porter & Bonneville, 1968.)

between two folds of the outer layer of the Schwann cell wall, while the lighter regions indicate contacts between two inner surfaces of these cell walls. The difference in brightness is, of course, related to the density of the heavy metal staining of different tissues and the resulting differences in electron opacity, rather than any real color or optical opacity differences in the living tissue. Figure 3.21 shows an electronmicrograph of one of the periodic interruptions of the myelin at a node of Ranvier.

A classification system based primarily on axon diameter and degree of myelinization has been adopted by most neuroanatomists as the basic taxonomy of transmission axon structure. The largest fibers are found in the peripheral nervous system. These large fibers are always covered with a myelin sheath and have been designated *A fibers*. The *A* fiber category is further subdivided on the basis of different peaks in the compound evoked potential (the superimposed responses of many neurons—see Chapter 8), produced by an electrical stimulus. The subcategories are usually designated A_α, A_β, A_γ, and A_δ. In each case, the categorization refers to that class of *A* fibers believed responsible for a given peak in the compound responses, the different peaks being associated with different

TABLE 3.3
Characteristics of the Various Classes of Nerve Fibers[a]

	A	B	s.C	d.r.C
Fiber diameter, μ	1–22	≤ 3	0.3–1.3	0.4–1.2
Conduction speed, m/sec	5–120	3–15	0.7–2.3	0.6–2.0
Spike duration, msec	0.4–0.5	1.2	2.0	2.0
Absolutely refractory period, msec	0.4–1.0	1.2	2.0	2.0
Negative afterpotential amplitude, percent of spike	3–5	none	3–5	none
Duration, msec	12–20	—	50–80	—
Positive afterpotential amplitude, percent of spike	0.2	1.5–4.0	1.5	
Duration, msec	40–60	100 –300	300–1000	
Order of susceptibility to asphyxia	2	1	3	3
Velocity/diameter ratio	6	?	?	1.73 average

[a] From Patton (1965).

conduction times resulting from diameter differences. The A_α fibers are characteristically faster conducting than the A_β fibers, and so on.

B fibers are also myelinated, and are of medium size, but occur only in the autonomic system. *C* fibers are very small unmyelinated fibers. The *C* fiber category is further divided into two subclasses: the *C* fibers found in the sympathetic nervous pathways of the autonomic system, and those found in the peripheral somatic nerves. Table 3.3 summarizes the properties of these three major classes of nerve fibers.

D. ANATOMY OF INTEGRATING NEURONS

Neurons in different parts of the body have shapes that differ from the prototypical form of the receptor or transmission neurons described in the previous sections. In the central part of the nervous system many cells have evolved specialized structures for the integration of information from several local sources, rather than the transmission of information from one distant place to another. Integration is a task of neurons that are generally called *interneurons,* i.e., neurons that are neither receptor nor motor units. But this is not necessarily so; both receptors and motor units are also capable of integrative action in many situations.

Figure 3.22 is a collection of drawings made from micrographs of central nervous system integrating neurons. All of these neurons show the extreme arborization typical of this sort of cell. The dendrites are abundant, large, and multiple-branched, while the axons (not needed for long-distance transmission) may be so small as to be almost indistinguishable from any of the many large dendrites. Another characteristic of this kind of neuron is that synaptic contacts occur at many locations on the cell body or the dendrites, rather than only at the terminal portion of the dendrites. Some central neurons can be seen to be liberally sprinkled with dot-like synaptic *boutons,* the microscopic bulbs typical of an interneuron synaptic junction.

Myelin sheaths are often very reduced on many central neurons. It is for this reason that the brain seems to be made of two different colored materials. The nuclei of the cerebral cortex are composed of interneurons that have minimal amounts of myelin, and thus these centers are seen in their natural pinkish-gray coloring. However, the neurons of the great transmission pathways of the brain that carry information to and from the periphery or among the various nuclei of the brain have myelin-covered axons. In this case, the myelin's whitish cast dominates the color scheme.

The reader interested in a more detailed consideration of the anatomy of interneurons and the problems involved in the study of this special class of neurons is referred to Horridge's (1968) volume on the subject.

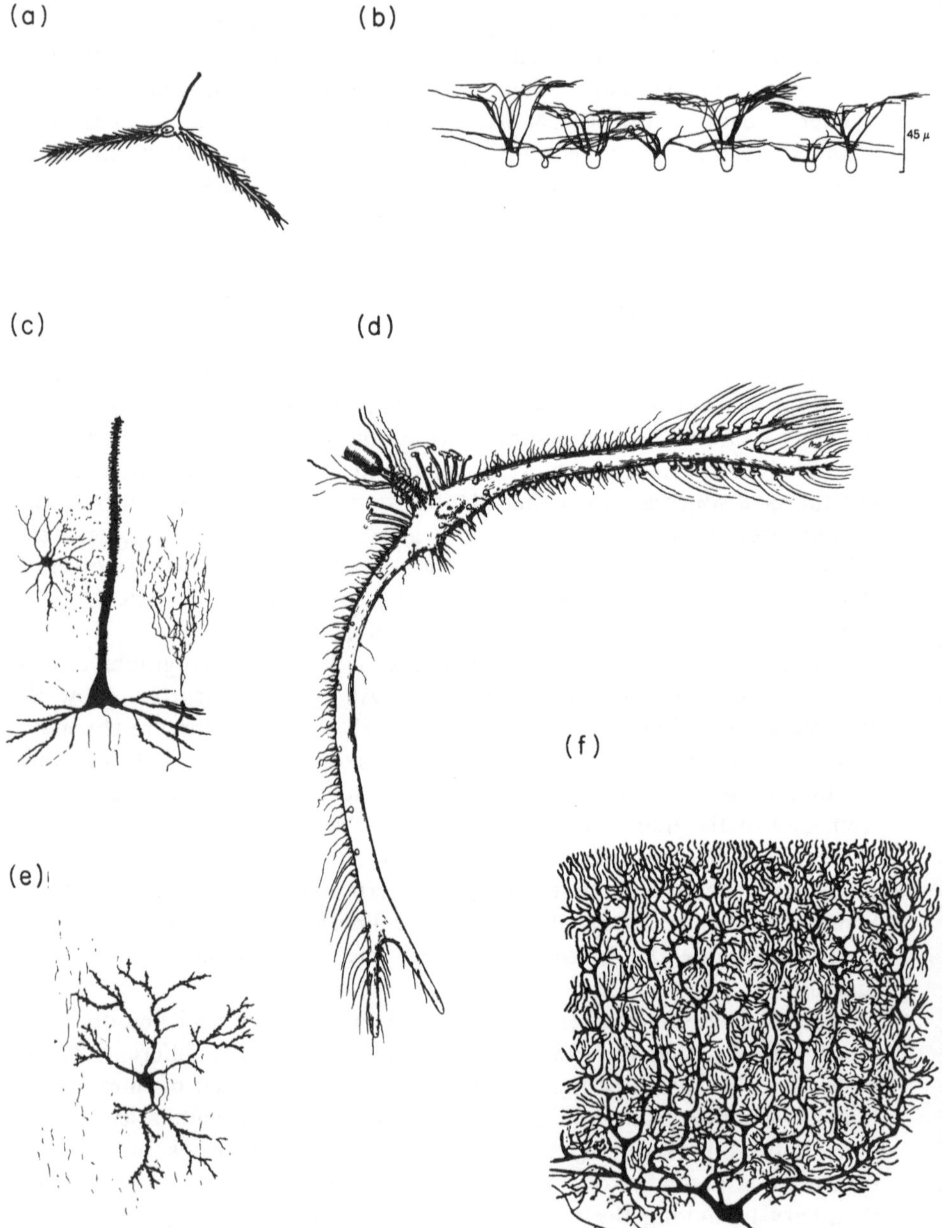

FIG. 3.22

E. ANATOMY OF THE PLASMA MEMBRANE

The final entry in this brief anatomical atlas will be a consideration of the structure of the cell wall, or plasma membrane, as it is more correctly known. A number of references have already been made to this extraordinary region of demarcation between the inside and the outside of the neuron. In later chapters it will be seen how the properties of the membrane are, in fact, the basis of almost all neural action.

At the level of microscopic analysis to which we now descend, it is not certain that we are in fact really dealing with structural entities of the same sort as the more macroscopic structures. The models to be described are moot, and the theories of molecular structure of just what it is that constitutes the membrane are far from common-sense explanations. Thus it is necessary to express an important caveat at this point. The membrane exists at the limit of resolution of even the most powerful electron microscopes available. Its dimensions are such that we have only a shadowy, indirect impression of what the membrane must actually be. While it is abundantly clear that there is certainly something different present at the interface between the cellular and extracellular fluids, it is to the realm of the indirect evidence of biochemistry, electrical properties, and ion transport that we must turn for details concerning the nature of the plasma membrane structure.

The last of the directly visible facts of the membrane is that it is a three-layered structure. Figure 3.23 is an electronmicrograph showing the tripartite structure, consisting of a central core and two outer layers. One of the outer layers is in contact with the internal cytoplasm of the cell, while the other is in contact with the extracellular fluid. The total thickness of the cell membrane is usually about 70 angstrom units. Biochemical assays performed on centrifuged components of cells, thought to be exclusively membranes, show that the membrane is composed primarily of a mixture of proteins and lipids. Bretscher (1973) describes how microchemical analysis of the protein components can be carried out with special detergents to enhance an otherwise weak tendency for the membrane components to separate when dissolved in water. These solutions can then be

FIG. 3.22. Drawings of a collection of neurons specialized for integration, rather than for transmission, to further illustrate the variability of neuronal anatomy, chosen from two important secondary sources (*The interneuron* by Horridge, 1968, and *The neurosciences: A study program,* edited by Quarton, Melnechuk, & Schmitt, 1967). (a) Bipolar cell from the median olivary nucleus (from Horridge, 1968); (b) retinal ganglion cells of the rat (from Brown, 1965); (c) cortical neurons from a kitten (from Scheibel & Scheibel, 1964); (d) Mauthner cell from a fish, showing many different kinds of synapses (from Bodian, 1952); (e) Golgi type II neuron from the kitten's cerebrum (from Scheibel & Scheibel, 1963); (f) Purkinje cell from the adult human cerebellum (from Purpura, 1967; after Ramón y Cajal, 1955).

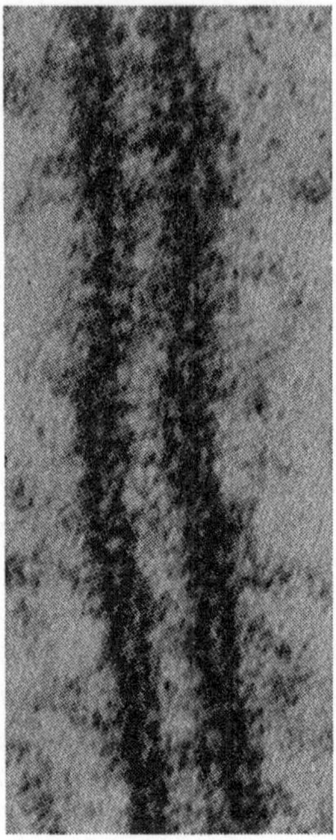

FIG. 3.23. Electronmicrograph of two neuronal cell, or plasma, membranes opposing each other across an intracellular cleft. Note the three-layered appearance of each cell membrane. (From Porter & Bonneville, 1968.)

placed on flat beds of polyacrylamide gels, on which the different proteins migrate at rates proportional to their molecular weights. The component proteins, thus, segregate themselves into a small number of subgroups with molecular weights of about 200,000, 100,000, and a range of lesser values, which can be viewed when the gels are stained. These values, however, hold only for the red blood cells with which Bretscher worked, and slightly different values would be obtained from different plasma membranes. Bretscher also pointed out that the lipid components are of three different kinds: phospholipids, neutral lipids, and glycolipids.

Until recently, the generally accepted view of the organization of these components into the membrane structure was based upon a description summarized by Davson and Danielli (1952) that assumes that the central core of the membrane is actually a double layer of lipid molecules. This central core was supposed to be covered by a monomolecular layer of protein on both the outer (extracellular) and the inner (intracellular) sides. The trilayered appearance produced in the electronmicrograph was assumed to be due to the selective staining of the protein with electron-opaque materials, to a density that was considerably different than that of the lipid.

In recent years the Davson and Danielli "theory" or "model" of the membrane has been challenged by a number of other researchers in the field of membrane structure. All of these alternative models face the problem of not only explaining the lipid-protein organization, but also of describing the pores or channels through which ions must pass. Some workers, most notably Singer and Nicholson (1972), have suggested that the membrane is only a liquid phase, and not an ordered molecular structure at all.

One of the newest models is that proposed by Bretscher (1973), who believes that recent data suggest quite a different organization than that proposed by the classic Davson and Danielli model. Bretscher believes that the main part of the membrane is actually a double layer of lipid, and that the portion of the bilayer adjacent to the interior of the cell is chemically different from that of the half adjacent to the extracellular fluid. He suggests that the membrane protein, rather than being uniformly distributed on either side of the lipid bilayer, consists of molecules that pass partly or entirely through the membrane, and that they indeed may be the ionic channels themselves.

The triple layering of the membrane, which appears in the electronmicrograph, is attributed by Bretscher to the fact that the lipid molecules are structured in an unusual way. The lipid molecule is assumed to be composed of a head and a double tail; the tails tend to be attracted to

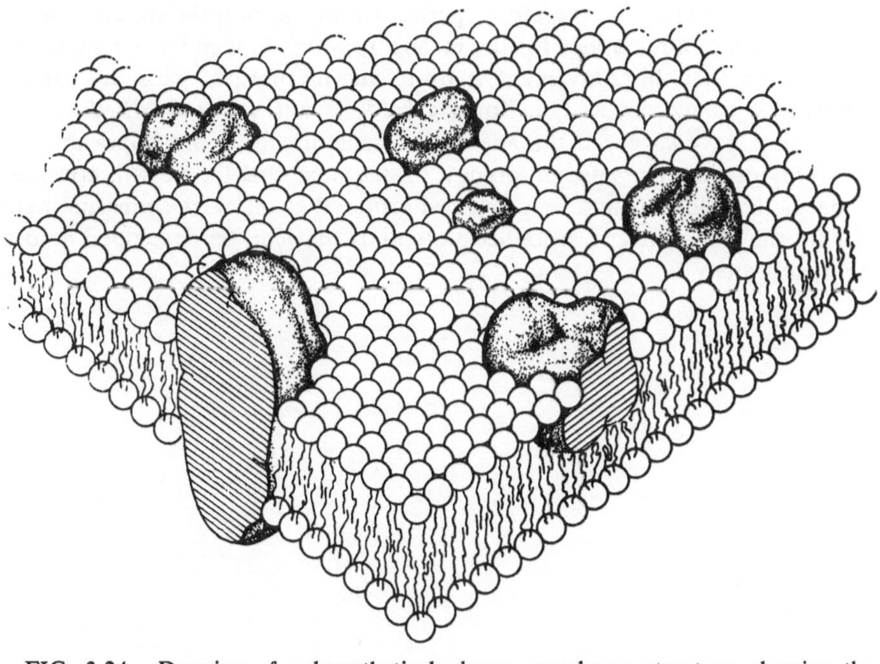

FIG. 3.24. Drawing of a hypothetical plasma membrane structure, showing the bilayer of lipid and the penetrating protein molecules. (From Singer & Nicholson, 1972.)

each other, and the heads tend to repel each other. The membrane is, therefore, formed when the lipids form a regular structure of two molecular layers that are arranged tail to tail (see Fig. 3.24). The outer electron-dense regions in the electronmicrograph are characteristic of the heads of these molecules, and the less dense regions in the middle represent the core composed of the tails. Many of the proposed features of the currently accepted model are summarized in Fig. 3.24.

It should be remembered, however, that there is no contemporary consensus on the exact molecular structure of the membrane, but only a set of competing models suggested on the basis of a not always consistent set of data.

The other main feature of the membrane to be explained is its selectivity to ions, a selectivity that varies from time to time. To achieve such a variable selectivity, the membrane must possess some transport mechanisms and/or pathways through which ions can pass under some conditions and not under others. This topic is discussed in greater detail in Chapter 5, but to preview an important point here, the "pore" or "channel" through which ions move might be better interpreted as some sort of chemical process, perhaps mediated by a penetrating protein molecule, rather than a real physical hole. A "pore" at the molecular level of the structure we are now considering is, therefore, perhaps more appropriately considered as something of a metaphor for the means by which ions move across membranes. Whether this movement is by means of physical gaps, by carrier substances, or by electrostatic forcing fields is not yet established.

We now have surveyed the anatomical structures that are important for our study of cellular neurophysiology. In the next chapter we will consider some methodological and instrumental issues that allow the measurement of both the transient and sustained potentials that can be developed across the membrane.

4
ELECTROPHYSIOLOGICAL TECHNIQUES

In the previous chapter we saw how electron and optical micrographs have vastly increased our knowledge of neuron structure. The present chapter will deal with some of the methodology that is helpful in examining the function of these same cells. The biology of the neuron (its low power levels and small size) requires that a specialized technology be used to detect and record any physiological response. As we saw in Chapter 2, because the neuron is generally quite small, microelectrodes must be used for detailed studies of membrane responses. The small size of the neuron also specifies other features of the recording apparatus. For example, in order to keep individual cells in contact with an electrode for extended periods of time, a mechanical framework is required that allows smooth control over the microelectrode position while holding the animal preparation in a rigid position. To locate and maintain a response from a single neuron, the microelectrode manipulator must be capable of controlled, repeatable movements as small as a micron. Furthermore, the low power levels of the neuronal response require that the electrochemical signals be amplified, just as their intrinsic rapidity demands that some means of storing a record of their occurrence for later, sustained examination must be present in the system.

The main purpose of this chapter is to review some of the modern developments in instrumentation and technique that meet these requirements. However, it is also useful to note that the equipment does more than simply meet certain biologically determined needs. It also constrains the sorts of observations that can be made and ultimately affects the theories that will emerge in neurophysiology, just as in any field of science. Perhaps this idea has been most eloquently summed up in the following words:

> In short, consciously or not, the decision to employ a particular piece of apparatus and to use it in a particular way carries an assumption that only certain sorts of circumstances will arise. There are instrumental as well as theoretical expectations, and they have often played a decisive role in scientific development [Kuhn, 1962, p. 59].

Kuhn might well have gone on to note that each decision about instrumentation also limits the experimenter in his sensitivity to the unexpected. Unless an instrument is able to probe some dimension of a response, there is no way that variations in the dimension can be observed or that they can be of any theoretical significance. This is precisely the reason that the introduction of a new measuring instrument almost always leads to the disruption of contemporary theory as totally novel observations are made. New dimensions, suddenly open to scrutiny, become extremely influential in theory building.

Specifically, it must be acknowledged that the technology introduced in this chapter limits and constrains what can be observed with regard to neurophysiology, just as it opens doors to new observations. In the enthusiasm and excitement of the initial introduction to these instruments, it is often forgotten that as fruitful as they have been, they have been profoundly effective in directing the attention of the neurophysiological research community to a relatively narrow range of experiments. Why has this been the case? It is mainly due to the fact that the working tip of the microelectrode is a point in space, and the cathode ray oscilloscope is mainly a plotter of time functions. Thus the response functions most easily detected are almost always functions of time at a particular point in space. This led to an enormous emphasis on the frequency patterns of single neurons, as one extreme example, and a relative neglect of other means of encoding information patterns. Only in recent years have such dimensions as the relative amounts of activity in spatially separated neurons come under investigation, as some researchers succeeded in simultaneously placing more than one microelectrode within a single preparation.

By expressing this caveat, it is not intended that the mass of data accumulated with the techniques to be described should be dismissed out of hand or their significance or impact minimized. They are all exceedingly important. However, it should be kept in mind that such constraints on our perspective may represent a source of distortion in the development of a global understanding of the biology of the nervous system.

Figure 4.1 is a block diagram of a typical electrophysiological recording system. The electrode shown in this drawing acts as an interface between the biological and electronic components of the system. It is mounted on

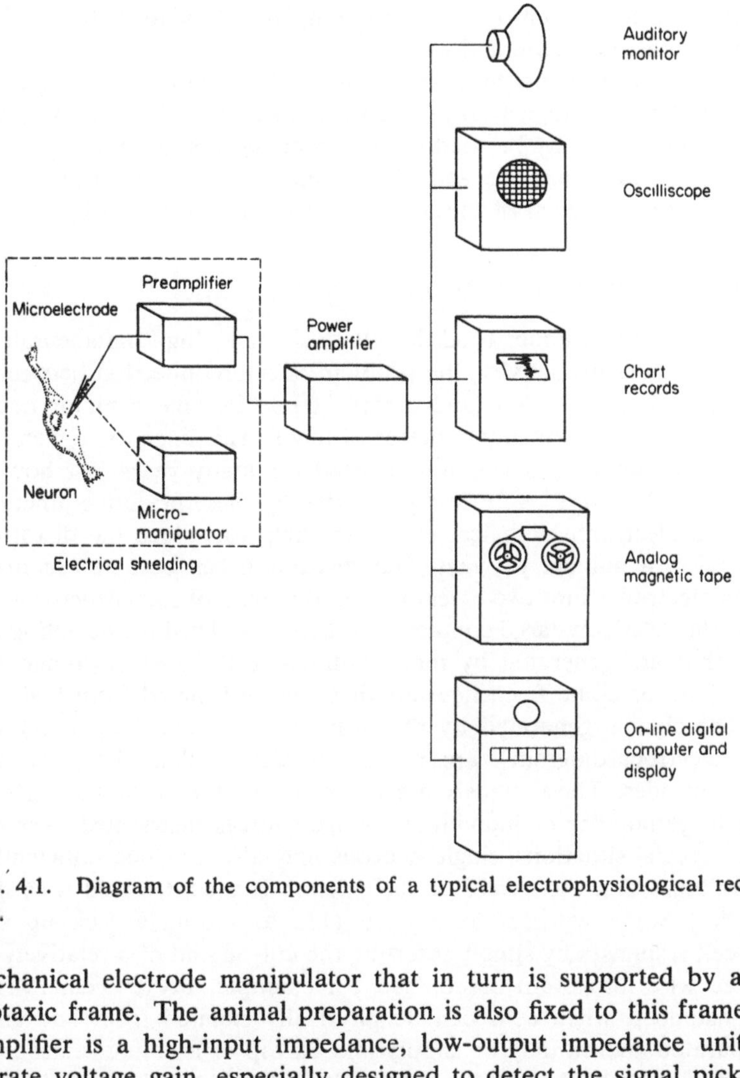

FIG. 4.1. Diagram of the components of a typical electrophysiological recording system.

a mechanical electrode manipulator that in turn is supported by a rigid stereotaxic frame. The animal preparation is also fixed to this frame. The preamplifier is a high-input impedance, low-output impedance unit with moderate voltage gain, especially designed to detect the signal picked up by the electrode, while draining only microscopic amounts of current from the neural elements. Even slight amounts of current drain could greatly distort the displayed signal. The main, or power, amplifier is responsible for production of the high power signals necessary to drive the display units, the devices that convert the detected temporal voltage patterns into readable formats.

In any neurophysiological recording system two or more of these functions may be combined into a single physical unit. Though this combination may result in a reduced number of separate physical boxes, at one place or another in the system each of the individual functions that have been mentioned must be accounted for.

A further stage of instrumentation that is becoming increasingly important in electrophysiological experiments is also shown in Fig. 4.1: an online computer that may be used both to produce simple or complex stimulus patterns and to analyze and display the results of an experiment. We shall now consider each of these parts of the system in detail.

A. ELECTROPHYSIOLOGICAL ELECTRODES

As soon as it was understood that there is something fundamentally electrical about nervous activity, the problem of interconnection between electrical measuring equipment and organic tissue became a major one. The earliest electrodes were simple pieces of metal, and since there were inadequate display devices, it was not realized for many years just how badly signals from the nerves were being distorted by these primitive junctions.

Modern electrophysiological electrodes help to reduce the distortion of the signal in a number of ways. Our discussion can best be organized by dividing electrodes into two categories on the basis of size. *Macroelectrodes* include the relatively large contacts that can be utilized for detecting neural signals that are generated by many different cells, are distributed over a wide region, or come from neurons that can be isolated from their neighbors. The signals generated by the brain (electroencephalogram) or the heart (electrocardiogram) are typical of signals that allow the use of macroelectrodes. These signals are the result of the combined activity of a very large number of individual voltage sources distributed over a wide area. In special situations, single neurons may also produce sufficiently isolated voltages that the details of their action can be detected with macroelectrodes. Some workers have been able to routinely pick up several single-cell responses by simply inserting the cut-off end of a relatively thick, insulated wire into brain tissue (see, for example, Ito & Olds, 1971). In other instances, when cells can be physically isolated from one another, the separated neuron may be simply laid on top of a macroelectrode. (See Hartline, Wagner, & Ratliff, 1956; Uttal & Kasprzak, 1962). But in general, to achieve sufficient signal amplitude and to observe the details of individual neuronal responses requires the use of *microelectrode* techniques.

1. Macroelectrodes

A macroelectrode can be as simple as a wad of cotton soaked in an appropriate salt solution, a metal hook, or a metallic button pasted to the skin or immersed in a conducting solution. In the latter case, the electrolytic solution has the function of reducing the resistance of the interface between the tissue and the electrode, so that most of the biologically generated voltage will be developed across the electrode, rather than wasted across the interface resistance.

The electrolytic solution also performs another very important function in electrodes that are exposed to moderately high current densities or prolonged immersion. Most metallic electrodes (especially metals like silver and zinc) *polarize* under conditions in which electrical currents are sustained in a single direction. Polarization is a phenomenon similar to the normal action of a battery, in that a constant electrochemical voltage comes to be maintained by the electrode itself on the basis of changes that occur on its surface. Polarized electrodes exhibit relatively high impedance to low frequencies and also have an intrinsic reverse potential opposing the biologically generated current. These distorting actions reflect a situation in which the electrode takes on significant electrical properties of its own rather than remaining inert. Polarization is also due in part to the electrolysis of the fluids in which the electrode is immersed. Gas bubbles formed from the electrolytic breakdown of the solution can gather on the electrode and contribute to the battery action, as well as increasing the interface resistance.

Polarization may also be due, in part, to the depletion of ions from the active region of the electrodes. When charge carriers are removed in any manner, the effect is to increase the internal resistance of the interface between the electrode and the specimen. For most applications in which electrodes are used for stimulation, polarization is not a significant problem, since the stimulating currents are so large. However, when tiny potentials are recorded from a nerve cell, polarization potentials can be equal to or greater than the signals themselves. The increased resistance, coupled with even the small capacitance of the electrode-preamplifier wiring, can reduce the overall frequency response of the electronic system by substantial amounts (i.e., the range of electrical frequencies that will pass through the amplifier is reduced).

Polarization is prevented by coating the electrode with one of its own salts or by isolating the metal from the fluid by a salt-solution bridge. The reader is referred to Silver's article in Donaldson (1958) for a more detailed review of the chemistry and prevention of polarization on electrodes.

2. Microelectrodes

The second type of electrophysiological electrode, the microelectrode, is designed to be small enough to penetrate the interior of the neuron without affecting its function. A microelectrode can, therefore, record the potential difference between the inside and the outside of a single cell, if a reference electrode is placed in the extracellular environment. Microelectrode tips can be drawn as small as $\frac{1}{10}$ of a micron. Eventually cells punctured even with these tiny electrodes do become less responsive, but a skilled electrophysiologist can often keep a microelectrode-penetrated cell alive for periods of up to several days. Apparently the initial puncture itself does less harm than is done by subsequent tissue movements around the electrode tip. These movements may enlarge the puncture wound to a size at which the cell's self-healing properties no longer work.

The earliest microelectrodes (see the discussion of Ling and Gerard's work in Chapter 2) were fine glass tubes, drawn by hand to submicron diameters after being heated to the temperature at which glass exhibits viscous flow. The fine glass tube, or *micropipette,* is not really the electrode, because glass is a good insulator. The micropipette must be filled with an electrolytic salt solution, which serves as the true electrical connection through the tiny hole at the end of the micropipette. Because of the very small diameter of the micropipette, capillary action exerts a substantial force that resists the flow of this electrolyte into the tube. It is usually necessary, therefore, to soak it for extensive periods in the electrolytic solution to fill the tube evenly. A more effective procedure involves three steps. The micropipette is first boiled in alcohol, then soaked in distilled water.

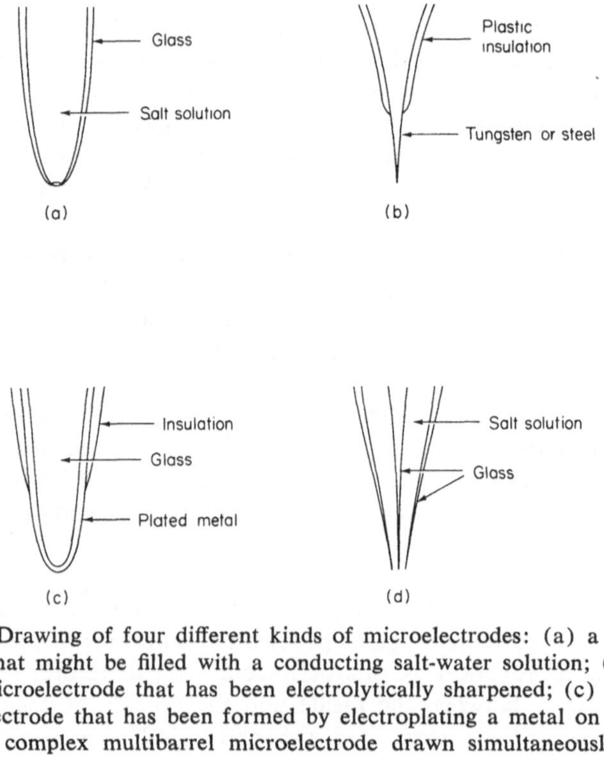

FIG. 4.2. Drawing of four different kinds of microelectrodes: (a) a simple glass micropipette that might be filled with a conducting salt-water solution; (b) an insulated metal microelectrode that has been electrolytically sharpened; (c) an insulated metal microelectrode that has been formed by electroplating a metal on a glass substrate; (d) a complex multibarrel microelectrode drawn simultaneously from two glass tubes.

The electrolyte is subsequently transferred into it by diffusion, by simply placing the micropipette in a jar filled with KCl or NaCl solution.

Though the electrolyte-filled glass microelectrode is still the best choice in many intracellular recording situations, metallic microelectrodes are increasingly used for extracellular recording. One way to construct a metallic electrode is to fill a glass micropipette with a low melting-point metal such as indium. An alternative method is to sharpen electrolytically a strand of relatively coarse wire by passing an electric current through it while it is suspended in an etching solution. By regulating the current rate it is possible to regulate the sharpness gradient of the point. Hubel (1957) and Spinelli, Bridgeman, and Owens (1970) discuss techniques for sharpening tungsten. Electrolytically polished electrodes of platinum and its alloys have also been successfully used.

New electrodes have been designed that are specific to the action of a single ion species. For example, Neild and Thomas (1973) describe a chloride-sensitive microelectrode constructed from a glass-insulated, chlorided silver wire, which can be used intracellularly. Thomas (1970) describes a similar, sodium-sensitive glass electrode specific to that ion.

Figure 4.2 shows several types of microelectrodes, including one of an ingenious multiple-barrel design.

B. STEREOTAXIC INSTRUMENTS AND MICROELECTRODE MANIPULATORS

A high degree of mechanical precision is required to place a microelectrode accurately and to maintain it in a stable relationship to the neuronal tissue under examination. Stability and control of this sort require a massive mechanical apparatus that is sufficiently rigid and sufficiently inertial to be free of unwanted mechanical motion. It is because of this necessity for size and mass of the accessory apparatus that only a very few experiments have been conducted in which several microelectrodes were operating simultaneously in the same preparation. At present, it appears that a dozen or so electrodes is the largest number that have been used at once in a single experiment.

The rigid framework used to hold both the animal and the microelectrode in position is referred to as a *stereotaxic instrument,* a defining term that is based on another function it serves. The location of some of the centers and nuclei in the brain or nervous system is relatively constant from one animal to another, and therefore brain atlases can be prepared that give the three-dimensional coordinates for structures referenced to some particularly prominent external landmark, such as the opening to the ears or a suture in the skull, for an animal of a particular species and size. (See, for example, Luparello, 1967, for the guinea pig; Jasper &

Ajmone-Marsan, 1960, for the cat; Karten & Hodos, 1967, for the pigeon; and Snider & Lee, 1961, for the Rhesus monkey.) Stereotaxic instruments are marked with graduated scales in three dimensions so that a microelectrode tip can be placed with high precision deep within the brain, without massive surgical damage to the animal. Precision micrometer screw drives are usually mounted on the stereotaxic instrument to provide the necessary degree of control and freedom from backlash. A photograph of a modern stereotaxic instrument is shown in Fig. 4.3.

Accuracy of placement with a stereotaxic instrument is good, but the variability between animals is such that post-mortem histological verification of the actual electrode location is usually desirable. The recording point can be marked with a heavy electrical current after the experiment, or with dyes injected electrophoretically through a micropipette.

The most precise control is required for the advance of the microelectrode itself. Mechanical micrometers still are the most suitable for this sort of high-precision movement, but often they are not coupled directly to the electrode holder. Rather, the actual connection is made through a miniature hydraulic system in order to reduce the transmission of movement from the manipulator to the electrode. Even very slight vibrations can alter the relationship between a neuron and a microelectrode so that a vigorously responding cell might be inadvertently lost by a gentle tap. Such waste can be minimized by this fluid coupling.

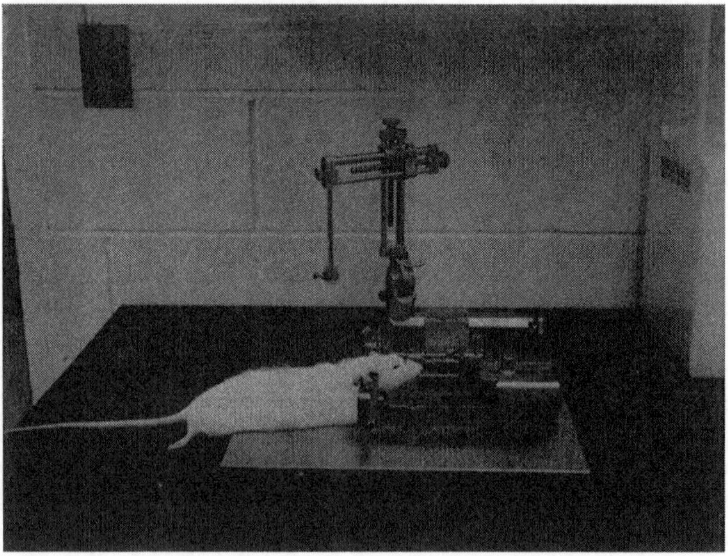

FIG. 4.3. A simple, but typical, stereotaxic instrument and microelectrode manipulator.

A new development has been the attachment of electrical motors to the micromanipulator. In this way the electrode may be moved automatically into or out of the nervous tissue in steps as small as 1 micron without the danger of a slip of the hand destroying an otherwise stable preparation.

In some instances it has been found useful to use micromanipulators that allow three-dimensional control of the tip location, rather than the single-dimensional thrust possible with the sort of instrument just described. Three-dimensional micromanipulators have the property of mechanical demagnification; that is, they convert relatively large motions to smaller ones by systems of gears and levers or hydraulic plumbing. While this type of micromanipulator is usually designed for microsurgery, it has proven useful in some electrophysiological recording situations.

C. PREAMPLIFIERS

The purpose of the electrode-amplifier-display chain is to reproduce and record the amplitude and time course of the neuroelectric signal under investigation, with as high fidelity as possible. But connecting electronic equipment to the delicate neuronal electrical source without distorting the waveform is not simple. In this section several possible sources of distortion are considered that could be produced by the preamplifier and the electrode-preamplifier interface. We will also consider ways in which this sort of distortion can be minimized.

Microelectrodes have considerable internal impedances. A microelectrode with a tip diameter of 1 micron may have an internal resistance on the order of 10 to 20 million ohms, and smaller electrodes may have internal resistances as high as 10^8 ohms. The internal resistance of the electrode and the input resistance of the preamplifier to which it is connected, along with the unavoidable stray capacitance of the wiring, therefore, form a resistance-capacitance voltage divider as shown in Fig. 4.4. Two problems are thereby created.

The first problem is one due to the internal resistive component of the electrode itself. The entire neuronal potential under study is effectively applied across the two ends (marked A and B) of the voltage divider shown in Fig. 4.4. But only a small portion of this voltage may be detected. This occurs because the part v of the total applied voltage V that is sensed across a resistor in such a series circuit is proportional to the ratio of the individual resistance value r and the total resistance R, a relation that may be expressed as

$$v = \frac{r}{R} V. \tag{4.1}$$

Therefore, if the input resistance of the preamplifier is smaller or even comparable in magnitude to that of the electrode, only a portion of the

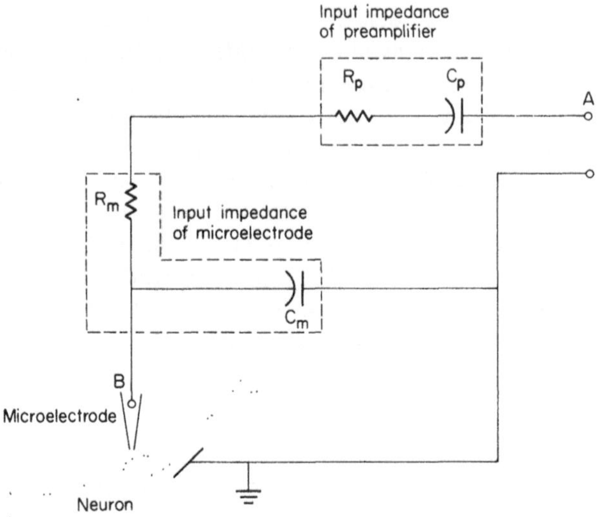

FIG. 4.4. Drawing of the electric circuit of a neuron, the recording electrode, and the recording preamplifier.

total neurally generated voltage will be "seen" by the amplifier. Since neural potentials are so small in comparison to the ambient electronic noise levels, this can be a serious handicap. In order to sense a large portion of the neural signal, the input impedance of the preamplifier should be very high compared to that of the electrode.

Preamplifiers with input impedances as large as thousands of millions of ohms (gigaohms) are now commercially available. Such units develop most of the signal voltage across their inputs, rather than across the electrode, even when they are attached to relatively high-resistance microelectrodes. High-input impedance preamplifiers were originally constructed with vacuum tube cathode-follower inputs or special high-input impedance electrometer tubes, but now that high impedance semiconductor devices have become available, all-solid-state, high-input impedance preamplifiers are commercially available from a large number of manufacturers. These solid-state preamplifiers often use field effect transistors (FET), a form of transistor that is wired very much like one of these special vacuum tubes and has the same feature of an unusually high input resistance.

Another advantage of high preamplifier input impedances is that the current drawn by the electrode-preamplifier circuit remains very small. If it were larger it could interfere with the normal metabolic processes of the cell and thus artifactually distort recorded data.

The second major problem associated with input impedance arises from the combination of the resistive and capacitive components of the electrode and preamplifier impedance, as also shown in Fig. 4.4. Even a very small input capacitance, when combined with the typically very high input resistance of a good preamplifier, can produce an electrical circuit with a very long resistance-capacitance (RC) time constant. The result of this long time constant is to reduce the high frequency sensitivity of the amplifying system. Therefore, the recorded biopotential will be seriously distorted by the inability of the preamplifier to follow the rapidly changing signal. Rise times will be artifactually elongated and very fast signals will not be passed at all by the electrode-preamplifier circuitry. Many preamplifiers, therefore, in addition to being designed for high input impedance, also contain special circuits whose purpose it is to reduce the time constant by minimizing input capacitance.

Older amplifiers accomplished the reduction in RC time constant by laboriously minimizing the actual input capacitance, either by placing the first stage of the amplifier very close to the tip, or by elaborate shielding. However, Amatniek (1958) has described an ingenious method for reducing the effective input capacitance of a preamplifier in an entirely different manner. He suggested that the input capacitance be neutralized by balancing it with an equal but opposite capacitive effect. In such a device, the input signal is simply inverted by an amplifier with a grain of 1. This inverted signal is applied to a second capacitor, which is adjusted by the experimenter to be equal to the input capacitance of the preamplifier. Thus there are two capacitors, equal in size and connected in parallel, but with opposite polarities of applied voltage. The net result is that the effect of the input capacitance is reduced to a vanishingly small value, even though the capacitance is still present. The advantages of neutralization of input capacitance are offset, in part, by an increase in the noise level of the output signal. However, this is not a serious impediment, considering the relatively large signals produced by high impedance intracellular electrodes of the sort requiring input capacity neutralization. Where extracellular electrodes may generate only 100 *micro*volts, intracellular signals may be as large as 100 *milli*volts.

In certain cases it is useful to design preamplifiers to have differential or bipolar inputs rather than a single-ended input. The need for the differential input arrangement is dictated by the fact that some biopotentials cannot be referenced to any external ground or common potential level, but exist only as voltage differences between two different parts of the tissue. A good differential input amplifier should also possess good *common mode rejection,* that is, the ability to reject signals that are common to both of the input electrodes. Satisfactory common mode rejection, however, requires balanced inputs and thus is not generally available with the idiosyncratic microelectrodes. If common mode rejection is good, a differential preamplifier will not pass any significant amount of the 60-Hz voltages, for example, that might be picked up from overhead lighting and other electrical equipment. If this "pick-up" is too great, metal screens or sheets can be used to shield specimen materials from most common sources of interference.

Preamplifiers usually do not have substantial voltage gain. They are primarily impedance matching devices coupling the high impedance of the electrode to the low input impedances of the main power amplifiers, in order to maximize power transfer from one to the other. They may thus produce an output signal that is only slightly larger in voltage magnitude than the original signal, but many times more powerful in current driving capacity. This low-level voltage capability means that many preamplifiers can be entirely battery operated, another feature that can contribute to the reduction of noise injected into the neurophysiological record by the electronic system.

Amplifiers in general can be separated into two different kinds on the basis of the sort of coupling that exists between successive stages. The first kind of amplifier is said to be *direct current* (DC) coupled from one amplifier stage to the next. With this sort of coupling even constant voltage signals can be amplified and passed through the system. Direct current coupling means simply that no series capacitors are used in the interstage circuitry, and constant voltages will thus pass as easily as changing ones. A number of problems are associated with the design of DC amplifiers, because they do allow DC and other slowly changing signals to pass. These slow signals may include such biopotentials as the resting potential of a cell membrane, potentials defined by the position of the eye, slow surface changes on the brain, and similar signals of long or continuous duration. These are the signals that require the use of DC amplifiers, but there may also be present spurious voltages with similar low-frequency characteristics that can distort or even obscure the desired signal.

Interfering voltages to which DC amplifiers are sensitive include amplifier base-line drift and electrode polarization effects, in addition to spurious electrophysiological signals. Amplifier base-line drift results from the amplification of tiny, gradual, long-term fluctuations in the values of component parts of the amplifier. In a DC coupled amplifier, a slight change in a resistor as a result of a temperature change can produce a minute voltage shift early in the train of amplifying stages that can be amplified at the output to a substantial voltage fluctuation. Such interference may completely obscure a low-frequency neural signal. Similarly, slight fluctuations in the characteristics of the electrode, such as the smallest amounts of polarization, can be amplified and result in large constant output signals.

Simpler, less noise-sensitive *alternating current* (AC) coupled amplifiers are usually used for the study of signals such as spike action potentials, electrocardiograms, and evoked potentials that are relatively brief transients, rather than for examining constant or long-duration voltage shifts. AC coupled amplifiers have capacitive circuits inserted between the several amplifying stages. These capacitors, by virtue of their high impedance to low-frequency signals, tend to block steady voltages and pass only signals that vary to some degree. By placing a bank of capacitors and resistors of various sizes on a ganged multiple position switch, it is possible to select the band-pass characteristics of the amplifier, since the resistance-capacitance combination determines what frequencies will be passed. AC coupled amplifiers thus have the advantage of allowing the investigator to select the optimum frequency range for the particular signals under investigation. It is possible to adjust the lower bound of the band-pass range to reject low-frequency shifts due to electrode shifts, for example, or the upper bound to reject signals due to high-frequency noise in the preamplifier. For 1-millisecond-long spike action potentials, a band pass from 300 to 2000 Hz would be desirable. For evoked brain potentials, a range of 3 to 300 Hz would be preferred.

D. POWER AMPLIFIERS

The purpose of the preamplifier is to replicate the tiny signal picked up by an electrode in a slightly more powerful form, while introducing as little additional electronic noise as possible. Low noise level is usually best achieved when only a modest amount of amplification occurs in the preamplifying first stage. However, the voltage and current capacities of the preamplifier are then usually insufficient to drive display and recording equipment, upon which the recorded signals are presented to the experimenter. Usually, therefore, another stage of power amplification is required to provide adequate input power levels to those displays. While in some modern equipment the actual power amplification components are distributed (some of it may be constructed into the preamplifier or into the display equipment), the function of power amplification must still be present some place in the overall system.

Power amplifiers usually give a substantial voltage boost, but probably more important, they provide sufficient current-driving capability to force even fairly massive indicators, like the stylus of a pen recorder, to new positions in short periods of time. This allows even fairly rapid varying signals to be traced. Emphasis has, therefore, been placed on development of high-power amplifiers with short time constants for rapid response. Depending on whether an oscilloscope or a pen recording display is used, the current requirements of the power amplifier may vary substantially.

E. DISPLAYS AND RECORDERS

Two types of displays are generally used in neurophysiological research. The first is the *oscilloscope,* a device designed to make graphic plots of rapid voltage functions. The oscilloscope is a combination of amplifiers and a *cathode ray tube* (CRT), an evacuated glass envelope with a source of electrons (the electron gun) at one end and a fluorescent screen at the other end that glows when hit by the electrons. Figure 4.5 shows the components of a typical CRT.

The electron source is a heated filament similar to that found in a radio vacuum tube. The heating action produced by a current passing through this filament boils off the outer orbital electrons of the atoms of the filament metal (usually tungsten). These electrons are then attracted by a positive potential on a secondary accelerating electrode, and/or repelled by a negative potential on the electrode source itself, toward the face of the CRT. The diffused bundle of electrons emitted by the filament is focused and collimated by placing an appropriate voltage on a special set of focusing electrodes and by masking the electron pathway with metal shields. The fluorescent brightness of the face of the oscilloscope is controlled by means of a grid electrode interposed in the electron beam pathway. The position of the electron beam on the face of the CRT is determined by magnetic or electrostatic deflection of the beam of electrons as they move from the electron source to the fluorescent material. Often both techniques, electrostatic and electromagnetic deflection, are used simultaneously in a single tube. When electrostatic deflection is employed alone, two sets of deflection plates are used—one for horizontal and one for vertical positioning. Similarly, electromagnetic control requires two sets of coils, but since the coils may be external to the CRT, a simpler construction technique can be applied within the evacuated tube itself.

A simple CRT is capable only of producing a spot of light on the fluorescent face at a place determined by voltages on the horizontal and vertical deflection plates. But obviously with such a point plotter it is possible to draw any figure desired by judiciously programming the sequence of signals to the deflection plates or coils.

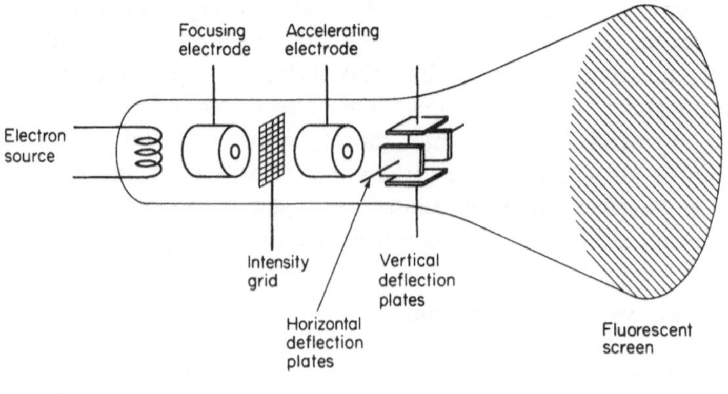

FIG. 4.5. Drawing of a cathode ray tube used in an oscilloscope.

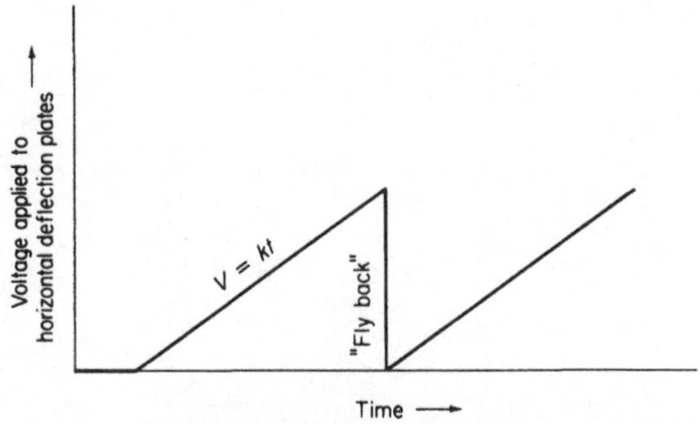

FIG. 4.6. Time plot of the voltage pattern that must be applied to the horizontal deflection plates of a cathode ray tube to allow the device to plot spatial equivalents of the pattern of a temporally varying electrical voltage.

For example, by applying a ramp voltage (the function $V = Kt$ where t = time as shown in Fig. 4.6) to the horizontal deflection system and applying the output of the bioelectrical power amplifier to the vertical deflection system, one can plot the variation in voltage of a bioelectric potential as a function of time. The ramp voltage, increasing linearly in time, displaces the beam of electrons across the CRT by a distance that is proportional to the elapsed time. Thus, the CRT plots a pattern in space in which the horizontal axis corresponds to time and the vertical axis corresponds to the voltage from the preparation.

Most modern oscilloscopes, like the one shown in Fig. 4.7, have the ability to hold off the beam of electrons and permit it to sweep out the horizontal time base only after some trigger signal (like the stimulus) releases it. Since such a system can also be used to limit the number of times the ramp voltage traces the function to a single sweep, this is a useful feature if one wants to photograph a single event for detailed examination. In other situations, repetitive sweeping helps to show the variability or constancy of the signal under investigation, but in all of these applications, it is important to trigger the release of the sweep at a constant, known point in time.

An important factor in the selection of a CRT for a particular application is the nature of the phosphor or fluorescent material on the viewing screen. Very short-persistence phosphors can be selected to minimize the amount of time a spot will glow after the bombarding electrons are removed. Short persistence is advantageous for those applications in which elongated persistence would cause interference between sequential displays. Long persistences are useful in retaining single events for prolonged observation. Sperling (1971a, 1971b, 1971c) has discussed many of these considerations in detail.

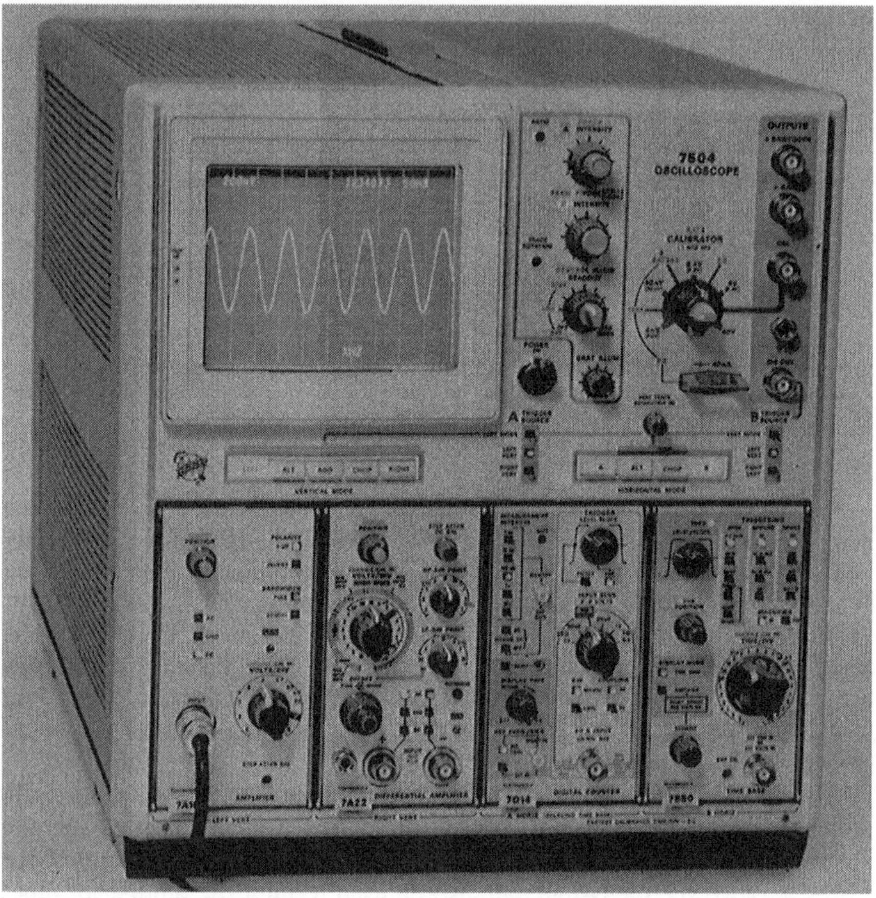

FIG. 4.7. A modern oscilloscope. (Photograph courtesy of the Tektronix Corporation, Beaverton, Oregon.)

Some "storage" oscilloscopes are specially designed to retain the trace of an applied signal indefinitely, until it is intentionally erased. This prolonged storage is accomplished by setting up a capacitive trace between the usual phosphor and a special additional grid electrode located immediately behind the phosphor, and then flooding the screen with a diffuse beam of electrons. Where the capacitive trace is present, a dot will be continuously brightened.

Another modern type of storage oscilloscope is essentially digital in nature. The analog signal is first converted into a digital form (see below) and then stored in some digital storage device, perhaps a small magnetic core memory. The signal is then retrieved and reconverted to analog signals for display on the CRT surface. Devices of this sort must be intentionally erased to rid them of stored information, since the digital record never passively decays.

The screen length of a typical research oscilloscope is no greater than 5 to 8 inches. To observe a long-lasting signal, the trace speed (as defined by the rate of rise of the ramp voltage) must be kept very slow. Under these conditions the individual signals cannot be resolved, since even complex transients would appear as merely a vertical line. As a compromise, a device known as a strip chart recorder may be used. Strip chart recorders do not have the same high-frequency capabilities as an all-electronic oscilloscope (the best optical ones do not pass signals with higher frequencies than several hundred Hz), but they do combine moderate resolving power with long record length. Strip chart recorders are usually galvanometric devices in which an applied signal mechanically moves a pen writer or light source a distance proportional to the magnitude of the input signal. In some other similar units, electric spark or thermal styli that can write only on specially sensitized papers are used.

F. COMPUTERIZED AUTOMATIC DISPLAYS

If computational equipment is inserted between the power amplifier and the oscilloscope or chart recorder, it is possible to include in the display an analyzed and summarized presentation of the experimental results, rather than a simple plot of raw data. This method is becoming increasingly popular in neurophysiological laboratories. For a full discussion of these techniques, the reader is referred to Uttal (1968, Chapter 8). The present brief discussion presents several data analyses and display methods that promise to be of particular significance to neurophysiologists.

One method is the application of simple statistical displays, such as the *amplitude, post-stimulus time,* or *interval histogram.* Figure 4.8a is a histogram displaying the distribution of amplitudes of individual neuronal responses, recorded from photosensitive neurons activated when the tail of a crayfish was stimulated with light. The amplitude histogram is a simple classification scheme in which the amplitude of a spike action potential specifies into which amplitude bin each signal will be sorted. In this case, 64 categories of amplitude were used, a value that happened to be dictated by the ability of the printer that was used to display these data.

Figure 4.8b is a reproduction of a *post-stimulus time* (PST) histogram, which shows the time distribution of intervals of the firing of a single cell in the auditory nerve after a click stimulus. In a PST histogram, data are categorized in terms of the time between the stimulus and the evoked responses. The time measurements themselves are thus classified into bins, the size of which depends upon the ability of the analysis method to

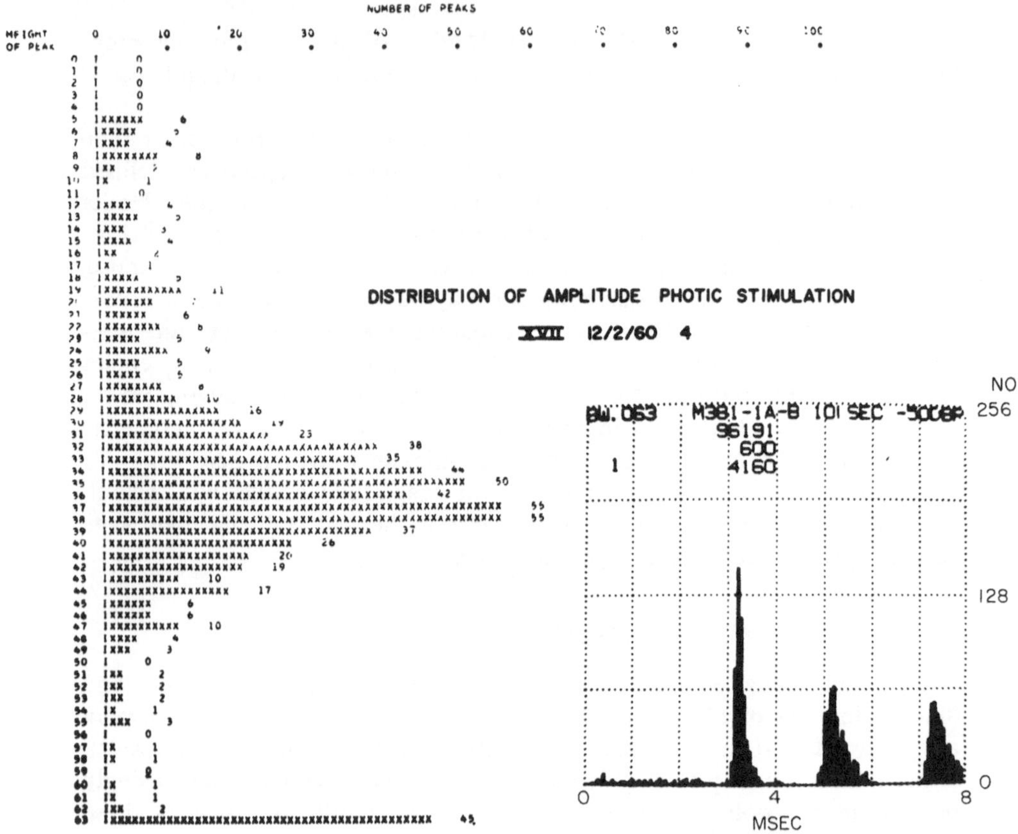

(a) (b)

FIG. 4.8. (a) Sample printout of an amplitude histogram obtained from neurons in the ventral nerve cord of the crayfish when the caudal photoreceptor is stimulated with visible light (from Uttal & Kasprzak, 1962); (b) Sample poststimulus time (PST) histogram from the auditory nerve of a cat (from Kiang, 1965).

partition time. Interval histograms plot the intervals between successive responses. For neurophysiological experiments, 1 millisecond is usually considered to be an adequate level of precision. However, in at least one experiment, measurable amounts of temporal interval irregularity were found to be on the order of tens of microseconds (Horvath, Halick, Peretz, & Miller, 1963).

In some experiments three-dimensional display of data is required to present adequately the details of a complex response pattern. One example is in the plotting of the receptive field of optic nerve or visual cortex neurons. Projections of three-dimensional data can be plotted on a two-dimensional oscilloscope quite easily if the oscilloscope is controlled by a computer. A sample of a projective plotting technique is shown in Fig. 4.9. In this case the quality of the display is enhanced by the suppression of hidden

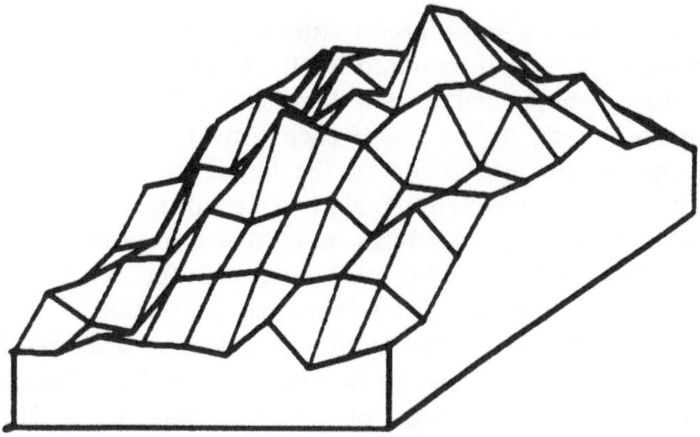

FIG. 4.9. Two-dimensional plot of a three-dimensional surface, representing the receptive field of a small group of adjacent neurons in the visual cortex of a rat. (From Gourlay, Uttal, & Powers, 1974.)

lines, a rather difficult programming task. Though this sample receptive field happened to be monophasic, both increases and decreases from the spontaneous rate could have been indicated. The shape and the extent as well as the valence of the receptive field can thus be made apparent in a graphic form. *Isometric displays* of this sort are a powerful means of displaying data, because they summarize in a single glance the results of many linear traces, representing multidimensional data, on the face of an oscilloscope.

In some types of bioelectric signal analyses, such as the amplitude histogram, it is necessary to measure exactly the amplitude or shape (time course of amplitude) of the signal. If data are subsequently to be analyzed by a digital computer, these analog (continuous and graded) amplitudes must be converted into digital (discrete and numerical) codes. The device that accomplishes this transformation is called an *analog-to-digital converter*. Several kinds of analog-to-digital converters have been developed, all working on the basis of a comparison between the signal and some voltage reference level.

In some of the earlier converters, a comparison was made between some input signal and a steadily increasing ramp voltage with a constant slope. During the time the comparison ramp voltage was rising, a counter tabulated the number of equally spaced pulses generated by an appropriately timed clock. When equality of amplitude was detected between the input signal and the increasing ramp voltage, the counter was turned off. If the ramp comparison voltage was linear and the clock intervals precise, the clock pulse counter would then contain a digital number that accurately represented the analog signal amplitude.

The difficulty with a ramp comparison type analog-to-digital system is the exceptionally long time required for the completion of the comparison. This is further complicated by another factor. In order to have equally spaced sample points, the period of each conversion has to be as long as the time required for conversion of the largest possible amplitude. This means that the rate at which signals can be coded is very low, and for signals that changed appreciably in amplitude during the comparison interval, a sizable error could be introduced into the calculation. In other words, the coding used in this scheme of conversion is uneconomical because each increment in the amplitude scale requires another unit of time. This may be considered to be classification by exhaustive inquiry, rather than systematic exclusion of possible values by a deductive test.

A satisfactory analog-to-digital conversion technique that does use a more logical approach is the method of *successive approximations*. This type of converter utilizes a much more economical code. Rather than counting each amplitude step, it essentially plays an electronic parody of the parlor game "Twenty Questions," classifying the signals into smaller and smaller groups with each successive "question." Thus only a few "questions" or successive approximations are required to identify a given amplitude, in contrast to the exhaustive testing of each possible amplitude that is characteristic of the ramp comparison procedure.

The sampling frequency capacity of an analog-to-digital converter is also limited by the time taken for a signal conversion. However, a complete successive approximation conversion may occur in as little as 1 microsecond. Nevertheless, a signal with a frequency high enough so that it varies at all during the conversion time will be inaccurately encoded. Auxiliary input devices, called *sample-and-hold circuits,* have therefore been designed for use with the analog-to-digital converter to help overcome this limitation. A sample-and-hold circuit samples the magnitude of an applied voltage at the beginning of the conversion and holds a constant record of that initial value until the conversion is complete. This is accomplished by means of a storage capacitor, the charge of which follows the input signal during the preconversion period, but which is electrically disconnected from an input signal as soon as the converter starts to perform its operation. When the capacitor is disconnected, it is not able to discharge through any other current pathway, and the initial signal level is maintained.

Often a single analog-to-digital converter may be called upon to process signals arriving on a number of separate lines from separate sources. Thus, it is important to be able to select one of these lines for conversion at any given time. The process of selecting and connecting only one of a number of possible signal lines is called *multiplexing* and can be accomplished either by relay contacts or by transistorized switches. An advantage of relays is that they have very low input impedances when the switch is closed and very large impedances when the switch is open. Thus there is very

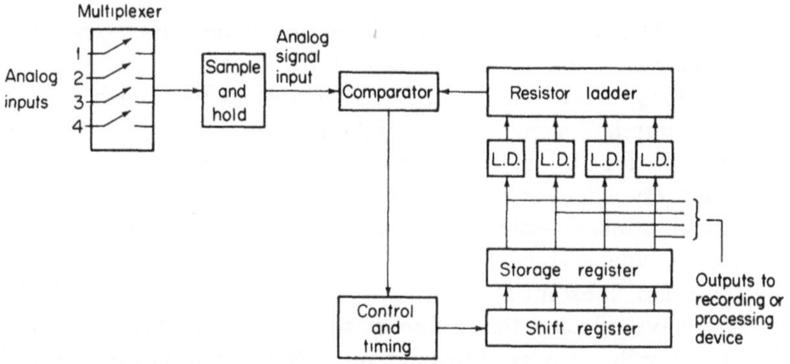

FIG. 4.10. Components of a complete analog-to-digital converter system. L.D. indicates a ladder driver. (Diagram from Uttal, 1968.)

little distorting electrical interaction between the several signal lines. Unfortunately, relays take a rather long time to switch among the alternative signals; a good relay multipexer may require 1 or 2 milliseconds to switch from one line to another. Higher speeds of operation can be achieved by solid-state multiplexers, which are capable of changing inputs in 1 or 2 microseconds.

Solid-state devices have certain limitations. A great deal of sophisticated electronics design is required to maintain the linearity and fidelity of a low-level neurophysiological signal being passed to the analog-to-digital converter through a solid-state switch. Similarly, there is a tendency for solid-state switches to cut off incompletely when open and to have a significant signal loss when closed. Adjacent switches can thus interfere with each other's operation.

Figure 4.10 depicts a complete analog-to-digital conversion system consisting of a multiplexer implemented with electromechanical switches connected to a sample-and-hold circuit, which in turn feeds the input to an analog-to-digital converter. A solid-state multiplexer is organized identically, except that the electromechanical switches are replaced by solid-state switches.

G. STIMULATORS

All the equipment that has been discussed so far is designed to detect, amplify, and display the tiny electrophysiological voltages generated by neurons. However, many experimental situations also require some sort of stimulating apparatus to elicit a neural response under controlled conditions. Over the years there has been a substantial change in the sort of stimulators used in neurophysiological experiments. In the early days almost all stimulators were impulsive types; that is, they were designed to produce the briefest possible stimulus. Each stimulus pulse could vary only in amplitude. The impulse was considered to be instantaneous, and little consideration was given to the pattern of the stimulus in either space or time. One distinguished scientist referred to this as the "flash of lightning and burst of thunder" school of stimulation. The stimuli used to elicit neural responses during this period were flashes of lights produced by stroboscopes or shutters, clicks of sound, mechanical taps to the skin, and impulsive electrical stimuli—that almost universal stimulus—applied to any place in the nervous system.

Over the years, substantial changes in the philosophy of neurophysiological research have affected the design of stimulators. Perhaps the most important single milestone occurred about 1960 when Hubel and Wiesel (1959) and Lettvin, Maturana, McCulloch, and Pitts (1959) discovered that stimuli with various features in time and/or space (for example, dots, lines, corners, and edges moving in particular directions at particular speeds), were much more effective than simple impulses in eliciting neural responses in the visual system. This was so even when the patterned stimuli were considerably less energetic than the impulsive ones.

As this trend toward more elaborately patterned stimuli continued, stimulator technology changed in some unusual ways. At first it seemed that there actually was a tendency toward simplification. The flashers and clickers of the past were replaced not by more elaborate stimulators, but by less complex ones. In their many distinguished studies (see, for example, Hubel & Wiesel, 1962, 1965), Hubel and Wiesel have almost always used a simple light source projecting an appropriately shaped image, which is simply moved about manually within the visual field of the animal. The work of Lettvin's group was done with an ingenious, though simple, stimulator that used magnets to move a stimulus about on an aluminum sphere placed in a frog's field of view.

However, in recent years a number of investigators have been forced by the increasing complexity of the experiments they wished to do to turn to more elaborate apparatus. Some, like Spinelli (1967) and Uttal (1969), have used computers to generate elaborate patterns of visual stimuli.

A full discussion of the many types of physical stimulators that have been used in electrophysiological research is beyond the scope of this book. The variety of photic, mechanical, acoustical, and chemical stimulation procedures is enormous. We cannot consider them in detail here, but the interested reader may want to look at another book by the author (Uttal, 1973, Chapter 2) for a more complete discussion of various means of stimulating neural activity. In this section only one kind of stimulator will be discussed—the electrical pulse stimulator. This one was chosen for two reasons.

First, electricity is the most general of all stimuli that can be used to activate any part of the nervous system. It is the one stimulus that allows the neurophysiologist to bypass the receptor and thus generate a known pattern of neural responses, rather than a pattern defined by the transductive mechanisms of the receptor. This is possible because electrical stimuli have the highly desirable property of producing one spike action potential for each electrical stimulus pulse, when applied to axons.

Second, the use of electrical stimuli complicates the use of the sensitive amplifying and recording equipment already discussed in this chapter. It is necessary, therefore, to understand how some of these complications can be overcome.

The major difficulty in simultaneously using electrical stimulation and high-gain preamplifiers in the same system results from the discrepancy between the amplitudes of the stimulus and the neuroelectric potentials. Stimulator voltage outputs may be as high as 100 *volts* before they can be effective in eliciting neural responses. However, the preamplifier is tuned to pick up signals that are at best measured in units of millivolts, and often only in microvolts. Thus the preamplifier is exposed during each stimulus pulse to a very much larger voltage than it can handle. Under these conditions most amplifiers overload and are momentarily blocked in their ability to pass any input low-level signals. Blockage of this sort is due to the fact that the stimulus voltage and the neuroelectric signal are closely coupled to each other electrically; that is, they are referenced to the same common potential. If the two potentials can be decoupled, or *isolated* from each other, this problem can be overcome. Potential isolation can be achieved in a number of ways, all of which involve some sort of electrical transmission from a grounded to a "floating" or ungrounded circuit. Simple transformers can be used as the simplest expedient, but transformers do not have good isolation or waveform preservation properties.

Two more recent developments have very much improved the degree of isolation that can be obtained. One innovation is a radio frequency system. This device, developed by Schmitt and Dubbert (1949), makes use of a very small radio transmitter that sends a signal across an air gap to an equally small tuned receiver. The net effect of the transmission of the signal across the gap between the two coils is to provide a voltage on the output that is free of any reference to ground except that leaking through parasitic capacitances.

An even greater degree of isolation can be achieved through the use of optical isolators. These systems substitute a voltage-controlled light source and a light-sensitive resistor for the transformer or radio frequency transmitter-receiver. When an input signal activates the unit, the light source, usually a photodiode, glows. The light then activates a photodetector, which changes resistance in a way that controls the magnitude of the current passing through the output circuit. The effect of this transmission, once again, is to isolate the stimulus from ground reference.

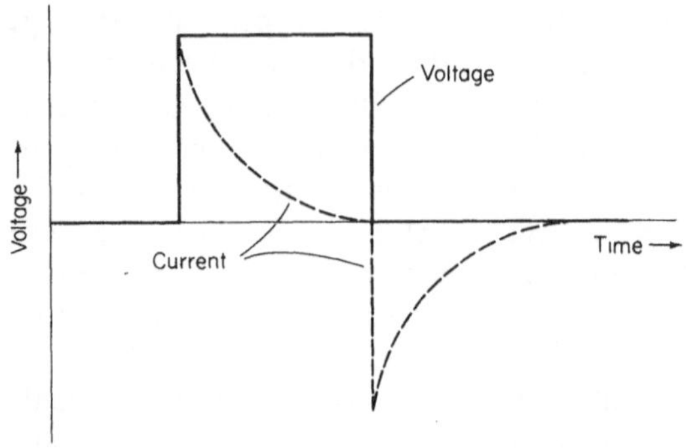

FIG. 4.11. Diagram showing the effect of applying a constant voltage current to a load (e.g., the skin) with substantial capacitive impedance. The generated current is essentially the derivative of the applied voltage. (From Uttal, 1973.)

The second major difficulty encountered with electrical stimuli revolves around the fact that the resistive and capacitive properties of the stimulated tissue can so distort the stimulus that a very constant voltage will produce a very non-constant current. This effect is mainly due to the charging of the capacitive components of the tissue in such a way that the applied voltage waveform is differentiated. Only the rise and fall (high-frequency components) of the waveform are passed when a current is so differentiated. The distortion can result in an actual stimulus current waveform like that shown in Fig. 4.11, even though the applied voltage stimulus is rectangularly shaped.

The manner in which this difficulty is overcome also depends upon special electronic circuitry. Circuits can be designed so that they produce constant current stimuli, rather than constant voltage outputs. In a constant current stimulator, the voltage applied to the tissue is not constant, but varies in the manner required to maintain a constant current through the tissue. Since most investigators believe that the applied current is a more satisfactory measure of stimulus amplitude than the voltage, this is a desirable feature. In any event, a constant current can be quantified more precisely than a variable voltage.

We now have considered a number of the important technical procedures and instruments that are important in cellular neurophysiological research. This equipment, as noted at the outset of this chapter, not only constrains and limits the universe of investigation, but also directly influences the theories that emerge to describe findings. Nevertheless, the fruitfulness of these instruments has been so great in the last 20 years that at least tentative answers can be given to some of the very important questions concerning the action of neurons. In the remainder of this book we shall consider the state of current neurophysiological data and theory.

5

THE CHEMICAL BASIS OF NEUROELECTRIC ACTIVITY: THE RESTING POTENTIAL AND A TAXONOMY OF ACTION POTENTIALS

The ultramicroscopic membrane that surrounds a neuron and demarcates it from neighboring cells does more than merely contain the cytoplasmic gel. Fluctuations in the permeability characteristics of the membrane are the source of all neuroelectric activity. The membrane accomplishes extraordinary functions because of its ability to pass some ions into or out of the neuron selectively, while barring the movement of others, and to individually vary the permeability to various ions over time. This variable selective and differential permeability to ions at the molecular level is thus the ultimate source of all responses at the behavioral level.

It must be emphasized that the following introduction to membrane processes is based on an admittedly incomplete model of ionic transport and equilibrium mechanisms. Although some of the basic notions are well established (for example, sodium and potassium ions are definitely known to be the primary ones involved in the ionic balances underlying both the resting and action membrane potentials), other important concepts (such as the detailed nature of the sodium-potassium pumping mechanism) are not yet fully understood. The material discussed in this chapter is presented, therefore, within the context of a developing theory of nerve membrane action that is acceptable, fruitful, and useful, but does not yet fully explain all the involved molecular and ionic processes. The reader should bear in mind that the "facts" presented and the model itself are highly susceptible to revision or even outright rejection as new information becomes available.

The following discussion is intended simply to introduce some basic concepts and a vocabulary that can be used in the more detailed discussion of action potentials in the chapters that follow. It is not intended to be complete. The reader interested in a more complete discussion is referred to Cole (1968), Aidley (1971), or Tosteson (1969), or for a highly mathematical treatment, to Plonsey (1969).

Electrochemical potentials exist across a membrane when the concentrations of charged ions differ on either side of the membrane. However, if the membrane is permeable to any of the involved ions there will be forces acting to alter those concentration differences. Some of these forces are passive, in the sense that they draw upon the potential energy of the concentration difference itself to reduce that difference. On the other hand, active forces are also at work across the membrane. By an active force, we mean a process that draws energy from the metabolic processes of the cell in a way that may actually increase the concentration difference and thus the resulting electric potential across the membrane. Let us consider these two categories—passive forces and active forces—in the following paragraphs.

A. PASSIVE FORCES

1. Diffusion

If two solutions, each containing a different concentration of a solute, are separated from each other by a membrane that is at least partially permeable to that solute, then molecules of the solute can pass by diffusion from one solution to the other. Random Brownian motions of the molecules of the solute totally account for this process and no metabolic energy is expended. The degree of diffusion is exactly that predicted by simple statistical-mechanical principles, i.e., on the simple statistical basis of random thermal motion, concentration differences tend to average out. In the initial state, when high original concentration differences are present, a low degree of entropy (disorder) is said to exist; that is, concentrations are very nonrandom (there is more solute on one side than on the other). If the membrane is able to pass molecules of the solute, the random processes lead to higher degree of entropy (greater disorder) of the molecules (the solute is considered to be more randomly distributed and in a state of greater disorder when the concentrations are equal). Since the ideal membrane we are discussing is assumed not to act in a selective direction (ions flow equally well in both directions), this increase in entropy occurs simply because more molecules would be expected randomly to cross the membrane from the high concentration side than from the low concentration side. Passive diffusion resulting from purely statistical distributions of molecular motion, therefore, can only result in equalization of the solute concentrations in the two solutions.

2. Osmosis

Assume that a similar semipermeable membrane partitions off two different concentrations of a common chemical solute in a common solvent. Assume also that the membrane is impermeable to the larger solute molecules, but will pass at least some of the smaller solvent molecules. Under these conditions the solvent will migrate from the side of low solute concentration (high solvent concentration) to the side of high solute concentration (low solvent concentration) in a way that also tends to equalize the two concentrations. Osmosis may therefore be considered to be a force that works primarily in the direction that reduces the solute concentration difference of the two solutions by transport of the solvent. Osmosis, however, is probably not simply diffusion of the solvent. The energy involved in osmotic pressure is usually greater than that predictable from simple diffusion forces, in spite of the fact that Brownian motion is presumed to be the driving force for both processes.

3. Potential Driving

If the solute is ionized, as it always is in cases involving dissolved salts, then, as we noted earlier, an electrical potential difference between the two sides of a semipermeable membrane can also exist simply due to unequal net charge arising from concentration differences across the membrane. For example, if there are more positively charged cations on one side of a membrane than on the other, the side with fewer has a net negative potential *relative* to the other, both because of the ratio of cations and the ratio of anions left behind.* This potential difference will tend to move some of the positively charged ions from the side where they are in excess to the other side of the membrane. Ion movements due to potential differences of this sort are said to be running *downhill,* since they reduce large concentrations and potential differences to smaller ones. Passive potential driving can, therefore, also be seen to be forcing concentration differences across a membrane toward equality, just as diffusion and osmosis do.

* The reader might receive the impression from this discussion that large numbers of charges are involved. In fact the total numbers involved in neural membrane potentials are very small—only one in every 100,000 or so of the charged ions inside a cell moves inward or outward—and the charge distribution effects are mainly localized within 20 angstrom units of the membrane. These facts should be kept in mind when one begins to consider the absolute rather than the relative magnitudes of ion flux.

The potential difference across a membrane, it is important to note, depends also upon the characteristics of the membrane as well as the concentration differences between the two solutions. Furthermore, it also depends upon the mobility of the ions themselves, and thus differences in the nature of the ions of the solute can result in major differences in transmembrane potentials.

B. ACTIVE FORCES OR PUMPS

If all the membrane forces acting on ions were passive, ion concentration differences on either side of a cell membrane would quickly be equalized and no transmembrane potentials would be maintained. Fortunately for organic life, there are also other forces acting in the opposite direction. Passive forces are counteracted by active metabolic processes in living cells, and these drive the ions back "uphill" against the action of passive forces. Since this requires the expenditure of energy, uphill pumping action must be metabolically powered.

The exact nature of ionic pumps that power the nerve membrane potential has been studied intensively in recent years by biochemically oriented neurophysiologists. It has been established that the pumping process that is mainly responsible for the production of concentration differences, and thus membrane potentials, in neurons is a double-barreled one, pumping sodium out of the cell and potassium into it.

It now appears that the two pumping actions are closely linked and may be parts of a single, complex chemical cycle. Hodgkin and Keynes (1955) and Caldwell, Hodgkin, Keynes, and Shaw (1960) showed that if the external environment of a cell is depleted in potassium, then as the pumping of potassium decreases due to a shortage of K^+ ions, the active pumping of sodium also decreases, even though there has been no change in the supply of available Na^+ ions. This result suggested a hypothetical model involving a chemical carrier substance. Assume that a single carrier in the cell membrane might be responsible for the transport of both Na^+ and K^+. When this carrier, X^-, combines with Na^+ at the inner edge of the membrane, it forms a compound we might call NaX, which for some biochemical reason, not yet understood, but obviously energy-consuming, has a tendency to move to the outer edge of the membrane. This movement of the compound NaX is in a direction counter to the substantial potential driving force for Na^+. The transport of a compound carrier substance (NaX) is further aided by the presumed neutralization of the ionic charge on the cation, so that the potential force does not have to be overcome. When it reaches the outer edge of the membrane, local chemical conditions change and two things happen to NaX. First, the bond between the Na^+

and the X^- anion breaks, releasing the Na^+ into the external environment. Second, the X^- anion is altered into a new carrier substance (Y^-), whose peculiar properties predispose it to combine with an ion of potassium. When it does so, this new compound, KY, under the influence of some equally unknown biochemical force, tends to move toward the interior of the cell membrane. There it, too, decomposes, releasing K^+ into the cell interior. The free Y^- radical in turn changes into an ion of X^-, which becomes available for recombination with an ion of Na^+. The cycle then repeats itself. A diagram of this hypothetical double-barreled pumping cycle is shown in Fig. 5.1.

This model has been challenged in recent years. It has been shown, for example, that more Na^+ ions are extruded from a cell than the number of K^+ ions injected into it by the active pumping process (Thomas, 1968; Carpenter & Alving, 1968). The exact amount is still uncertain; some researchers have suggested that two Na^+ ions are extruded for each K^+ ion injected, while others have suggested a three-to-two ratio. Thus a perfect exchange of one Na^+ for one K^+ does not seem to be the case, and the line representing the Na^+ extrusions in Fig. 5.1 should be stronger than the one representing the K^+ injection. Vigorous research aimed at answering this question is now underway in many laboratories. Particularly useful in this study is a newly discovered and highly specific inhibitor of the sodium pumping action—the chemical known as ouabain. Somehow ouabain manages to turn off the sodium pump without affecting the potassium pump.

Even though the Na^+ and K^+ pumping actions may not be on a one-for-one basis, they are certainly at least loosely coupled in some manner. The

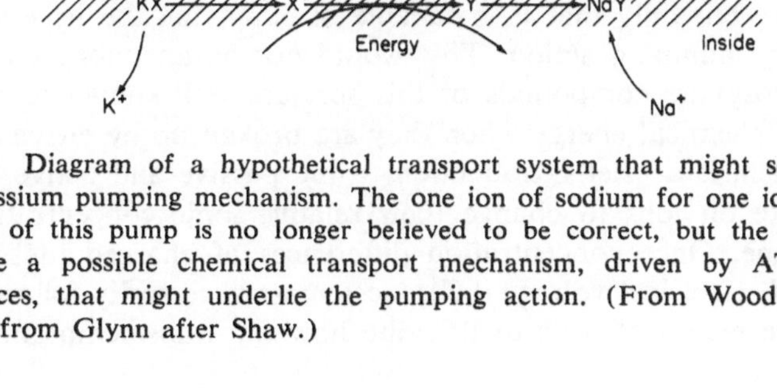

FIG. 5.1. Diagram of a hypothetical transport system that might serve as the sodium-potassium pumping mechanism. The one ion of sodium for one ion of potassium aspect of this pump is no longer believed to be correct, but the figure does conceptualize a possible chemical transport mechanism, driven by ATP-powered energy sources, that might underlie the pumping action. (From Woodbury, 1965, as modified from Glynn after Shaw.)

hypothetical double-barreled pumping model shown in Fig. 5.1 does not, unfortunately, emphasize another difference between the two barrels of the pump. The channels or pathways through which the Na^+ and the K^+ ions move are most probably not the same. Hille (1970) has reviewed a considerable amount of evidence, particularly with reference to pharmacological effects on the transport mechanisms involved, which supports the idea that the sodium and potassium ions pass through quite distinct channels in the membrane.

Hille notes, for example, that two highly toxic substances, tetrodotoxin from the puffer fish and saxitoxin from shellfish, have the ability to block the passage of Na^+ ions without affecting the passage of K^+. Quite to the contrary, tetraethyl ammonium ions are capable of closing the K^+ transport channel without affecting the one conveying Na^+. Hille and others claim that these diverse pharmacological effects suggest the presence of two distinct channels through which the Na^+ and K^+ ions travel, at least under the pressure of passive forces. It is of course possible that the difference in active pumping rates of "NaX" and "KX" may also be due to the nature of the channels through which these molecules pass.

The question of whether the pump is a chemical transport mechanism of the sort described in Fig. 5.1 is thus closely related to the issue of the nature of the channels through which the ions pass. Keynes (1972) has suggested that the rates of transmission of these ions through the membrane precludes any mechanism that requires mobile carriers of any appreciable molecular weight.

Whatever the outcome of this controversy, it is clear that the whole idea of the Na^+–K^+ pump does require that in some way these ions be forced uphill against the passive forces. This raises the question of the nature of the metabolic source that provides the necessary driving energy. Although the precise nature of the chemical cycle is not yet known in full detail, it appears that a substance known as adenosine triphosphate (ATP) may be the source of the energy required for the pumping action. Addition of ATP to preparations in which the neuronal membrane potential can be measured results in an increase in pumping action. This does not mean that ATP is the mysterious X^- or Y^- ion in the hypothetical transport mechanism, but only that it is capable of supplying the necessary energy to drive the pumping action. This would not be an unexpected result at all, for phosphate compounds of this sort are well known to be effective sources of chemical energy when they are broken up by enzymatic action.

This, then, is a brief introduction to the passive and active forces that can operate on ions to change their relative ionic concentrations across a membrane. Once concentration differences of charged ions have been established, electrical voltage differences must necessarily follow. The purpose of the next section is to describe how the transmembrane potentials result from the ionic distributions. We shall seek also to explain the origin of the normal resting potential of the cell. The resting potential is the non-zero electrical voltage measured across the membrane as the result of the interplay between the uphill pumping action and the downhill passive forces.

C. ESTABLISHMENT OF THE RESTING POTENTIAL

1. Concentration Potentials: A General Approach

Let us first consider how an electrical potential is produced across a membrane by a difference in the concentration of solutions on either side of the membrane. In the idealized case to be discussed in this section, it is important to remember that we will be considering the potential due to only a single solute. This potential is exerted across either a perfectly permeable membrane or across a liquid junction (that is, a junction between two contacting solutions without a demarcating membrane, a situation that can exist only momentarily). We will not yet be concerned about the specific details of the real neuronal membrane—a considerably more complicated system.

Equation (5.1) defines the membrane potential produced by two concentrations of a single salt solution separated by a perfectly permeable membrane, or by a liquid junction. This equation is a function of several factors that contribute to the membrane potential, some reflecting the nature of the ions and some reflecting membrane characteristics:

$$E = \frac{RT}{F}\frac{u - v}{u + v}\ln\frac{C_1}{C_2}, \qquad (5.1)$$

where R is equal to the universal gas constant, T is the temperature in absolute degrees, F is an electrical charge value that depends upon the ions that are involved, u and v are the mobilities in the solvent of the anion and the cation, respectively, and C_1 and C_2 are concentrations of the salt on either side of the membrane. The terms u and v, the ion mobilities, are affected both by the ion species and the nature of the solvent.

Equation (5.1) holds for the case in which the membrane is perfectly permeable to both the anion and the cation; thus this potential difference can only occur for a moment before it quickly declines to zero. If the membrane does not allow any flow of ions, it is said to be impermeable, and the concentration differences C_1 and C_2 will remain constant, and the terms u and v in that formula become zero. The factors u and v are also effectively zero when the flow of ions across some less than perfectly impermeable membrane has arrived at an equilibrium condition such that an equal number of ions are moving from one side of the membrane to the other.

To determine the potential of a single ion species that has different concentrations on either side of a semipermeable membrane, we can apply the expression known as the Nernst equation:

$$E = \frac{RT}{F} \ln \frac{C_1}{C_2}. \tag{5.2}$$

Equation (5.2) is used for evaluating the concentration potential produced at equilibrium conditions (i.e., when the net ion flow is equal to zero). Also remember that in this case we are dealing with a single ion species and not the full ionic system of the neural membrane.

2. Concentration Potentials across a Real Neuronal Membrane

Although the calculation of the potential due to solutions of a single salt or ion across an idealized membrane is relatively direct, calculations of the combined effects of several salt solutions across a semipermeable membrane are not so easy. A number of second-order interactions between the various factors complicate the prediction of the combined effects. For example, ion concentration differences that account for the individual electrical potentials are themselves affected by the potentials produced by other ions. Furthermore, the permeability of the membrane to a particular ion species is a function of the total potential difference, as well as of the individual ion type. Equilibrium potentials therefore become complicated interactive functions of several factors that feed back on each other.

However, it is possible to form a general picture of the result of these interactions by considering the behavior of the individual ions involved in the equilibrium process across a typical nerve cell membrane. The primary constituent ions of neuronal cytoplasm and extracellular fluid are Na^+, K^+, Cl^-, and certain organic ions. The large organic ions are, in general, unable to pass through the semipermeable membrane, and therefore do not contribute to the potential calculations except as they may be involved in cellular metabolism and the powering of the active pumps. The total cell membrane potential difference, therefore, is dependent primarily upon the mobilities and concentrations of the three inorganic ions, Na^+, K^+, and Cl^-.

The active sodium-potassium pump drives sodium out of the cell's interior into the extracellular fluid, producing a deficiency of this cation inside the cell. Since the internal sodium concentration is thus depleted, compared to its concentration in the extracellular fluid, that portion of the internal potential due solely to the sodium ion concentration must be at some negative value compared to the extracellular fluid.* The passive forces—osmosis, diffusion, and potential driving—are in opposition to the active

* Since the extracellular space is, for all practical purposes, infinitely large, it is therefore very easy to place an electrode in some part of the tissue that is electrically common to the extracellular fluid. The blood, or tissues rich in blood, is usually used as the reference against which all intracellular voltage measurements are made.

pumping process, which produced the relative internal depletion. Since the inside of the cell will be negative, with reference to the outside, for the positively charged sodium ions, and the concentration of sodium in the outside fluid will be greater than that on the inside, the three passive forces will all force some sodium back into the interior of the cell.

The potassium active pumping action, on the other hand, drives potassium into the cell interior, increasing the concentration of this ion compared to potassium concentrations outside the cell. If only the potential produced by this ion is measured, the interior of the cell will be positive with regard to the extracellular fluid. The passive forces acting on the potassium ions will force potassium to the outside of the cell, contrary to the passive effect on sodium.

The pumping processes for Na^+ and for K^+ approach a steady-state condition in which the number of ions being pumped in or out of the cell is equalled by the number of ions moving in the opposite direction under the influence of passive forces. At these steady-state conditions, the potential due to concentration differences of the ion is exactly balanced by an equal and opposite potential: the equilibrium potential produced by the pumping action.

The equilibrium potential for each ion species is determined by the magnitude and direction of the pumping force and by such factors as membrane permeability and overall membrane potential. The equilibrium potential is typically about +60 mV for Na^+ and about −90 mV for K^+, since the concentration potentials for Na^+ and K^+ are about −60 and +90 mV, respectively, across a typical membrane.

In addition to the inequality of the sodium and potassium pumping action, another factor contributes to the net difference in positive charges inside and outside the cell. The permeability of the cell membrane to passive efflux of K^+ is 50 times larger than its permeability to passive influx of Na^+. Thus, there is a much larger return flow of K^+ in the direction opposite to the pumped flow out of the cell than of Na^+ into it for both equilibrium and nonequilibrium conditions. This is an additional cause of the difference in the final charge concentration when ionic flow equilibrium is reached. A difference between the respective passive return flow and the pumped flow prevails at all membrane potentials until equilibrium is obtained. At that point the resting potential is established, and the net flow of each ion is zero. At the resting potential that reflects equilibrium, the active Na^+ pump is perfectly counterbalanced by the passive influx of Na^+, and the active K^+ pumping acting is perfectly counterbalanced by the passive K^+ efflux.

This, then, brings us to the main point of this discussion. How do the individual potentials summate to produce the total membrane resting potential that is measured between an intracellular and an extracellular electrode? The answer to this question is quite involved, but we can understand

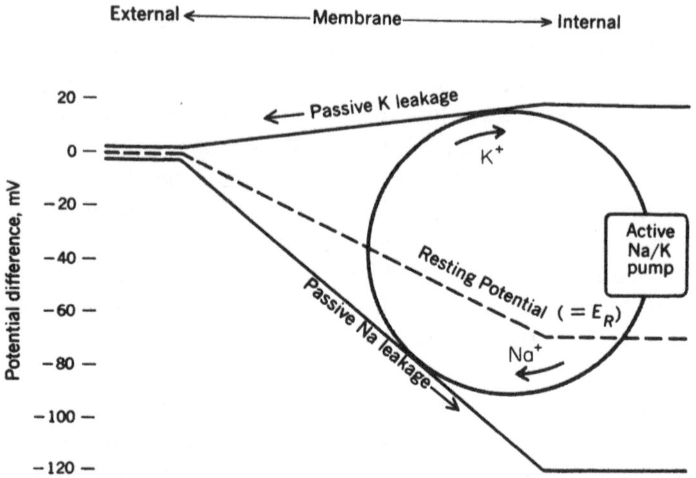

FIG. 5.2. Diagram of the potential levels, passive forces, and active forces that are involved in establishing and maintaining the membrane resting potential. (From Katz, 1966.)

the general nature of the process by considering a few simple interpretations of some very complicated concepts.

The total passive potential driving force operating on each ion will be the sum of its individual concentration potential and the total membrane resting potential. Thus the total driving force on Na^+ (given a typical —70- or —80-mV resting membrane potential) will be about —140 mV, and the total passive driving force for potassium will be about +10 mV. As we shall see in Chapter 7, the resting potentials produced by the potassium and sodium concentration differences almost completely account for all of the amplitude of the explosive neuronal spike action potential.

An entirely different set of forces operate on the chloride ion. There is some suggestion that there is a Cl^- pumping action too, at least in the squid giant axon (Keynes, 1963), but its purpose is unclear, since the Cl^- potential measured across most cell membranes is pretty much the same as the cumulative membrane potential itself.

The cumulative or resting membrane potential is the result of a complicated series of equilibrium interactions within and among the different ion species, their pumps, the "downhill" forces, and several membrane factors, most notably the membrane's individual and specific permeability to each ion. Figure 5.2, depicts the various forces that act across a typical neural membrane, and some of the interactions between them.

At equilibrium conditions a useful approximation for the prediction of the resting membrane electropotential, as a function of the concentrations of individual ionic potentials and the membrane permeability to the respective ions, has been suggested by Hodgkin and Huxley (1952b):

$$MP = \frac{RT}{F}\left[\ln \frac{P_K(C_K)_o + P(_{Na}C_{Na})_o + P_{Cl}(C_{Cl})_i}{P_K(C_K)_i + P(_{Na}C_{Na})_i + P_{Cl}(C_{Cl})_o}\right] \tag{5.3}$$

where *MP* is the total membrane potential; *R, T,* and *F* have the same meaning as in Eq. (5.1); P_K, P_{Na}, and P_{Cl} are the membrane permeabilities to the three ions, respectively; and the *C* values are the concentrations of each ion either inside (*i*) or outside (*o*) the neuron. This equation holds only at equilibrium, and since the permeabilities are actually themselves functions of the membrane potential as well as the ion concentrations, it is not, in general, applicable to the system at other times.

In summary, because of the differential permeability of the nerve cell membrane and the presence of a metabolically powered sodium-potassium pump, a difference in the concentration of a number of ions develops across the membrane. These ionic concentration differences result in individual electrical potential differences, which are defined by the cumulative steady-state conditions and the factors specific to each ion. The sodium-potassium pump transfers ions of Na^+ out of the cell and ions of K^+ into the cell. Na^+ is typically found in concentrations 10 times higher on the outside than on the inside. K^+, on the other hand, is found in concentrations 30 times greater on the inside. Chloride ions are also forced out of the cell by the equilibrium voltages of the Na^+ and K^+ concentrations, and possibly also by an active Cl^- pump, so that there is a preponderance (although to a lesser degree than the sodium or potassium discrepancies) of chloride ions in the extracellular fluid, compared to the intracellular solution. Inorganic cations, unable to move through the cell membrane, are found in high proportions in the interior of the cell. The several processes result in an equilibrium condition that produces a measurable electrical voltage known as the *resting potential.* However, this equilibrium is not static. Although the membrane resting potential remains relatively constant, the resting equilibrium condition represents a balance of the influx and efflux of the several ion species.

This, then, is a brief introduction to the nature and origin of the cumulative resting potential—the millivoltage that is measured across the quiescent cell membrane. As cells perform their assigned functions, changes of one sort or another in the membrane potential occur. These changes, more or less transient, are called *action potentials* and are the actual mechanisms by which information is encoded and processed in the nervous system. We shall consider the variety of action potential types in the following section.

D. A TAXONOMY OF ACTION POTENTIALS

So far we have described how non-zero and constant electrochemical potentials are developed across semipermeable membranes by both active and passive forces that alter ion concentrations. Pumping actions lead to the maintenance of a cumulative, steady *resting potential* that is maintained when the cell is not actively involved in information transmission or processing. The second major category of nerve membrane potentials are those transient shifts from the resting potential called *action potentials*.

Bullock (1959) has presented a taxonomy of the variety of electrochemical actions that constitute the response repertoire of the neuron. He emphasizes the often overlooked but very important point that neuronal action is not limited solely to the spike action potential, the easily detected, high-amplitude response of the axon that was the almost exclusive object of neurophysiological research in the 1930's and 1940's. (Some wags have suggested that "axonology" might have been a more appropriate name for the science during that period, rather than neurophysiology.) The point to be made especially clear in this section is that there are many ways in which a neuron can meaningfully respond.

As our technology and expertise in the study of neurons have developed, it has been discovered that other forms of membrane responsiveness are present even within the limits of a single cell, and these play important roles in neural information and transmission. Bullock stresses the idea that the spike action potential is actually only one rather unusual and highly specialized response type that has evolved especially to serve the function of long-distance transmission over the elongated axons of the neuron. Between other parts of the cell, however, other forms of communication are used. Bullock's schema is presented in Fig. 5.3.

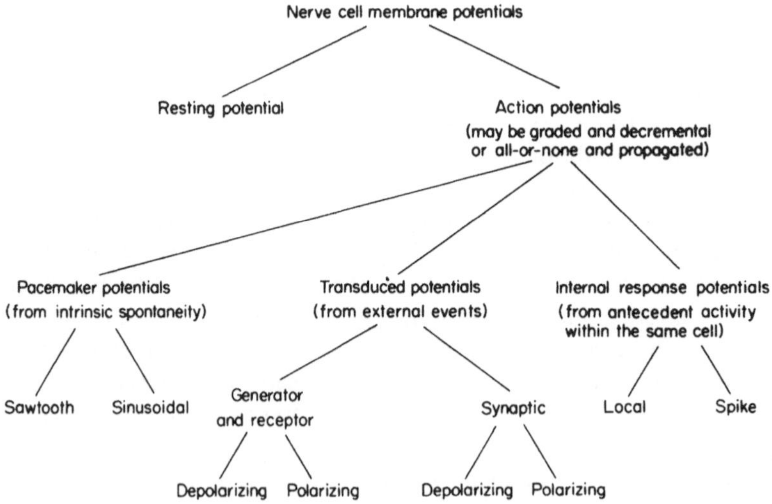

FIG. 5.3. Hierarchy of neuronal membrane potentials, emphasizing that the term "action potential" should be applied to a larger group of responses than solely the spike action potential of the axon. (From Bullock, 1959.)

Because of the very common usage of "spike" and "action potential" as synonyms, it must be re-emphasized that the term *action potential,* as used here, does not refer exclusively to the *spike* action potential, but to all the transient fluctuations described in this section. A more general and consistent definition of an action potential is: *any transient deviation from the resting potential that is involved in the information-processing activities of the neuron.* In general, deviations due to injury or to metabolic deficiencies are not considered action potentials in the sense used here. Following Bullock, three classes of action potentials are defined: pacemakers, transduced action potentials, and internal responses.

1. Pacemakers

Pacemaker action potentials include the periodic potential shifts initiated by an oscillatory or rhythmic mechanism within the neuron itself. These activities may serve as a master clock to trigger some other, more massive repetitive function. The critical condition defining a pacemaker action potential, however, is a periodic, intracellular response without identifiable external stimulation. The pacemaker function of the heart is perhaps the most familiar of these potentials, but pacemaker activity is also found in a number of other vertebrate and invertebrate neurons. Interestingly, von Baumgarten* has shown that it is also possible to set up a kind of pseudo-pacemaker activity (in addition to the naturally present pacemaker) in the nervous system of the mollusk *Aplysia* (the sea hare), by repeated stimulation with periodic stimuli. When the stimuli cease, the cell continues to respond at the appropriate interval.

2. Transduced Action Potentials

This class of action potential is composed of two subcategories. The first includes *receptor* or *generator potentials* produced when some form of external physical energy impinges on a specialized receptor capable of converting that physical energy to neuronal electrochemical energy. Receptor or generator potentials are graded and spread from one place to another in the cell only by passive electrical or electrotonic spread. They are the responses of highly specialized neurons whose main purpose is to amplify and convert the tiny amounts of physical energy involved in sensory activation. How this form of action potential is generated by the transductive mechanisms of the various senses is the subject matter of Chapter 6.

The second class of transduced potentials is the *synaptic potential.* Synaptic potentials are similar in many respects to receptor potentials but are generated in a very different way: They are produced as a result of a presynaptic neuronal response, rather than as a result of the transduction of external physical energy. There are many similarities between synaptic

* Personal communication, 1969.

and receptor potentials: Both are graded and nonpropagating, both may be summated and may either hyperpolarize or depolarize the cell membrane. The fact that both are graded potentials produced as a result of events that occur outside the cell, however, is the definitive criterion for placing them in the same class. In fact, if the postsynaptic membrane is considered as a specialized chemoreceptor for the transmitter substance, it becomes a perfect analog of sensory chemoreceptors.

3. Internal Responses

This type of action potential is generated by processes internal to the cell, unlike synaptic potentials and receptor or generator potentials. Internal responses may be initiated indirectly as a result of transduction processes, but the fact that they always have a prior intracellular cause distinguishes them from the former class. There are three types of internal responses, according to Bullock's classification system: local, spike, and the axon-hillock potentials.

a. Local potentials are graded and additive potentials produced within the cell by some previous intracellular activity. This type of action potential is a transient change from the resting potential, and it does not propagate to other portions of the cell but can, if of sufficient amplitude, trigger the propagating spike action potential. Though local potentials do not propagate, they can influence information transmission because they are additive and thus can alter the threshold for elicitation of the spike action potential by temporal or spatial summation. Local potentials can convey information to nearby portions of the same cell by passive or electrotonic spread. (See Chapter 7 for a complete discussion of this process.) The local action potential is sometimes considered to be one of the preliminary phases of the spike, rather than a distinct entity itself. However, since it does have slow and graded properties, and because not all local potentials result in a spike, Bullock's decision to place it in a separate category seems appropriate.

b. The familiar *spike action potential* is the self-propagating action potential for the conduction of signals over long distances along the axons. It is characterized by an all-or-none explosive responsiveness and a typical spike-like shape during its time course. The spike shape, of course, is due to the fact that the amplitude of the spike is quite large compared to the other types of action potentials and the duration is quite brief, typically around .5 msec. Chapter 7 also discusses the spike action potential.

The adaptive value of the spike is obvious. For long distances, some nondecremental form of transmission is required, and that is the role played by the self-propagating spike. Because energy is provided at each stage of the transmission by the local portion of the axon currently active, the response amplitude is a function only of local metabolic conditions and not of the characteristics of the energy that triggered the response. Until recently it was thought that spike action potentials could not be propagated across the membrane of the cell body or of the dendrites, but that graded and decremental potential shifts from the resting potential are the usual means of neuronal action in these regions. However, Llinás and his co-workers (1968a, 1968b) have demonstrated dendritic spikes in the cerebellar Purkinje cells of the alligator brain. The spike, thus, appears to be more general than simply an axonal response.

c. The *axon-hillock action potential* has much the same properties as the local potential, but according to Bullock, it is distinguished by its ability to convert the graded responses that are more typical of the neuronal cell body and dendrites into the spikes of the axon. Since other portions of the neuron also have been observed to perform this same conversion (for example, the ring of axonal membrane at the first node of Ranvier within a somatosensory Pacinian corpuscle), the notion of a specialized axon-hillock potential no longer has wide currency.

Bullock's categorization of the many modes of neuron action has specific implications for current theories of neural encoding of sensory, cognitive, and motor messages. If the neuron responds in so many different ways as information passes from one part of the cell to the next, it is clear that codes used to represent a given pattern of information cannot remain constant within the cell. Rather, at each place in the cell a different coding mechanism must represent the message pattern. As information is transmitted from cell to cell, the coding and recoding of information occurs again and again. At one point a graded action potential may represent the pattern, while at another, a train of spike action potentials may be the coding mechanism. The important implication is that there is no single or "best" neural code, and certainly no psychophysical test could tell how a given percept is encoded at any level of the nervous system. Rather, there are many different and equally useful neural "languages" used at different places in the nervous pathways.

This introduction to the nature of the ionic equilibrium conditions responsible for the production of the resting potential and the introduction to the nomenclature of action potentials have laid the groundwork for a more detailed discussion of the nature of the various kinds of action potentials. In accord with our plan to keep the thread of the discussion consistent with the biologically sequential series of events in the nervous system, the next chapter will consider the first step—the transduction processes in receptor neurons.

6
CELLULAR ASPECTS OF THE TRANSDUCTION PROCESS*

A. INTRODUCTION

Transduction, the process by which environmental stimuli are converted into neural signals, is a multistage process with important steps occurring at both the macroscopic and cellular levels. Before the actual conversion of energies (and energy conversion is the essential feature of transduction), each of the receptor organs carries out some nonneural or preneural modification of the incoming stimulus. The eye, for example, modulates the intensity of a photic stimulus by varying the pupil size. It also filters the light, since some wavelengths are selectively absorbed by the ocular media. Preneural modifications also occur in the mechanical receptors. The ear modulates auditory stimulus intensity by means of a system of membranes, muscles, and levers. Far more important, however, is the preneural conversion of the temporal patterns of acoustic energy to mechanical spatial patterns as a function of the hydrodynamics of the cochlea. This important alteration of dimensions is carried out on an entirely mechanical and totally nonneural basis.

Complex as these preneural modifications of the stimulus may be, they are not "neurophysiological" in the sense defined in this book. Only after a stimulus falls on a receptor and its physical energy is converted to the electrochemical energy of the cell do the real neurophysiological processes begin.

* This chapter is an adaptation and abbreviation of a considerably more detailed discussion of sensory transduction in a book on sensory coding by the author (Uttal, 1973). It is used with the permission of Harper & Row Publishers, Inc.

The transduction of physical to electrochemical energies that occurs within the receptor cell also consists of a number of distinct and separable processes. The *primary sensory action* may be defined as the specific interaction between the stimulus energy and some part of the receptor that triggers the actual energy conversion. From a certain perspective, the primary sensory action may also be considered to be a preneural process rather than a neural process itself. Nevertheless, it is the final and key preneural event that leads to the alteration of the resting membrane potential. Thus it is of special importance in our understanding of transduction. The primary sensory action must be defined and clarified in detail for each receptor type, because of the unique method found in each receptor. However, each of these individual methods serves the same general purpose: the production of some change in the properties of the membrane that leads directly to that deviation from the resting potential that has come to be known as either the receptor or generator potential.

B. RECEPTOR AND GENERATOR POTENTIALS—A DEFINITION

The infinitesimal amounts of energy involved in transduction make it clear that the essential parts of the process must occur at the molecular level. Some change in the permeability of the membrane, produced by a stimulus, presumably leads to a momentary fluctuation of the resting potential. This initial fluctuation is observed in all receptors in the form of a transient, graded shift in the potential measured across the membrane. This graded potential cannot itself be propagated to distant neural locations, but it can act as a precursor for signals that can be conveyed over long distances. Two generic terms have been applied to this intra-receptor fluctuation; it may be referred to as either a *receptor* or a *generator potential.*

The distinction between a receptor and a generator potential is based on an anatomical rather than a functional difference, depending on whether the receptor itself is a specialized terminal portion of a neuron with its own elongated axon, or whether the receptor is an axonless unit, synapsing almost immediately. A receptor potential is defined as a stimulus-produced, graded potential that occurs in a receptor that does not possess an elongated axon, and which, thus, cannot lead to a spike action potential within any part of that same cell. The influence of a receptor potential is exerted through the mediation of synaptic action, and a propagated spike action potential is produced for the first time only in a second-order neuron. A generator potential, on the other hand, is a graded potential produced in a receptor cell that has its own axon. The generator potential therefore is the direct precursor of a spike action potential that is produced within the confines of the same cell. Generator and receptor potentials appear to be universal intermediate mechanisms in receptor function, although they have not been unequivocally identified in all modalities.

Receptor and generator potentials are usually excitatory (that is, they increase subsequent neural activity), but this is not always the case. Arvanitaki, Takeuchi, and Chalazonitis (1967) report that both excitatory and inhibitory generator potentials can be produced in some large, olfactory receptor cells found in the subesophageal ganglion of certain molluscs, depending upon what kind of olfactory stimulus the animal encounters. Boeckh (1967) also discusses some olfactory receptors in the carrion beetle, which can produce either inhibitory or excitatory responses, depending upon the nature of the chemical stimulant. It is not clear whether an inhibitory potential that is produced by a sensory stimulus and that operates to reduce an ongoing amount of neural activity should be termed a "generator" potential. But there is no other generally agreed-upon term to describe such actions, and it does act to modulate, if not to generate, neural activity. Nevertheless, we must accept the general notion that both excitatory and inhibitory generator potentials can be produced by receptors. Inhibitory generator potentials of this sort may play an important role in establishing patterns of neural coding, even though their primary action is to reduce the amount of information flowing through the receptor.

It should be noted that there is no direct relation between *excitation* and *inhibition* on the one hand, and *depolarization* (a reduction in the resting potential) and *hyperpolarization* (an increase in the receptor potential) on the other. Excitatory receptor potentials are hyperpolarizations in the vertebrate retina, but depolarizations in most invertebrate photoreceptors, for example. In general, it seems that if spikes are produced in a second-order cell, then excitation is associated with depolarization. On the other hand, if second-order cells (like the bipolars of the vertebrate retina) do not themselves propagate spikes, then the direction of polarization is not associated uniquely with either excitation or inhibition of the next cell in the chain.

In general, inhibitory or excitatory responses are equally good candidates for the receptor potential. Some systems operate on the basis of considering a signal to be any deviation from some spontaneous level. Others operate through integrative circuits, which can actually enhance signal output at one point by inhibitory inputs at another. Furthermore, sometimes inhibitory effects can demarcate neural patterns more sharply by increasing the contrast between inhibited and excited regions.

C. TRANSDUCTION IN THE PHOTORECEPTOR

1. Primary Sensory Action in the Photoreceptor

The theory of the primary sensory action in vision, developed by George Wald and others over the last 20 years, asserts that the outer segments of rods and cones contain large, photosensitive molecules that break into two parts under the influence of incident light. The breakdown products resulting from the photodecomposition of these molecules, in some currently unknown way, induce a change in membrane polarization that can be considered to be the receptor potential. Wald has elucidated the mechanisms of the breakdown of the photochemical (which can be considered to be the primary sensory action in vision) and the sequential series of intermediary substances that are produced by spontaneous decomposition of the original photosensitive macromolecule. His work was mainly responsible for establishing the various steps in the photochemical "cycle" underlying the primary visual action, and most of the following discussion is based on his ideas.

For a number of years before Wald's discoveries, it was known that a curious violet chemical was involved in the visual process. Boll discovered the substance in 1877 and named it *rhodopsin.* The general idea of a hypothetical chemical cycle involving rhodopsin and its breakdown products was suggested by Selig Hecht in the 1930's. Wald (1968) states that it was under the influence of Hecht that he began a search for the chemical constituents that were only hypothetical concepts in Hecht's theories. In 1933 Wald made the initial, though not unexpected, discovery that vitamin A is present in large amounts in the retina. This discovery was unsurprising because vitamin A deficiency had long been known to produce night blindness. This discovery was immediately followed by the detection of another organic substance, which he called *retinene* (Wald, 1933), a chemical that was later determined to be an aldehyde of vitamin A. (An aldehyde is a chemical product produced by the oxidation of a precursor such as vitamin A.) Wald and others have now agreed that vitamin A should be called *retinol* and retinene, its aldehyde, *retinal,* to emphasize their very important roles in the visual process.

Over the years, Wald was able to flesh out the hypothetical notion of the photochemical breakdown cycle proposed by Hecht, to identify some of the specific chemicals (which were only algebraic expressions in Hecht's mathematical model), and to define the various steps of what is now known as the *rhodopsin cycle.* Wald, his students, and a number of other distinguished workers have identified the specific molecules and defined the chain of reactions that occur under the influence of incident light, in a way that makes the primary sensory action in vision perhaps the best understood of all the senses.

The essence of the now generally accepted Wald model is that the visual pigment found in each receptor cell of the eyes of a wide variety of animals consists of a prototypical, photosensitive molecule, composed of two essential parts or *moieties.* One part, the *carotenoid* (so named because pure suspensions often have the orange-red color typical of carrots), is a relatively simple structure. In fact, in almost all species, the carotenoid is a special isomeric form of retinal. The other part of the visual pigment is a much larger organic structure—a protein with a molecular weight of 30,000 to 40,000. Though it is now known that there are many different kinds of this large protein, Wald has coined the generic name *opsin* for it, regardless of the animal or photoreceptor in which it might be found, or its exact chemical structure.

Opsins differ from each other in two ways. Details of the organic structure vary from species to species, and the various opsins are therefore said to be *species specific*. But even within one animal, there is frequently more than one kind of opsin. In vertebrates one opsin combines with retinal to form the compound visual pigment found in the rods, while the color-coding trio of photochemicals observed in cones suggests that three kinds of cone opsin may exist. In some animals, the anatomical separation may be quite a bit more unusual than simply that found between different kinds of cells in the same sense organ. For example, two photopigments, each containing its own specific kind of opsin, are found at opposite ends of the body in the crayfish: one in the anterior eye, and the other in a simple photoreceptor cell in the tail.

The model of a composite molecule composed of a large proteinaceous moiety bonded to a small carotenoid is now accepted as the standard configuration of all visual photopigments. This model is the key premise of the Wald theory. Some details of the smaller structure, the carotenoid, will now be considered as an initial step in helping the reader to understand this important theory.

Retinol exists within the body in a balanced equilibrium between itself and its aldehyde, retinal. The structural formula of one kind of retinol, *retinol*$_1$, and one kind of retinal, *retinal*$_1$, are shown in Fig. 6.1. It can be seen that the two differ very little. Only the bonding of the terminal OH or O group differs. Retinol$_1$, however, is not an active participant in the visual process. Its main importance is in the role it plays as a biochemical precursor of retinal. The quantities of retinol$_1$, therefore, directly determine the availability of retinal$_1$; thus retinol$_1$ deficiencies in the diet lead only indirectly to visual inadequacies. The equilibrium between retinol$_1$ and retinal$_1$ is balanced so that the oxidation process producing retinal$_1$ is maintained only at the level it is used in the visual cycle and only, it is thought, in the retina.

A structural analysis of retinal$_1$ is further complicated by the fact that it can exist in several different *stereoisomeric* (three-dimensional) forms. All of the isomers of retinal$_1$, of course, have the same chemical formula, $C_{19}H_{27}CHO$, but they can differ in the three-dimensional pattern into which the constituent atoms are arranged. The physical shape of the molecule is a function of the angles between adjacent atoms at bonding points in the molecule. The various stereoisomers of retinal$_1$ differ in shape because of differences in bonding angles at carbon-to-carbon junctions. Visual

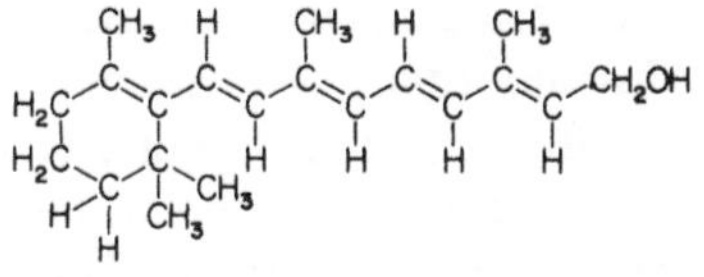

(a) All-*trans*-retinol$_1$, $C_{19}H_{27}CH_2OH$

FIG. 6.1. The structural formulae of three of the important molecules (all-*trans*-retinol$_1$, all-*trans*-retinal$_1$, and 11-*cis*-retinal$_1$) involved in the photochemical cycle of vision. See text for details. (Adapted from Wald, 1968.)

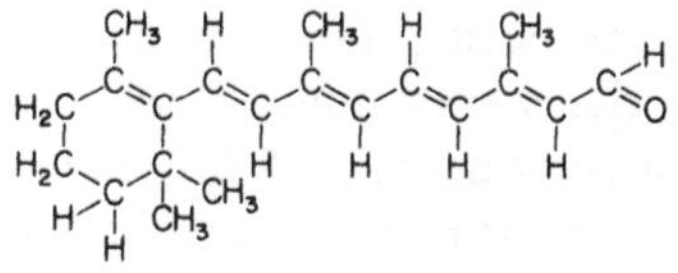

(b) All-*trans*-retinal$_1$, $C_{19}H_{27}CHO$

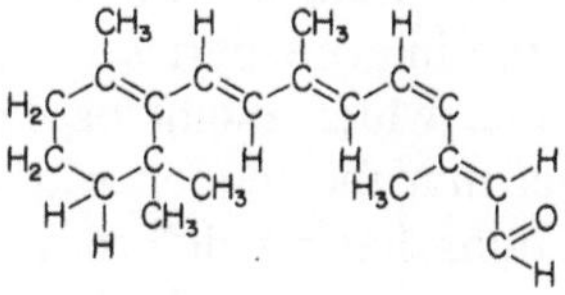

(c) 11-*cis*-retinal$_1$, $C_{19}H_{27}CHO$

photochemistry is based on the important fact that the chemistry of stereoisomers can differ even though their formulas are identical.

The two most important of the several stereoisomeric forms in which retinal$_1$ can exist are the *all-trans* and the *11-cis* shapes. The all-*trans* form is illustrated in Fig. 6.1b above. The shape of its tail is essentially straight, because all the carbon-to-carbon bonds in that tail are in the same direction.

The 11-*cis* stereoisomer of retinal$_1$ is illustrated in Fig. 6.1c. In this case there is a bend at the 11th carbon atom, which puts a kink in the tail. As Wald (1968) points out, the 11-*cis* configuration should not be expected to exist at all, since molecular dynamics suggest that it is intrinsically unstable. The extraordinary fact, however, is that 11-*cis*-retinal$_1$ does exist and does maintain this stable stereoisomeric form, *but only as long as it is kept in the dark.* This deceptively simple statement is the key to the whole story of the primary visual process.

There are two very important reasons for this key role. First, 11-*cis*-retinal$_1$ is the only stereoisomer of retinal$_1$ that is capable of forming a stable bond with the organic macromolecule, opsin. Second, the action of light on the rhodopsin molecule serves only to stereoisomerize 11-*cis*-retinal$_1$ into its all-*trans* form, thus forming an unstable substance that goes through the various stages of the visual cycle spontaneously without the necessity for any other external energy.

The net effect of these special characteristics of retinal$_1$ is that *the primary sensory action in vision can be specifically identified as the straightening or stereoisomerizing of the 11-*cis *to the all-*trans *form of retinal$_1$*. No other action of light is necessary once this triggering process occurs.

When 11-*cis*-retinal$_1$ is stereoisomerized by light and the combined photomolecule breaks down, a complex series of further spontaneous decompositions occur. The rhodopsin molecule, for example, does not immediately degenerate into all-*trans*-retinal$_1$ and opsin, but goes through a number of intermediate states, each of which has its own specific properties. These intermediate states occur very rapidly, and it was extremely difficult to determine their characteristics by the ordinary spectral absorption techniques used to examine the optical properties of the stable final breakdown products. A remarkable innovation introduced by Wald's group to this kind of research, however, has made it possible to take "temporal snapshots" of the intermediaries.

The technique, which might be called "freeze chemistry," takes advantage of the fact that the sequence of intermediary photochemicals involves chemical reactions having different temperature thresholds. That is, each of the processes can occur only at a successively higher temperature. Thus, the initial breakdown of rhodopsin into an intermediary called *prelumirhodopsin* will lead to a stable mixture of rhodopsin and prelumirhodopsin if the experiments are carried out at a temperature below —140°C. If the temperature is allowed to rise above —140°C, then prelumirhodopsin will spontaneously decay into another intermediary called *lumirhodopsin.*

At —40°C, lumirhodopsin further decays spontaneously into a substance called *metarhodopsin I* which, in the presence of H^+ ions and at temperatures exceeding —15°C, is converted into another substance known as *metarhodopsin II*. Metarhodopsin II is further decomposed into the final breakdown products, retinal$_1$ and opsin, through two alternative routes. Wald has shown that as the temperature rises to 0°C, in the presence of water, metarhodopsin II will spontaneously decompose into the final breakdown products; but it may also go through an auxiliary pathway involving another intermediary called *pararhodopsin.*

The whole process is diagrammed in Fig. 6.2. In addition to showing the sequence of breakdown products, this figure shows the temperature at which each sequential step is "unfrozen," as well as two other important sets of data.

First, the peak wavelength of the absorption curve of each intermediate is indicated. It can be seen that each of these chemicals absorbs maximally at a different point on the visual spectrum. This unique placement of the absorption peak is crucial to the identification of each breakdown product

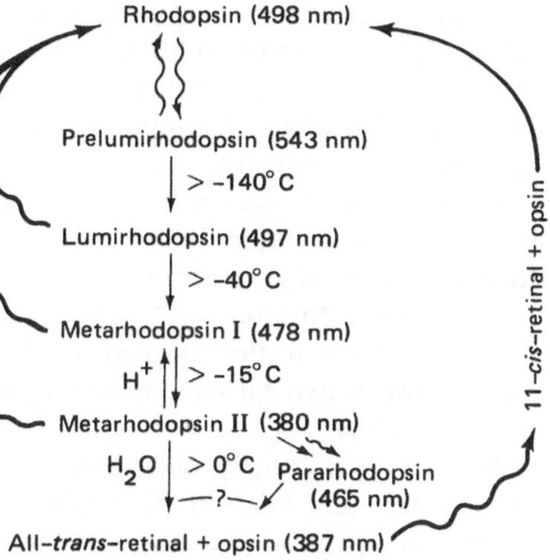

FIG. 6.2. Schematic representation of the complete rhodopsin cycle. Portions shown connected by wavy lines are driven by light, but those connected by straight lines occur in the dark. The wavelengths indicated in parentheses are the peaks of the absorption spectra of the particular intermediate breakdown products. The temperatures are thresholds for the next step in the reaction. Cooling the mixture below any of the thresholds can stop the reaction at any intermediate stage desired. (Adapted from Wald, 1968.)

and is the main currently available way of definitively showing that they are really separate, distinct chemicals.

Second, the shape of the arrow in each reaction in Fig. 6.2 indicates something about their energetics. Some are spontaneous and will occur in the dark (straight lines), while others are reactions that must be driven by light (wavy lines). It is important to note that there is only one light-driven breakdown reaction. This means that the visual response requires light only to initially trigger the entire sequence of dark reactions that leads to the free final breakdown products: retinal$_1$ and the protein opsin. The other light-driven reactions, interestingly enough, all move in the opposite direction—toward the regeneration of rhodopsin from the various intermediaries. Thus, the duration of the light stimulus contributes to the final equilibrium mixture of the involved substances.

The main advantage of the technique of freezing the chemical reactions is that processes normally lasting only a few microseconds, which are too brief to be examined at normal body temperature, can be fixed in time for a sufficiently long period so that measurements of their optical and chemical properties can be made.

It is not at all certain, of course, that each of the breakdown stages isolated with the freezing technique is not really itself a combination of several subprocesses with the same or similar threshold temperatures. Indeed, recent investigations have suggested that *lumirhodopsin* may be two substances (now known tentatively as *lumirhodopsin A* and *B*), but the details are not yet definitely established.

A complete understanding of the biochemistry of the visual photopigments is complicated further by the fact that, just as there are many types of opsins found in various species and two kinds of opsins (rod and cone types) found in some individual species, there are also two known types of retinol (retinol$_1$ and retinol$_2$) and thus two retinals (retinal$_1$ and retinal$_2$). The structural formulas of retinol$_2$ and retinal$_2$ are shown in Figs. 6.3a and b, respectively.

Retinal$_1$ and retinal$_2$ and their respective retinols are not simple stereoisomers of each other, but are distinct chemical structures with different formulas. Retinal$_1$, as noted earlier, has a formula of $C_{19}H_{27}CHO$, while retinal$_2$ has a formula of $C_{19}H_{25}CHO$. The critical structural differences can be observed in the benzene ring body of the molecule, where a double bond between two carbons in retinal$_2$ replaces a pair of auxiliary hydrogen atoms in retinal$_1$.

Retinal$_1$ had usually been assumed to be present in land vertebrates, marine fish, and adult frogs. Retinal$_2$ was assumed to be found in vertebrates that spend at least the early parts of their lives in fresh water, such as freshwater jawed fish, lampreys, and some amphibia. Recently, however, Schwanzara (1967) surveyed the pigments of a large number of freshwater fish and found that both retinal$_1$ and retinal$_2$ systems were present. She attributed this to the fact that freshwater fish live in different conditions of ambient lighting, and somehow this environmental fact has led to a natural selection of an optimal pigment for each environment.

As we mentioned earlier, an important complication in understanding the mechanics of color vision is that there appear to be several different kinds of cone opsin in some animals. Based on direct spectrophotometric measurements by Brown and Wald (1964) and Marks, Dobelle, and MacNichol (1964), it is known that there are probably three kinds of cone

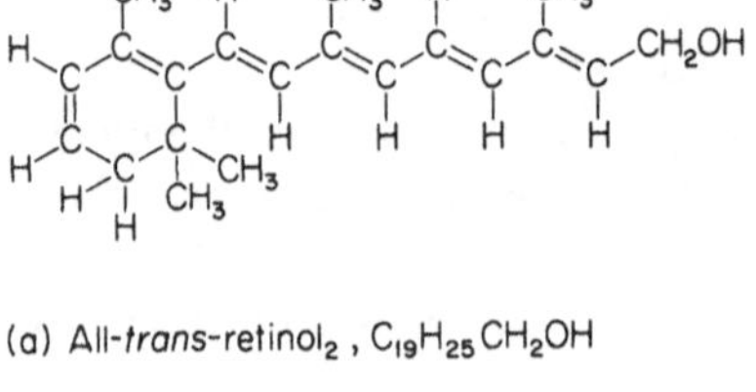

(a) All-*trans*-retinol$_2$, $C_{19}H_{25}CH_2OH$

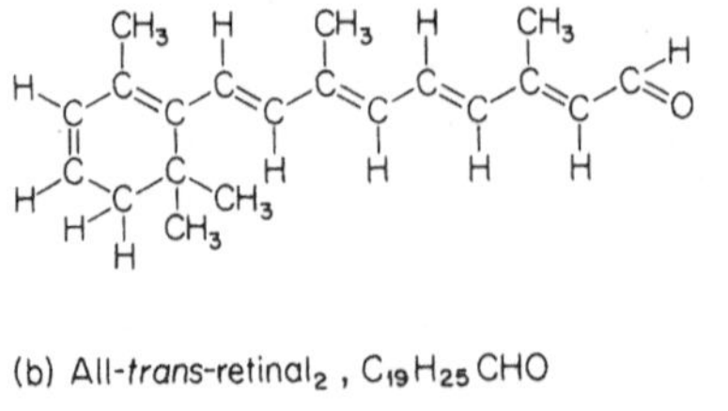

(b) All-*trans*-retinal$_2$, $C_{19}H_{25}CHO$

FIG. 6.3. The structural formulas for retinol$_2$ and retinal$_2$. (Adapted from Wald, 1968.)

TABLE 6.1
Possible Combinations of Opsins and Retinals That Might Form Photopigments[a]

	$Retinal_1$	$Retinal_2$
Rod Opsin	Rhodopsin $\lambda_{max} = 500$	Porphyropsin $\lambda_{max} = 522$
Cone Opsin R	$Iodopsin_1$ = erythrolabe $\lambda_{max} = 555$ or 570	
Cone Opsin G	$Iodopsin_2$ = chlorolabe $\lambda_{max} = 525$ or 535	Cyanopsin ?
Cone Opsin B	$Iodopsin_3$ = cyanolabe $\lambda_{max} = 445$ or 450	

[a] Adapted from Wald (1959)

pigments in the primate eye. Since the $retinal_1$ in each of these cone pigments would presumably be the same, the observed differences in absorption spectra must be due to the presence of three different kinds of cone opsin. Each cone has only one of each of these chromatically distinct photopigments, which Rushton (1962) has named *chlorolabe, erythrolabe,* and *cyanolabe.* The entire photochemical system has been summarized in Table 6.1.

2. Production of the Visual Receptor Potential

So far the discussion has concerned mainly the characteristics of the retinal, part of the large photosensitive molecule. The structure of this relatively small portion of a molecule of rhodopsin, and of the other photopigments, is the main determinant of the characteristics of the primary sensory action. However, the opsin moiety, a much larger part of the photosensitive molecule, is thought to be critical to the actual production of the receptor potential. The source of much of the material discussed here is once again the work of George Wald.

The bonding between the retinal and opsin moeities can be described by the following chemical equation:

$$C_{19}H_{27}HC{=}O + H_2N\text{-opsin} \rightleftharpoons C_{19}H_{27}HC{=}N\text{-opsin} + H_2O \quad (6.1)$$
$$11\text{-}cis\text{-retinal}_1 + \text{rod opsin} \rightleftharpoons \text{rhodopsin} + \text{water}$$

The key portion of this reaction is the substitution of the retinal molecule (with the exception of its oxygen atom) for the two hydrogen atoms that had been attached to the nitrogen atom on the opsin. While the oxygen and hydrogen form a free molecule of water, the opsin and the retinal are attached by a carbon-to-nitrogen double bond—a condition that can remain stable only if the retinal remains in the 11-*cis* stage. However, in addition to the availability of the double-bonded oxygen on the retinal molecule and the H_2N group on the opsin (as indicated in Eq. 6.1), an atomic side chain known as a *sulfhydryl group,* SH, is also required for the combination to occur. This bond also becomes available for chemical interaction when photodecomposition occurs. Under the action of stimulating light, the rhodopsin molecule fractures and "exposes" two of these highly reactive sulfhydryl groups. At a later stage in the sequence of the intermediate breakdown products, a hydrogen bond is also exposed; but apparently this occurs too long after the initiation of the actual excitatory process to play a very significant role in the actions to follow. Specifically, the opsin fragment with its exposed sulfhydryl groups is supposed by Wald directly to affect membrane permeability, either of the disc membrane or the cell membrane.

"Exposure" of the sulfhydryl groups is a somewhat ill-defined term, for these groups apparently are not part of the actual bonding reactions holding retinal to the opsin. Wald suggests (see Fig. 6.4) that the exposure is actually a function of the stereogeometry of the opsin moiety, rather than of unavailability due to a specific chemical bonding. Originally the sulfhydryl groups are buried deep within the body of the rhodopsin molecule. The structural changes occurring in the opsin molecules during photodecomposition bring them to the surface, where they become available for chemical interaction. Whether this geometrical change is a metaphor for a much more complicated catalytic or direct chemical reaction is hard to say. The important fact is that shortly after the initiation of the bleaching process, the opsin is no longer an inert part of the rhodopsin complex but is altered

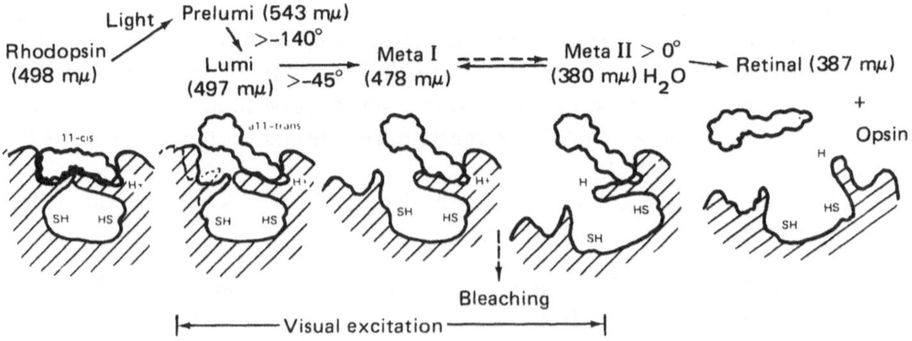

FIG. 6.4. Wald's schematic model of the decomposition of a rhodopsin-like photochemical, showing the various stereochemical stages in the breakdown of the molecule. Note the freeing of the —SH groups, a process that, he believes, may be associated with the production of the generator potential. (Adapted from Wald, 1968.)

to form a reactive structure capable of participation in other chemical processes, and by other means besides rupture of the C═N bond.

However, this model does not yet explain the production of the graded electrical receptor potential. Wald suggests another plausible mechanism that might be involved in the production of the receptor potential, which some investigators have suggested is observed as the *A* wave of the electroretinogram. If they are correct, the *A* wave may be properly referred to as the *late receptor potential* (LRP).

Wald's suggestion is that the exposed sulfhydryl groups on the opsin immediately begin to react chemically with either the outer cell membrane of the receptor itself, or with the more immediately available interior membranes of the lamella of the receptor outer segment, shown in Fig. 3.5 If the opsin in some way did change the permeability of either membrane, there would be a resulting depolarization or hyperpolarization as passive forces redistributed, in at least this local region, the ions whose concentration differences produced the resting potential. A local current sink or source would thus be produced that could lead to action at a distance—presumably at the synapse at the base of the receptor cell. Such a hypothetical system would explain one adaptive reason for the presence of the plate-like lamella in the outer segments of photoreceptors. The lamella might act as amplifiers of a sort. A local depolarization in one region might summate with those in others to produce a receptor potential magnitude far greater than that which could be produced by the single cell membrane.

Hagins, Penn, and Yoshikami (1970) have discovered that the action of the photobreakdown products is to reduce, rather than to increase, the permeability of the rod or cone membrane, and thus to reduce a continuous ionic current observed when receptors are in the dark. The decrease in permeability could lead to hyperpolarization of the membrane potential, and it is indeed hyperpolarization that is typically found in vertebrate photoreceptors when they are exposed to light. (Invertebrate photoreceptor potentials are more usually found to be depolarizations.)

There is a second process that may be involved in transduction. Brown and Murakami (1964) have shown that if the photically driven voltage (the electroretinogram) between the front and the back of a dark-adapted vertebrate eye is recorded with the use of very high stimulus energies, there occurs a rapid voltage shift, which seems to be associated with the initial stages of the photochemical breakdown. The reaction is very rapid, occurring with a latency of only a few microseconds.

This fast visual response, the *early receptor potential* (ERP), is dependent upon parallel alignment of the rhodopsin molecules in the outer segments of the intact eye. It will not occur in solutions of the photochemical in a test tube (Cone & Brown, 1967). Almost no other change in the physics of the situation seems to affect the amplitude or latency of the response. Since the signal is measurable only at very high stimulus intensity, it is not clear what role it might play in normal vision. Nevertheless, the ERP is of great interest in suggesting that there may be a direct electrical action produced by some of the steps in the breakdown of the photochemical, one not mediated by an intermediate membrane permeability change due to the presence of the breakdown products.

Cone (1967) feels that the ERP is a direct result of the redistribution of charges as the rhodopsin molecule goes through the intermediary stages of decomposition following the initial photoisomerization. Cone and Cobbs (1969) were able to evoke a distinctive pattern of the ERP at different times following a strong preliminary conditioning flash, and they believe that the shape of the ERP in these cases is characteristic of the presence of one or another of the breakdown products. Since the threshold for the ERP is at least a million times greater than that of the LRP, the ERP does not appear to be the direct antecedent of the receptor potential per se, but it probably will continue to be an important measure of the photochemistry of retinal pigments.

Obviously, the problem of transduction in the vertebrate visual receptor is enormously complicated. While no one has yet presented a complete explanation of the process, we are beginning to learn some things about the involved subcellular mechanisms. In summary:

1. The process is initiated entirely by the action of the light that alters the stereoisomeric state of the retinal portion of the photochemical.
2. The ultimate intracellular result of the breakdown in the photochemical is to produce a reduction in the permeability to sodium ions of the plasma membrane of the outer segment of the receptor cell, thus hyperpolarizing the cell.
3. This membrane permeability change can be recorded as the LRP, which is probably the same as the *A* wave in the electroretinogram.
4. The ERP, on the other hand, is an indication of the processes that are occurring within the pigment molecule rather than at the membrane.

D. TRANSDUCTION IN THE AUDITORY RECEPTOR

1. Primary Sensory Action in Audition

The primary sensory action in the cochlea is thought to be the shearing action on receptor hairs, which is created by movement of the tectorial membrane that lies along the top of the hair cells. This simple statement is ambiguous, however, for there are many ways in which such a mechanical shear could affect the hairs themselves. The response could be a direct function of the amplitude, velocity, or even acceleration of the deformation produced by the force of an acoustic stimulus. A detailed explanation of the mechanics of the auditory primary sensory action has come from the work of Georg von Békésy. Von Békésy received the Nobel Prize in 1961 for his work on the mechanics of the ear, and much of the following discussion of modern auditory receptor theory is based on his contributions.

Given that the ear is sensitive to mechanical forces, are the receptors then deformation, velocity, or acceleration transducers? Von Békésy (1951) approached this problem in a novel way. He recorded the cochlear microphonic potentials that were produced by a trapezoidal acoustic waveform such as that shown in Fig. 6.5a. If the cochlear microphonic (thought by some investigators to be the actual receptor potential) were dependent solely on the deformation, then the response should follow the trapezoidal shape of the stimulus itself, as shown in Fig. 6.5b. If it were a velocity detector, then it would be expected to produce a differentiated waveform, producing signals only when the basilar membrane is moving, like that in Fig. 6.5c. If, on the other hand, it were an acceleration detector, then

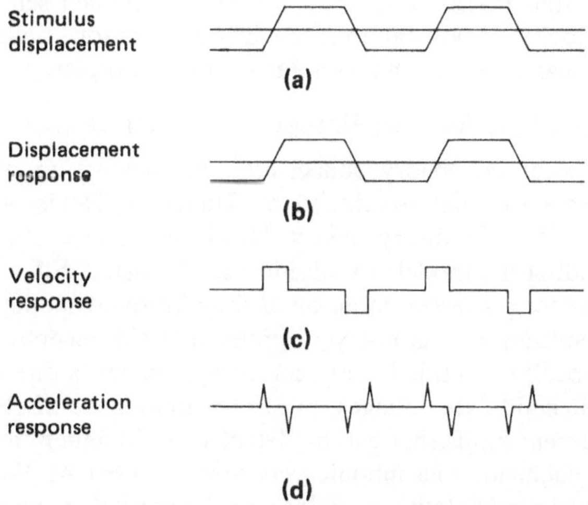

FIG. 6.5. Schematic representation of the possible observations that might have been obtained in von Békésy's experiment to determine the specific nature of the primary auditory sensory action. Shown are (a) the nature of a trapezoidally applied stimulus; (b) the cochlear microphonic response that would be expected if the primary sensory action was dependent only on the displacement; (c) the response that would be expected if the primary sensory action was based on a velocity-sensitive detection process; (d) the response that would be expected if it was an acceleration-sensitive response. Direct observations indicated that the waveform was most like (b); thus the hair cells act as displacement detectors. However, more recent evidence indicates that this may be true only for the outer hair cells. (Adapted from von Békésy, 1951.)

the waveform observed should follow the double differentiated shape of the stimulus as shown in Fig. 6.5d. Von Békésy's (1951) observations with "carefully prepared cochlea(s)" led him to conclude that the cochlear microphonic in this case is, in actual fact, trapezoidal like the function representing the position of the basilar membrane. Therefore, the transducer action is dependent upon the displacement, rather than the velocity or acceleration, created by the acoustic stimulus.

Recently, this conclusion has been challenged in part. Dallos and his co-workers (1972) have suggested that this generalization may hold only for outer hair cells, and that inner hair cells respond as velocity detectors.

Whatever is the critical aspect of the stimulus, we must also ask where these detectors reside. All investigators agree that some part of the cochlear hair cell is the acoustic transducer. Apparently, although this is not so certain, the hairs or cilia transmit the mechanical forces of the stimulus to a thickened plate at their basal end. This base plate is often presumed to be the actual mechanosensitive tissue and is thought to be specialized in some unknown way for the conversion of mechanical energy to electroneural membrane fluctuations. Though it is not known what the structural mechanism is that makes this particular part of the cell sensitive to deformation or velocity, it is clear that the primary sensory action in audition is the mechanical action of the stimulus on this base plate.

2. Production of the Auditory Receptor Potential

The most complete theory concerning the production of the auditory receptor potential is that developed by Hallowell Davis, a distinguished senior contributor to auditory theory. However, there is still some controversy surrounding his model. In addition to the usual difficulties in understanding the exact processes going on at the ultramicroscopic level of molecules and membranes, it is not yet agreed that the receptor potential has been unequivocally identified. The lack of agreement is due not to a shortage of possible candidate voltages, but to an abundance of electrical signals of several different kinds that can be detected in the inner ear.

When the cochlear microphonic was first detected by Wever and Bray (1930), it was immediately assumed to be the neurally transmitted acoustic signal. During subsequent experiments by them, however, it was shown that the cochlear microphonic is a most unusual nervous response. It has enormous high-frequency following capabilities (up to many thousands of hertz) and is so independent of body metabolism that it can be detected long after the death of the animal (Wever, Bray, & Lawrence, 1941). Since nerve impulse rates are known to be limited to about 1000 Hz, and mammalian nerves are known to be extremely sensitive to anoxia, the microphonic clearly could not be the result of a propagated nerve impulse pattern. For many years following their discovery, therefore, the cochlear microphonic was considered to be merely an artifact, perhaps of the orderly, quasi-crystalline structure of the array of hair cells. In recent years, as knowledge of sensory mechanisms has increased and the necessity for identification of a receptor or generator potential has become more apparent, there is renewed interest in assigning this role to the cochlear microphonic. Davis' (1965) theory is, therefore, important also in helping us to understand the nature of the auditory transduction process, even if the cochlear microphonic turns out not to be the acoustic receptor potential.

Davis' theory must be separated into two parts. The first presents a compelling theory of the production of the cochlear microphonic. The second part involves a somewhat less well-founded logical leap. Davis' assumption in the second part is that the cochlear microphonic is indeed the true receptor potential, and that it is directly responsible for the release of some transmitter substance at the synaptic junction between the hair cell and the dendrites of the cochlear nerve neurons. To understand fully the difference between the two parts of the theory, we must examine Davis' statements in detail and consider where observation becomes speculation.

In brief, Davis' theory states that the cochlear microphonic, or the *AC potential,* as it is often called, is generated by a gating action of the acoustic signal. A substantial constant (direct current) voltage difference between the interior of a hair cell and the fluid of the scala media is known to exist, and it is this DC voltage that Davis looks upon as the pool of energy tapped to produce the microphonic. Fig. 6.6 shows the general scheme of the endocochlear voltages within the inner ear. The two connecting scalae, the scala vestibuli and the scala tympani, are filled with a single fluid, the perilymph. The perilymph is continuous throughout the two scalae via a tiny opening, the helicotrema, located near the apex of the cochlea. Though these scalae are interconnected and have nearly the same potential with respect to the lymphatic fluid or blood, there is, surprisingly, a slight discrepancy in electropotential between the two. The scala vestibuli usually exhibits a potential 2 or 3 mV more positive than the +5 mV typically recorded in the scala tympani. This slight potential difference is probably due to local differences in the ionic composition of the perilymph in each case, but the source of the specific ion concentration is obscure. In any event, the potential difference between these two scalae is so small that it is not considered to play a significant role in the auditory process.

A different situation obtains in the scala media, however. There, a positive potential of +80 mV with respect to the blood is recorded with DC measuring systems. Another important potential difference is also present in this situation: the very large negative voltage typically found inside any nerve cell. In this particular case a —80-mV intracellular potential can be measured within the hair cell itself.

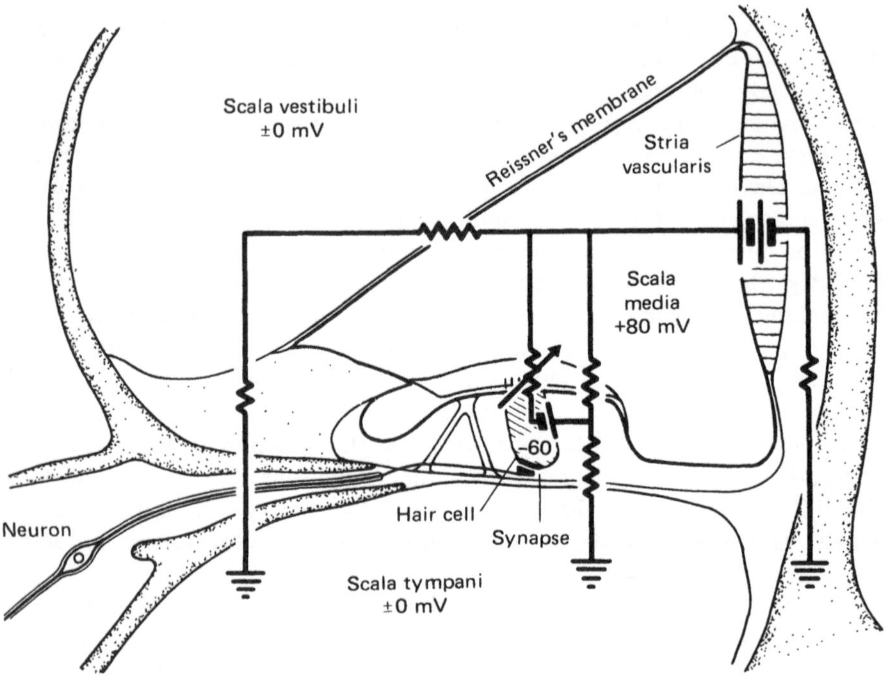

FIG. 6.6. Davis' "resistance microphone" theory of the generation of the cochlear microphonic, diagrammed in a way that emphasizes the voltage values found in the inner ear. According to Davis, the hair cell acts as a valve to gate the very large bioelectric potential difference between the inside of the hair cell and the scala media, to produce a current that is observed as the microphonic. See text for complete details. (From Davis & Silverman, 1970.)

The scala media potential and the receptor intracellular potential, therefore, sum to give a very large total potential across the hair-cell membrane; a voltage of 160 mV is often detected in this situation, an extraordinarily large potential for a single cell membrane and probably the largest transmembrane potential found in the human body.

Davis suggests that this very large potential difference produces a substantial DC resting current because of the finite impedance of the cell membrane. It is his further suggestion—and this is the key assumption in the part of his theory that concerns the origin of the cochlear microphonic—that the base plate underneath the hairs is capable, in some undefined way, of varying its impedance under the influence of mechanical deformation. As the impedance varies, the resting current is altered. In this manner, the substantial current flow that is driven by the large transmembrane potential can be modulated or gated in such a way that the total change in the DC current can be quite a bit larger than the energy that initiated the action. This process would be analogous to the processes involved in electron tube or transistor amplification of electronic flow, in which small signals control larger ones.*

There are a number of advantages to Davis' theory of cochlear microphonic generation. The theory meets the necessary criterion for energy amplication than von Békésy had shown to be necessary, and the microphonic clearly must be the end product of an energy-producing metabolic system. There are still some problems in this first part of Davis' theory (it is not exactly clear what mechanisms might underlie a change in "resistance" or "impedance" in the context of ion flow through the base plate), but most such criticisms are reductionistic details, which are not crucial to the essence of this part of his argument.

More severe criticisms of the Davis theory, however, may be directed at the second, more speculative portion. The explicit assumption in this second part is that the cochlear microphonic, produced as described in the first part of his theory, is, in fact, the receptor potential; that is, he assumes that the cochlear microphonic is specifically responsible for triggering the synaptic action, which leads to spike action potentials in second-order neurons. On the other hand, even Davis appreciates the weakness of this assumption, for he does state that the same processes accounting for modulation of the resting DC current may also, in some other way, increase the release of neurochemical transmitter substance.

However, a number of questions remain unanswered in this case. What, for example, is the role of the cochlear microphonic, if it is not the receptor potential? Is it merely a concomitant of and, in fact, irrelevant to the critical processes involved in neural activation; or does it play some role in transmitter substance release? Indeed, it is well known that microphonics of the same sort are produced by many other tissues of the body. Many of these other microphonics are sufficiently separate from any sensory transduction process to make it seem certain that they are not involved in neural processing.

Thus the real question concerning the identification of the cochlear microphonic with the receptor potential is: Is there any experimental evidence that it *alone* is sufficient to initiate synaptic activity? Davis (1965) felt there was none, and he himself looked upon this as the most speculative link in his argument.

Further speculation might suggest other possible links. The release of transmitter substance by the acoustic hair cell may be the result of some smaller voltage mechanism hidden in the relatively large cochlear microphonic. The release of transmitter substance might even be a direct effect

* It is interesting to note that a similar gating process has been suggested by Hagins, Penn, and Yoshikami (1970) for the vertebrate photoreceptor, and that a gating theory of nerve membrane potentials also has wide currency in contemporary neurophysiology.

of the mechanical deformation on some locus of the receptor cell membrane that does not require any kind of an electrical intermediary. This other locus, for example, might be the presynaptic portion of the membrane, and the presynaptic potential itself would then be the true receptor potential if this were so. For all practical purposes, however, it is unlikely that a separate membrane potential could ever be measured in the presence of the larger cochlear microphonic. No way has yet been found to disassociate a separate receptor potential, if indeed it does exist.

Davis is very much aware of such criticism. Nevertheless, within these limits, his theory is better structured and more plausible than any competing one.

E. TRANSDUCTION IN THE SOMESTHETIC RECEPTOR

1. Primary Sensory Action in Somatosensation

The wide variety of cutaneous and proprioceptive receptors are probably stimulated by the action of a number of different primary sensory actions. Hensel, Ström, and Zotterman (1951) have shown that neurons in the lingual nerve leading from the cat's tongue exhibit properties that suggest a direct sensitivity to temperature. Some of the fibers observed responded only when the temperature was maintained within a certain narrow region, while others seemed to respond only to transient changes in the temperature of the cat's tongue. Some fibers seemed to respond best to cooling and others to warming, while those that responded to ambient or steady levels also seemed to be separated into cold- and warm-sensitive groups. Quite complicated patterns of response were often observed among these steady discharge fibers, which are thought to differentially encode warmth and cold on the one hand, and warming and cooling on the other.

In general, however, we have little knowledge of the primary sensory actions that lead to thermal sensations or to such complex experiences as pain, itch, or tickle. Most of what is known about the primary sensory action in somesthesis is based upon research on mechanoreceptors such as the Pacinian corpuscle.

The identification of the primary sensory action of the Pacinian corpuscle is one that must be presented in a carefully qualified way. Experiments with the corpuscle in its normal state (that is, with the outer capsule intact) might suggest that the primary sensory action in somesthesis, unlike that found in the outer auditory hair cells, is the transient position change produced by a deforming stimulus. It might be concluded, therefore, that the Pacinian corpuscle is purely a velocity detector, since responses are elicited only at the onset and termination of the stimulus. If, on the other hand, the investigator prefers to exclude the nonneural modifications that must certainly be introduced by the elastic outer capsule, and considers only the neural aspects of the transducer, then he would want to look at the pattern of the generator potential after removal of the capsule. Under these conditions, the Pacinian corpuscle generator potential exhibits exactly the same properties as the cochlear hair cell: It responds relatively continuously to a continuous deformation. However, removal of the entire layered capsule, by far the majority of the mass of the corpuscle, does very little to the spike action potential response characteristics of the enclosed somatosensory neuron. Some other differentiating or velocity-sensitive mechanism apparently is also at work at the interface between the generator and spike action potential.

In this context it would have to be concluded that the receptor within the Pacinian corpuscle is a static deformation detector with regard to the properties of the generator potential, but a velocity detector with regard to the final spike action potential pattern. Figure 6.7 depicts the various steps in this multistage somesthetic process with and without the capsule.

2. Production of the Somesthetic Generator Potential

Werner Loewenstein's laboratory has provided the most up-to-date description of the production of the generator potential in a somesthetic receptor preparation. As in the case of the acoustic receptor, our knowledge of the underlying molecular details of this process is sketchy, but a fairly good description of the involved events has been developed.

It seems clear that the primary sensory action is a mechanical deformation of the nerve membrane itself. But this does not answer the question of how the membrane produces the generator potential. Measurements can be made of the generator potential produced by a given deformation, and the functional relationships between it and a stimulus can be elucidated; but it is not yet known how the graded generator potential is actually produced. No physiochemical theory has been advanced that is capable of explaining how a mechanical deformation might alter the permeability of the membrane to ion flow for the somesthetic receptors. Although it is simple to compute the variation in the resting potential once membrane permeability has been altered, we do not know why the permeability changes. The critical step in the production of the generator potential is still shrouded in the mystery of molecular phenomena. All we can say now is that ion redistributions do occur as membrane permeability changes, and these ion redistributions produce the depolarization called the generator potential.

We might speculate about the ways in which a mechanical deformation might alter membrane permeability. Are these specialized tissues stimulated to release substances that act chemically on membrane permeability? Are

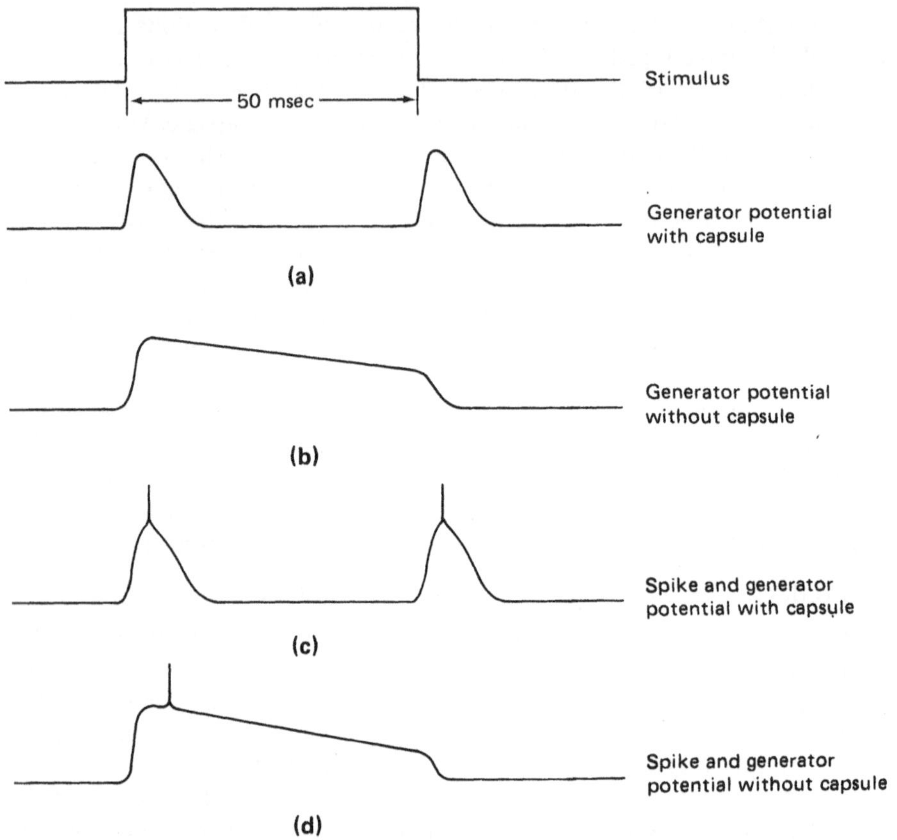

FIG. 6.7. Schematic representation of the effect of the removal of the outer multilayered capsule on the production of the generator and spike action potential patterns produced by a long-lasting stimuli applied to a Pacinian corpuscle. See text for details. (Adapted from Loewenstein & Mendelson, 1965.)

"pores" or "channels" enlarged or altered in shape in a way that allows some ions previously incapable of crossing the membrane barrier to do so? Are charge distributions altered so that the electrostatic forces repelling and attracting certain ions form new patterns? Are active transport mechanisms, which actively move ions from one side of the membrane to the other, brought into play? These and other explanations might be invoked to account for the actual transduction, but no one has yet developed a convincing theory. One reason it has been difficult is that the level of analysis is ultramicroscopic. It is not at all certain, for example, exactly what is meant by such concepts as "pores" or even "membranes" in a situation such as this, where structures of atomic dimensions must be hypothesized. (See the discussion on the theory of membranes in Chapter 3.)

While the exact nature of the conversion process at the molecular level cannot be clarified, Loewenstein and Rathkamp (1958) have at least established where in the membrane the various substeps in the transductive process occur in the Pacinian corpuscle. Their experimental setup is shown in Fig. 6.8.

Loewenstein and Rathkamp used a powerful technique, the pressure block, in order to determine the role of specific portions of the receptive tip of the axon. When a fine wire was used to compress the neuron at the first node of Ranvier (as shown in Fig. 6.8), the spike action potential disappeared, even though the generator potential was still present and normal. The effect of the block on the spike action potential can be seen in Fig. 6.9, which shows the spike before, during, and after the compression block. Compression blocks applied at more central nodes of Ranvier do not show a corresponding drop-out of the spike action potential. The conclusion drawn by Loewenstein and Rathkamp is that only the first node of Ranvier is specialized for the conversion of generator potentials into spike action potentials.

The next question, then, is: What is the site of origin of the generator potential itself? The pressure block was also used by Loewenstein and Rathkamp to answer this question, but in this case it was applied to the unmyelinated portion of the nerve ending at the tip of the axon. The result of this experiment was that the closer to the tip of the neuron the block was placed, the larger and more normal-looking was the generator potential. Figure 6.10 presents these results. The major conclusions to be drawn from this important experiment are: (*a*) it is the unmyelinated terminal end of the nerve fiber that is uniquely capable of converting mechanical energy into generator potentials, and (*b*) this capability is not localized. Rather, it is distributed over the entire receptive portion of the terminal, and the magnitude of the potential generator depends upon how much of the terminal portion of the neuron is stimulated.

An important general observation that emerges from the Loewenstein and Rathkamp experiment is that there is a high degree of specialization exhibited by different portions of the nerve membrane, even though these portions might be located within a few micra of one another. The most terminal portion of the axon can produce generator potentials but cannot be stimulated by any means to produce regenerative spike action potentials. The narrow collar of membrane at the first node of Ranvier, on the other hand, cannot be stimulated directly to produce generator potentials. Yet this collar is the only portion of the nerve that seems to be capable of converting the generator potential to the regenerative spike for long-distance communication. More central portions of the axon, while perfectly capable of propagating spikes, cannot convert the graded generator potentials into spikes unless those graded potentials are abnormally large.

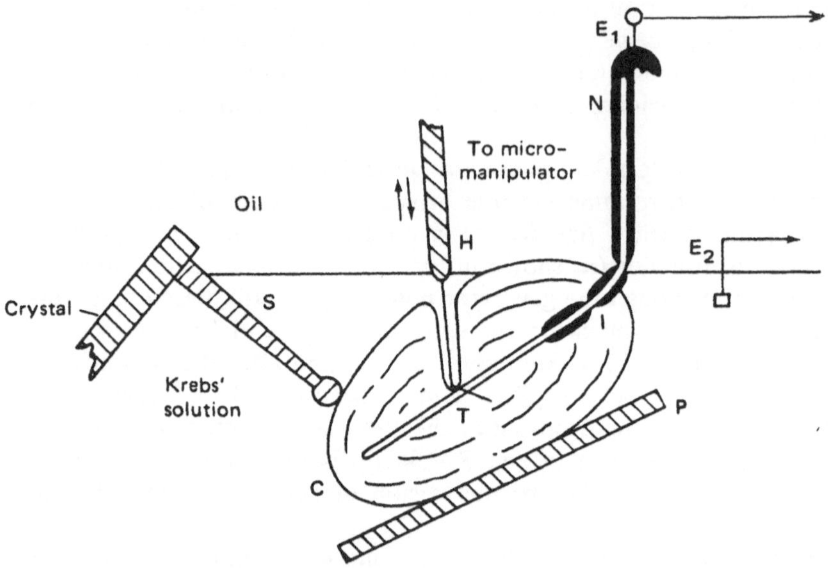

FIG. 6.8. Schematic representation of a preparation of a Pacinian corpuscle and the recording techniques used for the analysis of the transductive process. S = stimulator; C = Pacinian corpuscle; P = glass plate; H and T = wires used to block the receptor; E_1 and E_2 = electrodes. (From Loewenstein & Rathkamp, 1958.)

FIG. 6.9. Loewenstein and Rathkamp's pressure-block experiment, which demonstrated that the first node of Ranvier inside the Pacinian corpuscle is the critical point for the conversion of the generator potential into the spike action potential. Arrows indicate the point of presentation of two brief mechanical stimuli. In (b), a pressure block is applied at the point indicated by the horizontal line. Only the generator potential is seen in this case, but both before and after the block [(a) and (c)], the stimuli produce a pair of spikes. (From Loewenstein & Rathkamp, 1958.)

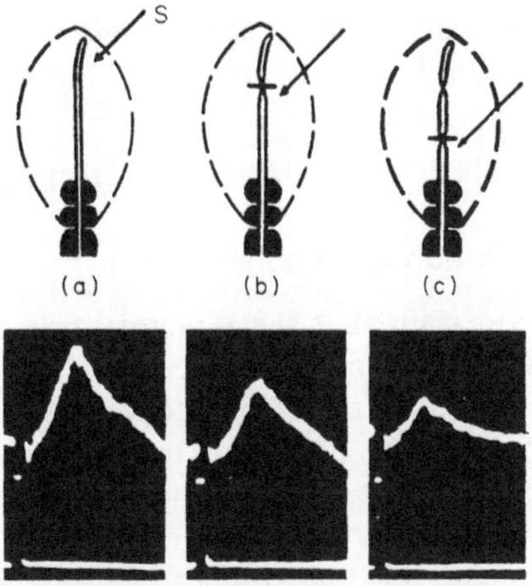

FIG. 6.10. Loewenstein and Rathkamp's pressure-block experiment, showing that the amplitude of the generator potential is a function of the available membrane at the terminal of the neuron. Arrows indicate the point of application of a single stimulus. Horizontal lines indicate the point of the mechanical block. In (a), (b), and (c), progressively smaller portions of the terminal are involved, and the generator potential is progressively reduced. (From Loewenstein & Rathkamp, 1958.)

There is an immediate temptation to compare this process to the mechanical transduction occurring in the cochlea. There are indeed fundamental similarities in the processes, but there are also dissimilarities that would strain any such analogy. The very high positive endocochlear potential in the scala media has no counterpart in the somesthetic receptor system. Thus, the notion of an impedance-reducing mechanism capable of gating exceptionally large biopotentials, as suggested by Davis for the cochlea, has no equivalent in the Pacinian corpuscle. Furthermore, the specialized subcellular microstructures of the cochlear hair cell—the hair, and the cuticular base plate (or the equivalent structure of cells of the vestibular epithelium)—have no corresponding components in the neuronal terminal within the Pacinian corpuscle. The special role thought to be played by the cuticular base plate, therefore, would have to be performed by some other membrane mechanism, and no specialized substructure has yet been found in the end-organ of these cutaneous receptors.

It is not at all certain that the picture emerging from the work of Loewenstein and Rathkamp for the Pacinian corpuscle is generalizable even to all other somesthetic mechanoreceptors. However, there is a strong presumption that it is and that the corpuscular sheaths, varied in shape as they are, do not play a direct role in the differential encoding of any other cutaneous or proprioceptive signals. What they do, according to Loewenstein, is preneurally modulate the stimulus. Thus, the notion that the transducer function is actually an evolved sensitivity of the terminal portion of the anatomically undifferentiated neural membrane itself seems to be the best explanation for most somesthetic receptors.

F. TRANSDUCTION IN THE OLFACTORY RECEPTOR

1. Primary Sensory Action in Olfaction

Although it seems certain that the olfactory primary sensory action is a chemical interaction between the molecules of the stimulating odorant and the constituent molecules of the olfactory epithelium, there is no convincing theory concerning the details of the process. Molecules differ in several ways, and it is not yet certain which difference is crucial to their distinctive reactions with chemoreceptors.

The problem of identifying the primary sensory action in olfaction is complicated by the fact that some apparently antithetical hypotheses may not be quite so far from each other as they initially seem. Superficially competing theories often really represent what ultimately turn out to be only alternative ways of looking at chemical reactions. For example, Amoore's (1952, 1964) theory of stereoisomeric shape coding, to be discussed in detail later in this section, seems at first glance to be quite different from some of the binding-type theories; but it may actually be based upon ideas of shape, which are in fact derivable from the fundamental nature of the chemistry of molecular binding. Moreover, conventional chemical interactions (specific reactions that occur between specific molecules) do not explain why many different chemicals with similar shapes, but dissimilar structures, produce similar olfactory experiences. These points can be made more concrete by considering some of the modern theories of olfactory primary sensory action.

a. Davies' "puncture and penetration" theory. Davies (1965, 1970) has suggested a model of olfactory primary sensory action he calls the "puncture and penetration" theory. His main idea is that molecules of odorous substances are absorbed by some mechanism into the lipid surface of the olfactory receptor cell membrane, but only temporarily. After a brief period, these molecules may either escape outward from the membrane, a process most properly referred to as *desorption,* or diffuse through it. Because of the molecular properties of the membrane, however, the hole that is left after the odorant leaves cannot be immediately filled, and there is a temporary gap through which Na^+ and K^+ ions can diffuse under the influence of passive forces. According to Davies, the primary sensory action is thus the absorption of the odorous stimuli by the membrane.

According to this theory, the distinctive responses of different olfactory stimuli is a result of their specific diffusion properties. Davies postulates that the quality of a sensation will depend upon the relation between the diffusion rate of the stimulus molecules away from the hole and the healing rate of the hole.

b. "Optical" theories. Over the years a number of investigators have attempted to link the infrared or ultraviolet optical absorption or emission properties of chemicals to the primary sensory action. Early theories suggested that the key link was simply the direct absorption of infrared energy by the receptor tissue from the odorants (or vice versa). However, this notion was disqualified as a possible mechanism by Ottoson (1956), who showed that no smell could be induced by a substance through a membrane, even if the membrane was transparent to its known infrared absorption or emission wavelengths. Direct chemical contact between the stimulus and the receptor is definitely required for olfactory perception.

The most important of the modern optical theories is Wright's (1964) postulate of "vibrational" coupling. He believes that the molecular basis of olfactory discrimination lies in the fact that each individual chemical structure possesses a specific vibrational pattern, which is only incidentally correlated to its infrared spectrum. Wright believes that the primary sensory action in olfaction is a coupling of the nearly resonant frequencies of vibration of substances in the olfactory pigment with those of the odorant stimulus molecules. Thus, substances with identical vibration patterns should smell alike.

However, recent experiments (Russell & Hills, 1971; Friedman & Miller, 1971) have shown that *enantiomeric* forms of some chemicals do not always smell alike. (Enantiomeric molecules have identical structures, except that they are mirror images of each other; they have identical rotational and vibrational energy patterns.) The enantiomeric forms *R*- and *S*-carvone, for example, smell like oil of spearmint and caraway, respectively. These experiments were carried out with particular attention to the control of impurities. By interconverting between the two enantiomeric substances, the smell could be systematically changed, even though the chemical formula and the vibrational characteristics remained constant. These controls overcame earlier criticisms of impurities as a source of error in experiments with enantiomeres, and, as the issue of the journal *Nature* in which these two experiments were published put it, these data are "death" to any theory that attributes olfactory specificity to the optical or vibrational properties of the macromolecule.

c. Amoore's stereochemical configuration theory. At the present time there simply is no universally accepted theory of olfactory primary sensory action. The one that has received the most attention recently is the development by Amoore (1952, 1964) of a suggestion made by Moncrieff (1949), now referred to as the stereochemical configuration theory of olfaction. "Stereochemical" is a term that refers to the three-dimensional shape of molecules.

Amoore's theory is based upon a number of indirect pieces of evidence. His original approach was based on an assumption that, at first glance, might appear to be a somewhat farfetched oversimplification. He simply assumed that if there were specific receptor sites, there would have to be a finite, perhaps even a very small, number of distinct basic smells. This notion is supported by the fact that specific anosmias (smell "blindnesses") are known to exist. Such specific insensitivity suggests the presence of a set of specific receptors, one type of which may be missing and so produce the deficit.

Most ingeniously, Amoore turned to a literature search to determine what basic smell names were used by chemists more often than any others. These, he thought, might in some linguistic manner reflect the biological characteristics of a set of basic smells. The literature search did turn up a surprising fact: A large proportion of smells reported by chemists could be accounted for by only seven smell names: *camphoraceous, musky, floral, minty, ethereal, pungent, putrid.*

Amoore continued his literature search and examined the stereochemical shape of a large number of odorants that were routinely described as having one of these seven smells. He found that there was indeed a rough correlation of shape and odor. Each of the seven basic smells, except for pungent and putrid, was associated with one of the prototypical shapes shown in Fig. 6.11. Camphoraceous smells were highly correlated with chemicals whose stereoisomeric shape was ovoid. Musky smells were characterized by oval disks; floral smells, by mace-shaped molecules; minty smells, by oval wedges; and ethereal smells, by rod-shaped molecules. The two exceptional smells, pungent and putrid, were not consistently of one shape or another; but Amoore observed that pungent smells were very often positively charged molecules (indicative of an electron deficiency somewhere in their structure), and putrid-smelling chemicals were very often negatively charged molecules (indicative of a surplus of electrons somewhere in their structure).

The next step in the development of Amoore's theory required him to make a speculative leap that was both ingenious and, as it turned out, treacherous. He assumed that if the molecules of odorants did have such specific shapes (or charges), then the way in which they would be captured by the olfactory receptor might depend upon similar specificity in the shape of the receptor sites. Figure 6.12 displays the set of five shaped and two charged receptor sites, which Amoore thought might be sufficiently shape-specific to particular basic odorant types so that one type of odorant would fit in better than any of the others.

It would be well to mention at this point that this notion of shaped receptor sites, which the reader can see emerges as a sort of "key-and-lock" theory, might be considered to be a sort of metaphor. Receptor site "shape" might be nothing more than a description of a region with a very specific

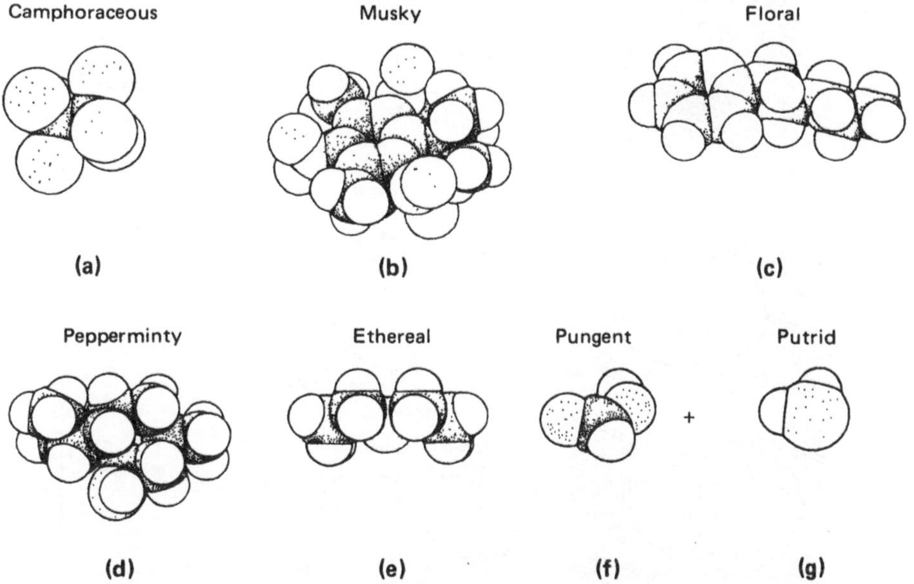

FIG. 6.11. The stereogeometric forms of the seven basic smell substances suggested by Amoore, and the names commonly associated with these stimuli. Five of the basic smell chemicals are individually shaped. Two are based upon the electrical charge of a relatively small molecule. (From Amoore, Johnson, & Rubin, 1964.)

sort of chemistry, just as molecular "shape" of the odorant might be a metaphorical description of a specific form of stimulus chemistry. The main significance of Amoore's theory, then, might lie not so much in the exact nature of a site as in the general notion that there exist small localized spots that have heightened chemical receptivity to one or several types of odorous molecules. These localized spots on the microvilli of a single receptor cell need not be exclusively of a single type. Thus, while a given receptor spot may be quite specific, it would still be possible to have receptor cells that were broadly tuned to a variety of odorants.

In order to test his theory, Amoore carried out a number of interesting experiments, which have modified and elaborated the original hypothesis (Amoore, 1965; Amoore & Venstrom, 1967). In one experiment, laboratory-synthesized molecules of the appropriate shape were sensed by expert observers in an attempt to determine whether the hypothetical association of smell with molecular shape held true for synthetic molecules. Generally, the answer was affirmative.

In another experiment, Amoore (1965) developed two precise ways of measuring shape similarities—a step considerably beyond a simple subjective, qualitative categorization of a given molecule as "ovoid" or "rod-shaped." The first of these techniques involved the manufacture of solid models of the molecules and Plexiglas "tank" models of the hypothetical

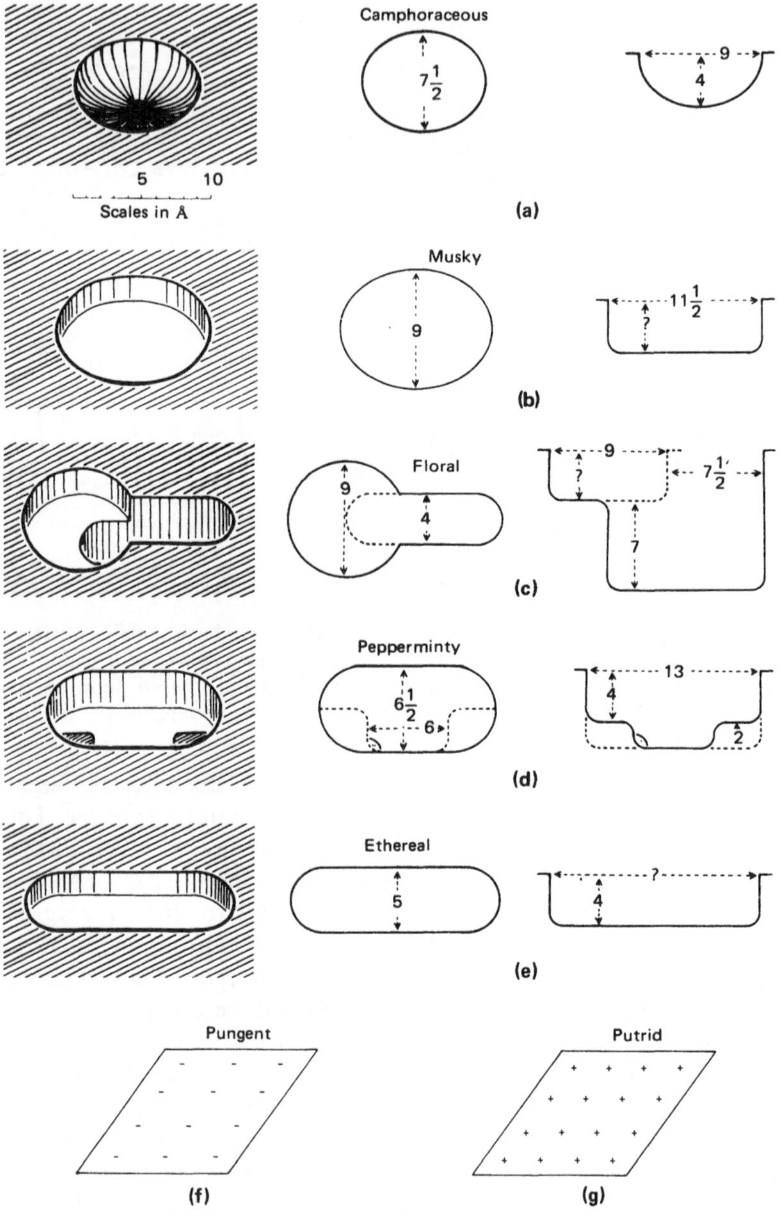

FIG. 6.12. Amoore's suggested shapes and dimensions for the five shape-sensitive receptors and the two charge-sensitive sites on the olfactory receptor. (Adapted from Amoore, 1962.)

receptor sites. The model of the molecule was then immersed in the water-filled receptor site model and measures were taken of the amount of water displaced by the model molecule. The complementarity (C), or degree of fit of the molecule in the tank, was then defined as

$$C = \frac{V_d^2}{V_s \times V_m} \tag{6.2}$$

where V_d was the volume of water displaced by the model, V_m the volume of the model, and V_s the volume of the site. The better the fit, the closer the complementarity would approach 1, that is, a perfect fit. However, because all models could not be immersed completely, and because of the "knobby" shape of these molecular models, the actual complementarity figures varied from a high of 0.44 to a low of 0.03.

Unfortunately for his theory, these complementary scores did not correspond very well with an associated set of psychophysical tests of odor similarity. Many molecular shapes with similar complementarity scores smelled quite different from each other, and many with dissimilar scores smelled quite alike. In fact, as Amoore (1965, p. 629) says, for the ethereal and camphoraceous sites, the results were "hopeless."

It should be remembered that, in this particular case, a comparison was being made between well-known molecular shapes and hypothetical receptor sites, of which virtually nothing is known. For this reason Amoore tried an alternative method of comparing shapes, one that did not involve receptor shapes. Silhouette cutouts were made of the top, front, and side views of each test molecule. This time the silhouettes were compared not with a hypothetical receptor site, but with the silhouettes of a standard set of odorous chemicals. Table 6.2 lists the standard odorants chosen by Amoore.

Metrics of the similarity of a test and a standard molecule were then computed in the following way. Amoore first physically determined the center of gravity of the cutout silhouette. The radial distance from this center to the periphery was measured for each 10 degrees of arc. Comparisons were then made by subtracting each distance on the sample from an equivalent measure on the standard and computing the mean deviation $\bar{\Delta}$ of all the 108 angular radii (108 radii are present, because each of three silhouette views of a single molecule has 36 measured radii). The similarity index was then computed as the ratio

$$\text{Similarity} = \frac{1}{\bar{\Delta} + 1} \cdot \tag{6.3}$$

Correlations were then computed between psychophysical tests of odor similarity and these measures of shape similarity. Relatively high degrees

TABLE 6.2
Set of Standard Odorants That Will Produce the Seven Basic Smells (According to Amoore)

Standard odor	Representative compound
Ethereal	1,2-dichloroethane
Camphoraceous	1,8-cineole
Musky	15-hydroxypentadecanoic acid lactone
Floral	*d*,1-β-phenylethylmethylethyl carbinol
Minty	*d*,1-menthone
Pungent	formic acid
Putrid	dimethyl disulfide

of correlation between odor and shape similarity were found for 107 different substances.

These geometrical analyses required a major modification of the original stereoisomeric theory. Amoore still believes, on the basis of the silhouette similarity measures, that molecular shape is the basic discriminating characteristic of olfactory chemicals. But he has backtracked with regard to the notion that equivalently shaped receptor sites account for differential receptivity. Simple three-dimensional geometry of receptor sites seems inadequate to explain the selective action of molecules of a given shape. For the most recent statement of his ideas, see Amoore's (1970) book *Molecular Basis of Odor.*

If a key-and-lock notion cannot be used to explain different sensory effects of molecules of varying shapes, the possibility must again be raised that shape alone is an epiphenomenon merely reflecting some more fundamental characteristic of the interactive forces that hold molecules together; and it is that more fundamental property, rather than shape, that must ultimately explain receptor specificity.

In summary, it seems most likely that the primary sensory action in olfaction is a direct chemical interaction of odorant molecules with more or less specific receptor sites on the surface of the olfactory epithelium. The detailed nature of this chemical reaction is, for all practical purposes, still unknown. With the rejection of the key-and-lock aspects of Amoore's theories, the single compelling fact remaining is that the shapes of the molecules are highly correlated with psychophysical discriminations. Future research will have to unravel how this shape difference is discriminated by olfactory receptors.

2. Production of the Olfactory Generator Potential

Since the nature of the olfactory primary sensory action is unknown, the mechanism of the olfactory generator potential is equally obscure. We can only speculate that in some way chemical action leads to a change in receptor cell membrane permeability in the olfactory epithelium. This permeability change allows the ionic equilibrium to shift in such a way that a graded potential is produced, which initiates spike action potentials in the tiny axons of the olfactory nerve.

This simple speculative statement, however, is complicated by the fact that there is a great deal of controversy about whether or not the olfactory generator potential has actually been detected. Slow potentials are known to be generated at the olfactory epithelium by odorous stimulants. Nevertheless, it is not yet clear whether these slow potentials actually are the true generator potentials. Ottoson (1956) has made extensive studies of a graded negative potential, which occurs with moderate latency and a very slow time course after the application of a stimulus. Figure 6.13a shows a typical negative electro-olfactogram recorded with the apparatus shown in the inset. This signal is now referred to as the "Ottoson potential," in honor of its discoverer.

However, a positive potential of similar temporal characteristics has been observed under some conditions. Takagi, Shibuya, Higashino, and Arai (1960) observed that the polarity of the graded Ottoson potential is reversed with certain stimulants (see Fig. 6.13b). In addition, Takagi and Shibuya (1959) showed that there is often also a negative off-potential associated with the typical negative Ottoson or the anomalous positive on-potential, as shown in Figs. 6.13b and c. Figure 6.13d shows the associated spike activity in the olfactory bulb and its relation to some of these graded potentials.

A number of discrepancies prevent confirmation of any of these graded potentials as generator potentials. Takagi (1967) spells out some of the discrepancies in an interesting review of the problem. One of the most damaging is a report that showed that the Ottoson potential could be eliminated (by wiping off the mucous usually found on the olfactory epithelium) without any diminution of olfactory bulb spike action potentials (Shibuya, 1964). Takagi also suggests that at least some of the system of graded potentials may be no more than secretory potentials produced by the sustentacular cells of the olfactory epithelium when stimulating chemicals are present. If this were so, it would be a somewhat disappointing state of affairs, because no known candidate for the generator potential role would then be left.

Clearly, there is much more to be learned about the transductive process of olfaction, both with regard to the generator potential and its origin,

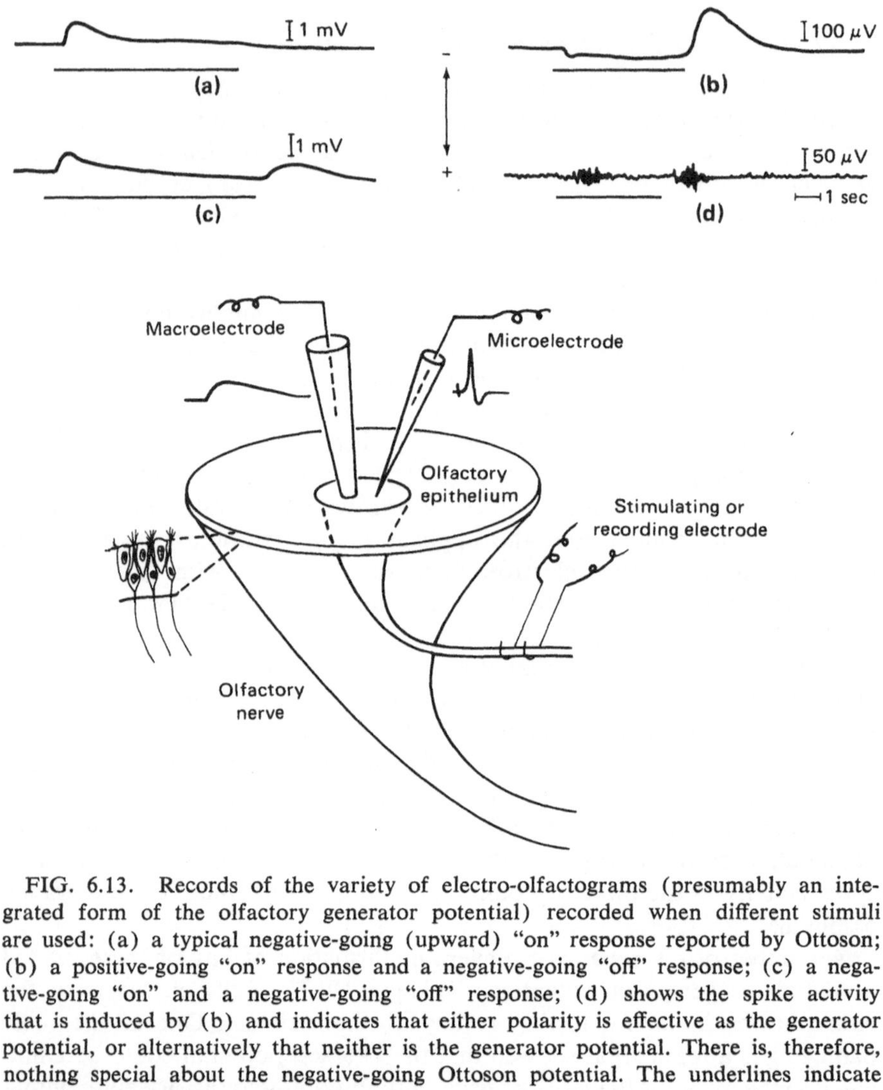

FIG. 6.13. Records of the variety of electro-olfactograms (presumably an integrated form of the olfactory generator potential) recorded when different stimuli are used: (a) a typical negative-going (upward) "on" response reported by Ottoson; (b) a positive-going "on" response and a negative-going "off" response; (c) a negative-going "on" and a negative-going "off" response; (d) shows the spike activity that is induced by (b) and indicates that either polarity is effective as the generator potential, or alternatively that neither is the generator potential. There is, therefore, nothing special about the negative-going Ottoson potential. The underlines indicate the stimulus duration producing the various potentials. The insert shows the recording apparatus. (Adapted from Takagi, 1967.)

and with regard to the mechanism for conversion of generator potentials to spike action potentials. Unfortunately, the latter conversion is probably a function of a region at the base of the olfactory receptor cell, where its axon emerges; and the olfactory axons are so small that technical difficulties probably will prohibit substantial progress in understanding action potential generation in the foreseeable future.

G. TRANSDUCTION IN THE GUSTATORY RECEPTOR

1. Primary Sensory Action in Gustation

Gustation is a sense that seems to be mediated by direct chemical interactions between molecules of the stimulus and the constituent molecules of the taste receptor cells. Yet the very directness of this chemical action makes our knowledge of the primary sensory action less complete than for some less directly activated modalities. The primary processes involved in taste are so microscopic that it is not yet even clear exactly where the site of the primary action resides. It is speculated that the significant chemical action occurs at specialized portions of the microvilli of the gustatory receptor cells.

Because of this uncertainty and difficulties with stimulus control, research in the area of gustatory transduction has concentrated on problems that are amenable to chemical analysis. For example, attempts have been made for many years to find chemical correlates of the four classic fundamental tastes—sweet, sour, bitter, and salty. It should be emphasized, however, that even the notion of a "fundamental taste" is an idea drawn from only the loosest analogy between visual color-mixing ideas and gustation. While this analogy may have some useful pedagogic value, the lack of a single dimension of variation, comparable to photic wavelength, makes it a weak one. This is particularly so in light of the fact that correlations between chemical structure and taste are incomplete, at best. Furthermore, taste scientists have generally found it difficult to create complex tastes from mixtures of the four fundamentals. There is apparently much more going on in the tongue than in the eye, where quality does seem to be a relatively simple additive mixture of a set of fundamental stimuli.

Pfaffmann (1959), in a comprehensive review of the taste literature through the 1950's, describes the tentative, inconsistent nature of the associations that have been made between taste and chemical structure. Sourness, for example, seems to be a function of the pH or acidity of a solution. The greater the number of hydrogen ions present, the greater the sourness of a substance. But to confuse the issue, not all acids are sour. Amino acids are sweet, and picric acid is bitter.

A sensation of sweetness also results from stimulation with a wide variety of very dissimilar substances. Typically, organic molecules are the ones that taste sweet, but so do heavy metal salts of lead and beryllium. This probably accounts for the high incidence of lead poisoning in children in poverty situations. Lead salts from painted walls are used as a dangerous substitute for the sucrose or lactose of candy or milk.

Dzendolet (1968) has suggested that a common feature of sweet-tasting substances is that they have some molecular bonding arrangement that makes them proton (hydrogen ion) acceptors. This mechanism would be approximated any time a hydroxyl ion (OH^-) or a hydroxyl-like structure is created by some chemical action. For example, Dzendolet believes that the metal ions of inorganic salts, which are created upon dissolution, become surrounded with water molecules, that is, become hydrolyzed. However, since the metal ions are positively charged, they tend to repel some of the hydrogen ions (protons) of the hydrolyzing water, and these ions become lost or trapped by other proton sinks, such as saliva. In some instances the loss of these protons leads to an OH^- group becoming sufficiently free from one of the repelled hydrogen ions to act as an acceptor of some other hydrogen ion. If this new hydrogen ion comes from the gustatory receptor sites, a sensation of sweetness is presumably evoked. Heavy metals such as beryllium and lead are presumed to produce sweet experiences in this same manner. Figures 6.14a and b display the initial hydrolyzing reaction of a typical metallic ion and the resulting repulsion

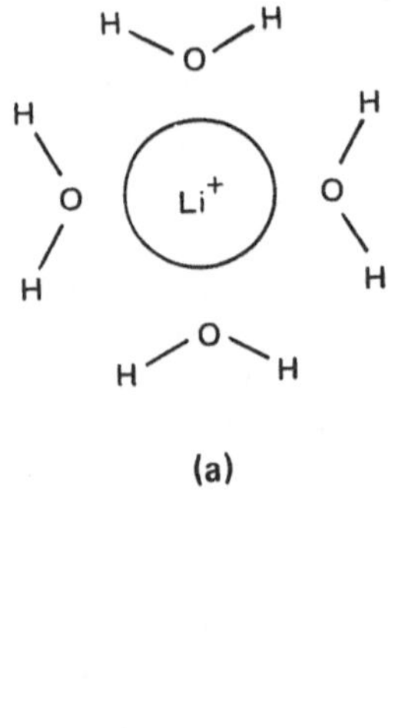

(a)

FIG. 6.14. Dzendolet's scheme of H^+ ion replacement, which may lead to specific sensitivities for sweet-tasting substances. He proposed that when ions like Li^+ become surrounded by water molecules, the positive charge of the Li^+ tends to force an H^+ ion away from one of the water molecules. At some subsequent time, the resulting OH^- molecule may become an H^+ acceptor. It is this attraction for H^+ ions that Dzendolet believes is common to all sweet-tasting substances. (From Dzendolet, 1968.)

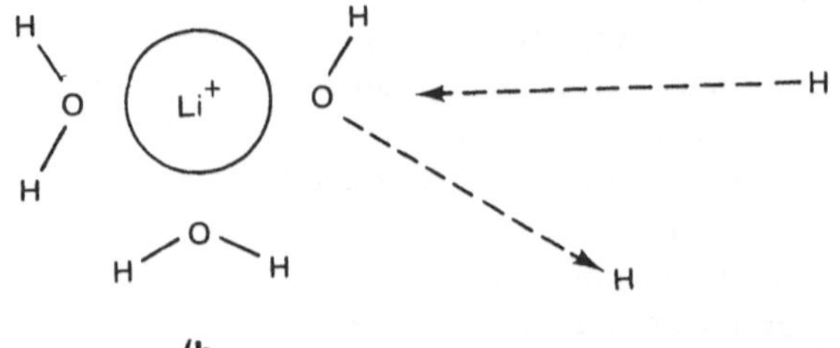

(b)

of a positively charged hydrogen ion from one of the water molecules. A similar process, Dzendolet suggests, operates for other sweet-tasting substances, such as sucrose, saccharin, and urea. This concept, however, is in disagreement with the idea of heavy sweet binding substances discussed below.

Many chemical classes contain substances reported to have a bitter taste; typically, as the inorganic molecules increase in weight, they taste increasingly bitter. Very often, bitter substances contain nitrogen.

Salty tasting substances are usually ones that ionize in solution. Saltiness seems to depend on the ionization of soluble salts, which are usually made up of a metallic cation and a halogen anion. The metals usually involved are sodium (N^+), potassium (K^+), lithium (Li^+), magnesium (Mg^{++}), and calcium (Ca^{++}). But the ammonia radical (NH_4^+) is also often found as the cation in salty-tasting substances. Typical anions include chlorine (C^-), bromine (Br^-), iodine (I^-), the sulphate radical (SO_4^{--}), the hydrocarbonate radical (HCO_3^-), and the nitrate radical (NO_3^-). Salty-tasting substances are of particular interest because they often include among their constituents some of the ions involved in the maintenance of the transmembrane resting potential. It would seem that the addition of these species of ions could be expected to influence ionic equilibria very directly, but this appears not to be the case.

Beidler (1967) has presented one of the most comprehensive contemporary views of the primary sensory action in gustatory transduction, based on what is known of the structure of the plasma membrane. His main idea is that the protein chains, which may possibly be part of the surface of the membrane, are linked together by several different kinds of bonds. Because of imperfect bonding of many sites on the unstimulated membrane, a wide variety of free bonding sites are available to be linked to any appropriately reactive stimulus substance. Beidler suggests that the membrane might also have locally charged sites or even a cumulative charge over its entire surface.

The net impact of the presence of this variety of binding types and sites is that many kinds of solutions can be bound in numerous ways at various places on the receptor surface. Beidler points out that not only charged ions are capable of this sort of interaction, but also many other nonelectrolytes such as sugars, steroids, and urea, both at the surface and within the membrane.

Beidler's notions, while they are extremely interesting and most certainly important in understanding the primary sensory action in gustation, are distressingly incomplete. No specific statement can yet be made about the particular types of bonds necessary to produce a given sensation. Simple correlations between chemicals and tastes remain elusive. Various acids, cations, and large organic molecules bind in ways that seem specific to each particular chemical, and little of a truly general nature seems yet to have been elucidated.

A further complication is the fact that not all chemicals serve to excite a reaction when they are bound to the membrane. Some seem to inhibit sensory activity, and some seem to play no direct role but merely alter the charge distribution on the membrane—an alteration that might have substantial indirect effects on the binding of other, more directly effective agents. In general, but again not exclusively, anions tend to bind to the charged regions on a membrane in an inhibitory manner. Cations, on the other hand, generally tend to excite. Perhaps the essence of Beidler's current ideas about the primary sensory action can be best summed up by quoting him directly:

> The relative number of anionic and cationic membrane sites, as well as the concentration and effectiveness of each kind of cation and anion in the taste solution, determine not only the magnitude of the taste response but also its direction, whether inhibitory or excitatory [Beidler, 1967, p. 532].

But the exact nature of the specific primary action for the various taste qualities is yet to be worked out.

A step in this direction has been taken, however, by Dastoli and his co-workers. Dastoli and Price (1966) and Dastoli, Lopiekes, and Price (1968) have extracted a substance from the tongue of cattle that exhibits a binding action to various sugars in much the way predicted by Beidler, and in a manner that is highly correlated with sweetness psychophysics. The sweet-sensitive substance appears to be a protein with a molecular weight of about 150,000.* Dastoli (1969) reported that he was also able to extract a protein that seemed to be quite specific to bitter-tasting substances. The type of interaction between these proteins and the taste stimuli seems to be best described in terms of the weak electrostatic binding actions already discussed.

2. Production of the Gustatory Receptor Potential

Clearly, neither the site nor the mechanism of the gustatory receptor potential production process are fully understood. Some facts seem highly likely, however. The various binding actions described almost certainly lead to changes in the permeability characteristics of the receptor membrane. It must thus be assumed that an alteration in the permeability of the membrane to the specific ions involved in the maintenance of the resting potential occurs as stimulating chemicals bond to the membrane. As the Na^+ K^+, and Cl^- ions change their steady-state conditions across the membrane, a change in the resting potential develops. The effects of this change in polarization might be to release transmitter substance at the base of the

* This is a large enough weight to suggest that such a molecule might penetrate the entire thickness of the membrane, as shown in Fig. 3.24.

gustatory cell, or to affect directly the postsynaptic regions, thus leading to activation of the initial fibers in the gustatory nerves. Beidler suggests that the transmission of information contained in the receptor potential might also possibly be a result of a quasi-physical effect, such as the reorientation of the constituent molecules of the membrane. The effect might equally well be a simple electrical sensitivity of presynaptic regions supplying the transmitter substance. Again, we must simply admit that we are at the frontier of contemporary knowledge, and aside from these speculations, no more definite information is yet available with regard to the primary sensory action in gustation.

Ozeki (1970) has studied the ionic origin of gustatory receptor potentials. He believes that there is some fundamental difference in the membrane processes that are stimulated by sodium chloride and quinine hydrochloride, respectively. To arrive at this conclusion, Ozeki inserted KCl-filled glass microelectrodes into gustatory receptor cells and recorded the electrotonic potentials produced by a series of applied constant-current electrical test pulses superimposed on the receptor potential. The brief (100 millisecond) electrotonic potentials were purely passive electrical responses, merely reflecting the momentary resistance of the membrane. His findings suggested that the action of sodium chloride on the membrane was to reduce the electrotonic potentials by reducing the overall membrane resistance. This is shown in Fig. 6.15a by the reduction in size of the brief electrotonic potentials produced by the test pulses during the period of the receptor potential, compared to the unstimulated periods.

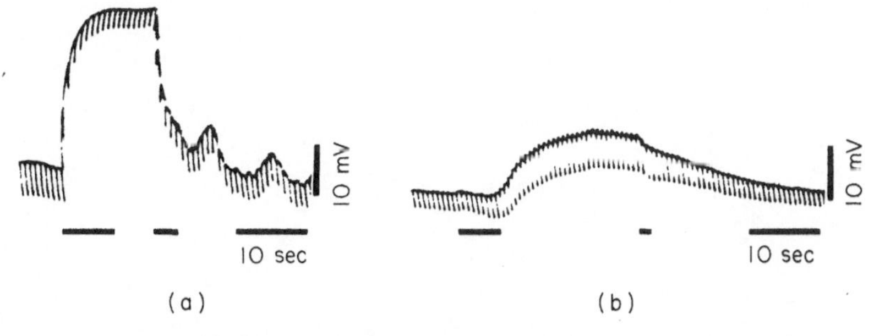

FIG. 6.15. Two receptor responses from a rat's gustatory receptor cell. The continuous curve is the receptor potential, while the small downward-going spikes are the potentials produced by a series of applied test stimuli. These spikes are, therefore, nothing more than the electrotonic potential produced by the action of the test current, and their amplitude is purely a function of the membrane resistance in accord with Ohm's law. (a) The response of the cell to NaCl; (b) the response of the cell to quinine hydrochloride. The small spikes decrease in size in (a) but increase in (b), indicating that the depolarizing membrane receptor potential in (a) is due to a decrease in membrane resistance, but in (b) it is more likely due to an increase in membrane resistance. (Adapted from Ozeki, 1970.)

On the other hand, the response to quinine hydrochloride was quite different. Figure 6.15b shows that membrane resistance actually increased when the cell was stimulated, for in this case the electrotonic responses to the test pulses were enlarged during the period of the receptor potential. On the basis of a voltage-clamp type of experiment, Ozeki concluded that this increase in membrane resistance during stimulation with quinine hydrochloride was due to a specific decrease in membrane permeability to potassium ions. However, the same sort of clamping experiment suggested to him that the decreased resistance observed during stimulation with sodium chloride was due to a specific increase in the permeability of the membrane to sodium ions. Thus, the membrane actions underlying the origin of receptor potentials produced by quinine hydrochloride and by sodium chloride seem to differ. Whether similar differences occur with the other basic taste stimuli is still moot, but this sort of differential effect seems consistent with the specific ionic binding site hypothesis of the primary gustatory sensory action proposed by Beidler.

The discussion of gustatory transduction completes our brief review of the initial stages in neural processing and transmission. The various receptors convert physical energy through one or another primary sensory action into a generator or receptor potential. This graded action of the receptor ultimately leads to a spike action potential—a signal that can be propagated over long distances in the nervous system. In the next chapter we consider the underlying ionic mechanisms of the transmission process.

7
THE TRANSMISSION PROCESS

A. INTRODUCTION

Now that we have discussed the ionic basis of the resting potential and the transduction processes that initiate neural activity, we can turn our attention to the transmission of information from one point to another in the nervous system. The activation of a receptor cell would be without biological value if the information pattern generated by the energy transduction were not transmitted to some remote destination. The transduced information *must* be decoded or integrated with information arriving over other transmission lines, if the organism is to respond meaningfully.

There are two ways in which information can be transmitted by individual neurons: "spikes" (propagated, nondecrementing, "all-or-none" signals), and more localized, graded, summatable potentials.

The spike action potential, with its special characteristics, is the mechanism *par excellence* for the transmission of neural information over long distances in the nervous system. Its particular utility lies in the fact that it is not reduced in amplitude or shape as it progresses from one place to another. Graded potentials, on the other hand, rapidly diminish in amplitude and change in waveform the further a measurement is made from their original source. This diminution is due to membrane resistance and capacitance. At best, graded potentials can be detected at distances no greater than a millimeter or two from their points of origin.

The reason for this difference in the action of the two potentials is simple to describe superficially, but difficult to explain in detail. On a very qualitative level, axons that are capable of sustaining the propagated spike action potential provide the energy necessary for continuation of the progressive, regenerative action at each point along their course. In contrast, the regions that have evolved specifically to sustain graded potentials generally do not provide any such energy; the spread of their effect is explained by the passive spread of current and voltage through the electrically conductive medium of the membrane and its internal and external fluids.

It is difficult to go beyond a descriptive statement of the specific properties of the membrane that account for the spike and the graded potential. The structural characteristics of the membrane, which allow one portion to propagate a spike and prevent another region from doing so, are clearly to be found at the molecular level and are not yet understood in detail.

What we do know is that in the case of both the graded and the regenerative spike potential, electrical current flow is mediated solely by ions. There are few free electrons in an electrolytic solution, and therefore no electronic conduction. Although there has been some speculation about some semiconductor effects in nerves, such effects have not yet been demonstrated in any way convincing enough to make them serious contenders for explaining membrane action. Charge transfer by ionic carriers, therefore, is the basic concept for explaining both kinds of transmission: passive spread of graded potentials and propagation of spikes by active energy-consuming processes.

Although the neuronal transmission of charge is mediated by ionic carriers instead of electrons, the same charging and discharging characteristics displayed by a metallic cable are exhibited by neurons. Elements with resistive and capacitive properties are present in the nerve membrane. Therefore, the same mathematical model that describes an electrical TV cable can be used to describe the passive spread of graded potentials along the axon. The numerical values of the time and space constants are quite different, of course, but the model is still valid.

The velocity at which the graded potential spreads depends upon the RC (resistance $\times$ capacitance) time constants at each link of the neuronal "cable." Although maximum speeds of a few tens of meters per second are characteristic of propagated spike action potentials based on the same sort of ion flow, decremental spread occurs much more slowly—often less than one meter per second. Furthermore, the spread of graded potentials along the nerve membrane is due primarily to a flow of ions forced from place to place by stationary potential fields, whereas propagated potentials are driven by potential sources that are sequentially triggered and move along the neuron, drawing energy from each sequential location. The spike, therefore, can be just as large at its destination as it was at its source, unlike the decremental spread that leads to progressively weaker signals the greater the distance from the source.

With these general points in mind, we now shall consider the specific attributes of decremental graded action potentials for short-range communication, and the propagation of spike action potentials for long-distance communication.

B. DECREMENTAL CONDUCTION

Let us assume as a sort of thought experiment that it is possible to perturb the resting membrane potential at one region of an axon, while holding membrane properties constant at all other regions. In such a situation, a *dipole* of voltage difference is said to exist between the perturbed region and those places that have maintained the original membrane potential. In this way an ionic current is produced, which passes through the cell's interior and external medium. Such a current can affect distant locations, even though there is no movement of the abnormal voltage perturbation itself (i.e., no propagation of the signal). This ionic current also conveys charge in directions dictated by the simple laws of electrostatic attraction and repulsion, and at speeds determined by the membrane resistance and capacitance, in accord with the "cable" model mentioned above. Similarly, the amplitude of the signal that can be measured at any given distance from the original perturbation will be a function of the shunting resistance across the membrane.

Figure 7.1 is a diagram showing some typical values of the capacitive and resistive elements of one nerve membrane cable model. The relatively long charging times of the capacitive elements in this resistance-capacitance combination is most responsible for slowing the spread of currents to other points along the nerve membrane. The amplitude of the potential at a distant point will, therefore, depend not only upon the distance, but also upon the length of time since the perturbation was applied. The voltage at several distances, plotted as a function of time, is shown in Fig. 7.2. After a sufficient period of time has passed, all locations along the nerve membrane will be charged to some maximum value, as defined by the shunting resistance.

Though this sort of localized, decremental, passively spreading potential is not propagated by an active process and affects only its immediately neighboring locations (the effect may spread in some cases as little as 100 μ), it is a very important means of neuronal transmission and information processing. The critical reason for its importance lies in the fact that decremental potentials can be added and subtracted algebraically by various portions of the neuronal membrane. Thus, although one local perturbation may be insufficient to produce an informationally significant effect, several can summate to create a cumulative potential large enough to trigger a significant response. Similarly, some decremental signals

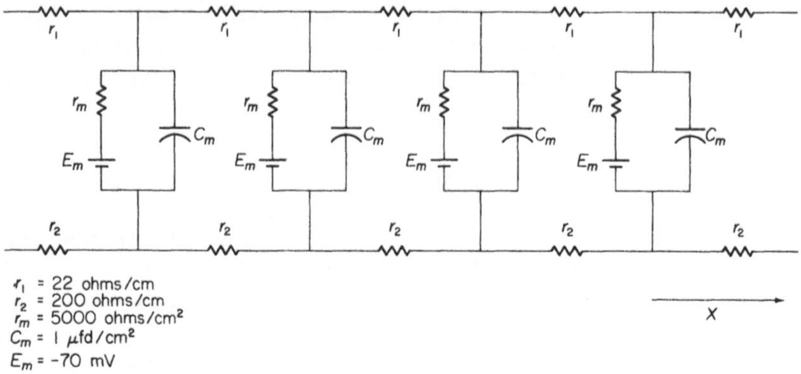

FIG. 7.1. Diagram of the cable model equivalents of a section of nerve membrane. Values are typical but can vary considerably, depending upon the species from which the neuron was obtained. (Adapted from Plonsey, 1969.)

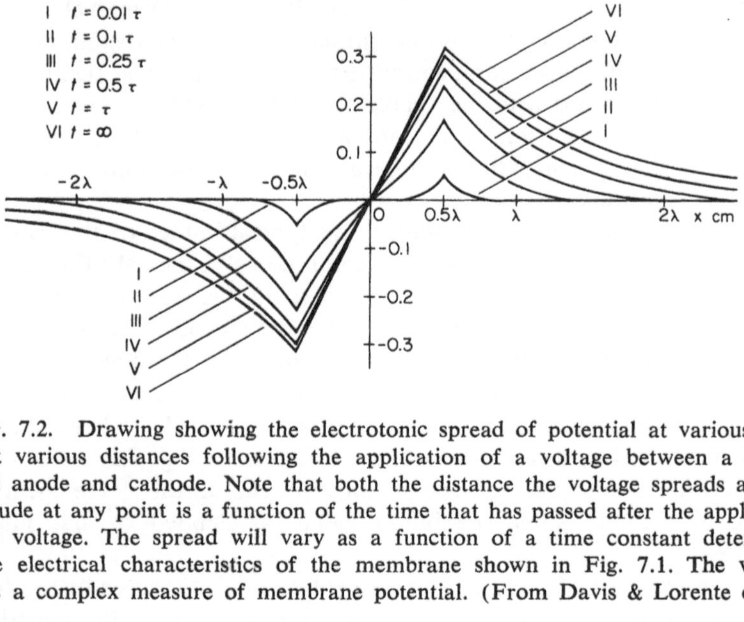

FIG. 7.2. Drawing showing the electrotonic spread of potential at various times and at various distances following the application of a voltage between a closely spaced anode and cathode. Note that both the distance the voltage spreads and the amplitude at any point is a function of the time that has passed after the application of the voltage. The spread will vary as a function of a time constant determined by the electrical characteristics of the membrane shown in Fig. 7.1. The vertical axis is a complex measure of membrane potential. (From Davis & Lorente de No, 1947.)

may subtract from or inhibit others, thus modifying the pattern in substantial ways.

Thus, the most important function served by the graded potential is that of integration. Its ability to add to or subtract from other, similar signals allows information to be combined, intermixed, and to interact in ways that vastly increase the repertoire of behavioral responses possible to organisms. The interaction is so complex that molar human behavior, as the most impressive example, appears to be even qualitatively different from the action of individual neurons, in spite of the fact that they must be based on the same physical and chemical laws.

On a more immediately researchable level, scientists can examine elemental interactions at the cellular level. While we cannot yet say how single-cell responses lead to human cognition, we can determine the values of the capacitive and resistive elements in the "cable model" describing a specific neuron. To do so involves the application of an artificial stimulus—an electric voltage—to a select portion of the neuron, which acts as an artificial local potential. By measuring the temporal and spatial properties of the membrane current response to this stimulus, we can estimate the value of the resistive and capacitive components of the membrane "cable."

It must be remembered that graded potentials are produced not only by synaptic action and receptor transduction, but also by the propagating spike action potential. As a spike is created at any point on a neuron, it becomes a source of passively spread "local" potentials. These local potentials, which also spread by means of decremental conduction, are probably the mechanisms through which sequential portions of an axon are successively ignited to produce the travelling spike action potential. Thus the graded potential plays an important role in propagation of the spike, in much the same way that it does in synaptic and transductive processes.

C. THE PROPAGATED SPIKE ACTION POTENTIAL

1. A Qualitative Description

Let us now reconsider some of the details of ionic equilibria across a membrane. When an electrical stimulus is applied to a membrane, it serves as an artificial local potential. This perturbation produces a new electrochemical balance as ions move about under the influence of the applied electrical force. Similarly, a change in permeability of the membrane or in the concentration of certain ions might also produce corresponding changes in the ionic equilibrium and thus in the level of polarization of the resting membrane.

Small changes in the equilibrium of the ionic system are usually reversible, but larger ones may not be. If the depolarization of some specialized membranes is carried to a critical threshold level, the restorative processes that would ordinarily reestablish the resting potential are no longer functional. Rather, once a specific membrane potential threshold is crossed, a spectacular event occurs: an almost explosive change in the potential across the membrane. What had perhaps been a moderately passive local depolarization now ignites into a rapid, massive depolarization of the membrane potential.

Figure 7.3 shows the results of the application of various levels and polarities of local polarizing potentials to a portion of a cell membrane specialized for spike conduction. Figure 7.3a shows that the application of an additional negative potential merely drives the resting potential to a more negative value—hyperpolarizing it—roughly in proportion to the applied voltage. Application of depolarizing positive going potentials of increasing magnitude, as shown in Figs. 7.3b, c, and d, however, results in depolarization commensurate with the applied potential, but only up to a certain threshold limit. When that limit is exceeded the spike action potential is triggered. The local membrane region suddenly shows a rapid

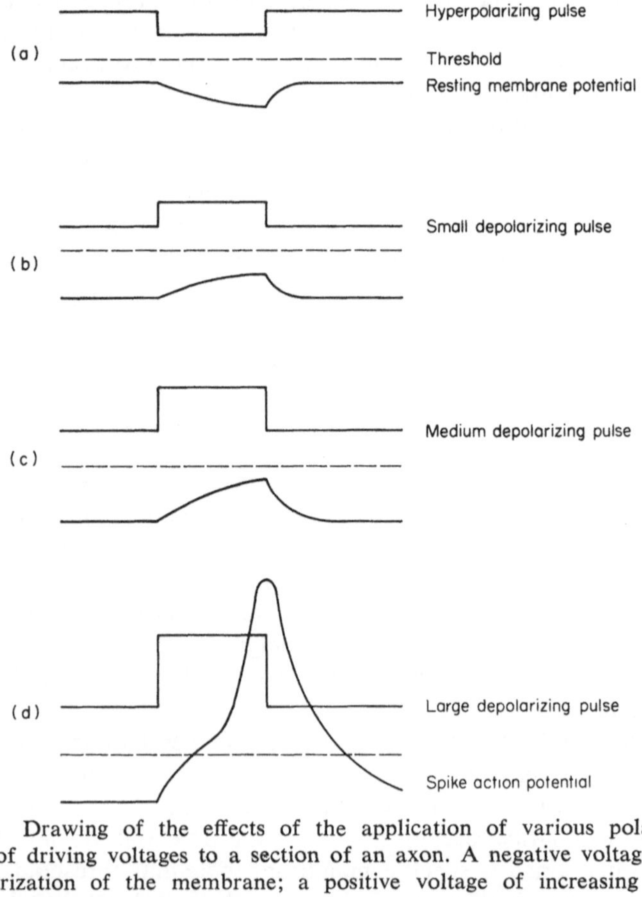

FIG. 7.3. Drawing of the effects of the application of various polarities and amplitudes of driving voltages to a section of an axon. A negative voltage produces a hyperpolarization of the membrane; a positive voltage of increasing amplitude succesively increases the depolarization until the threshold for the evocation of a spike action potential is exceeded.

depolarization far beyond the values that might have been produced passively by the applied voltage. In fact, the depolarization overshoots the point of zero transmembrane potential so that interior of the cell momentarily becomes positive with respect to its exterior. After a brief period, typically about .25 millisecond, the membrane potential begins to return to its resting level. The return is not immediate, however; following the spike there occur a series of after-potentials that are deviant from the original resting potential.

Figure 7.4 shows the shape of an entire spike action potential complex. A number of separate events can be identified: the initial local potential, the rise of the spike, the return of the spike, the positive (often incorrectly referred to as the negative) after-potential, and the negative (often incorrectly referred to as the positive) after-potential.

The positive after-potential occurs during the recovery period, before the return of the spike to its original resting potential level. The positive after-potential is associated with the *absolute refractory period,* a period of time after an axon has fired during which it is completely inexcitable, no matter how intense the stimulus. The negative after-potential is a much longer hyperpolarization of the nerve cell membrane, lasting for many milliseconds (up to 50) in the period following the spike. It is associated with the *relative refractory period,* a period following a spike during which an abnormally intense stimulus is required to produce another spike. Since the positive and negative after-potentials are continuous with each other, it must be assumed that some portion of the waveform of the spike must pass through the original resting potential level. A relatively normal, although brief, period of usual excitability between the two refractory periods might therefore be expected. This is exactly what happens. Figure

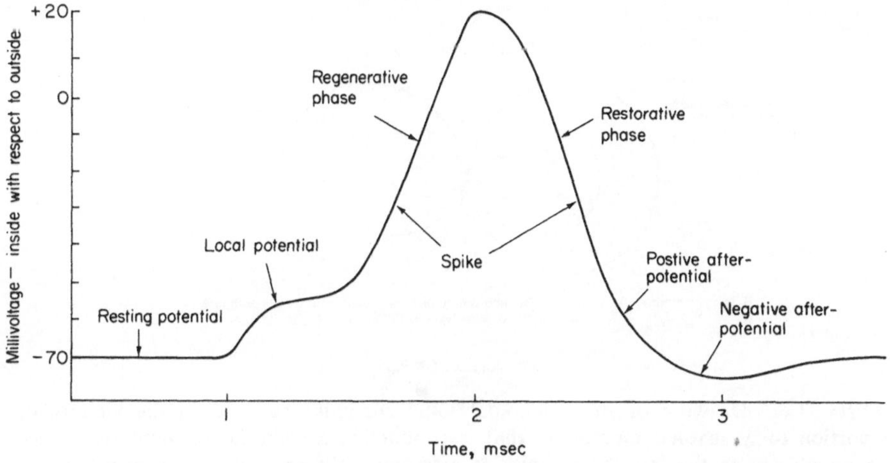

FIG. 7.4. Details of the various phases of a spike action potential.

7.5 shows the course of these periods of altered excitability following a nerve spike action potential in a typical *A* fiber. It is evident that at least a brief period of normal excitability is present.

The spike action potential itself can also be considered as a strong potential source. Currents generated by a dipole, composed of a spike and nearby portions of a cell membrane that are still at resting levels, can depolarize the membrane in the same way the original, artificial depolarizing stimulus did. This sort of spike-driven local current flow is depicted in Fig. 7.6.

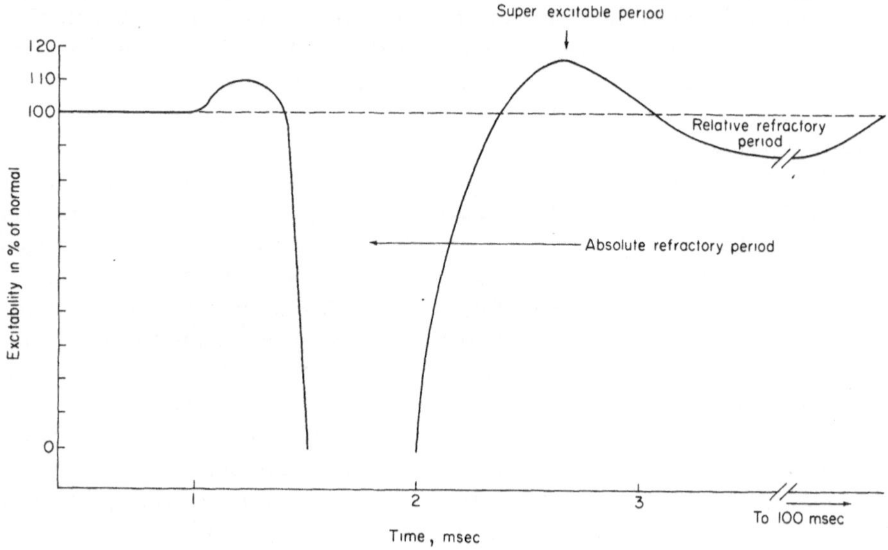

FIG. 7.5. Details of the various stages of excitability during the course of the spike action potential.

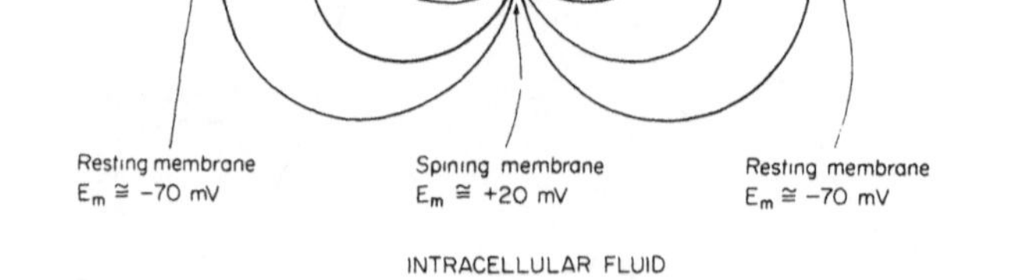

FIG. 7.6. Drawing of the solenoidal ionic currents produced in the vicinity of a portion of an axonal membrane that is conducting a spike action potential. Since the membrane to the left of the spike is refractory, the spike can only propagate to the right.

As the spike occurs, it stimulates or triggers a local depolarization of the next portion of the nerve membrane. This local depolarization can also lead to a spike, just as the original stimulus did. In this chain-like manner the spike action potential is propagated down the axon until it reaches a region that is incapable of spike production.

This qualitative description of the spike action potential is summarized in the following outline:

1. Triggering of the spike action potential occurs when specialized regions of the axonal membrane have been depolarized beyond a threshold value by some stimulus.
2. The spike action potential is self-propagating by virtue of its ability to sequentially depolarize successive portions of an axonal membrane. This is accomplished under the influence of a local current driven by the spike.
3. The amplitude of the spike action potential, once it has been triggered, is independent of the stimulus amplitude. (This is known as the *all-or-none law.*) The amplitude of the spike is, however, dependent upon local ionic and metabolic conditions.
4. The occurrence of a spike action potential leaves a series of refractory periods behind it. These refractory periods serve to maintain the directionality of the spike by preventing backward conduction. This extremely important characteristic contributes to stabilization of the entire nervous system.
5. The speed of spike conduction is a function of ionic factors, temperature, the diameter of the axon, and its degree of myelination, but it is totally independent of the nature of the stimulus.

2. The Ionic Basis of the Spike Action Potential

The preceding qualitative description of the spike action potential does not clarify the underlying ionic mechanisms accounting for this extraordinarily large neural response. An elegant explanation, however, has been developed, based on the ionic model and notions of the semipermeable membrane discussed in Chapter 5. It was established there that the best current explanation of the resting potential is one based upon the sodium and potassium ion concentration differences and equilibria across a cell membrane. We now add to those notions the general principle that fluctuations in the membrane resting potential (that is, action potentials) are mainly due to local, transient changes in the permeability of a membrane to these two particular ions.

Recall the qualitative description of the spike action potential. First, there occurs a graded depolarization of a membrane by the local potential to the point at which a threshold is crossed. Then an explosive depolarization of the membrane potential occurs. This is followed by an equally rapid return toward the original resting potential level and a series of after-potentials. An acceptable model of the spike action potential must explain a number of things about this process: (1) What is the nature of the threshold and the regenerative depolarization of the cellular response? (2) Why does the membrane potential completely reverse its polarity by the particular amount it does? (3) What is the nature of the restorative process accounting for the return of the spike action potential to its original level of negative polarization? (4) What ionic processes account for the after-potentials?

The currently accepted answers to these questions are based primarily on the work of Hodgkin and Huxley (see, for example, Hodgkin & Huxley, 1952b, and Hodgkin, 1958), who received the Nobel Prize in 1963 for their description of the ionic basis of nerve membrane activity. Other important contributions include Keynes (1951) and Cole (1968). The theory to which they have all contributed has a number of key points:

1. The state of the membrane potential at any point in time is primarily a function of the sodium and potassium ionic distributions present at any time.

2. The concentrations of the ions are functions of membrane permeability, as well as the sodium-potassium pumping action. Therefore, the potential across the membrane is also a function of the permeability of the membrane.

3. However, membrane permeability is *also* a function of the potential across the membrane. Specifically, for sodium and potassium, the greater the membrane depolarization, the greater its permeability to these two ions. This is a critical point, because it establishes a positive feedback cycle that almost totally accounts for the regenerative (positive feedback) process that results in the spike action potential. The greater the depolarization, the greater the permeability; and the greater the permeability, the greater the depolarization. This statement holds true until depolarization has peaked, at which point sodium permeability begins to decrease very rapidly again. It is this reversal and abrupt decline in sodium permeability at maximum depolarization and the effect of a slightly delayed, but similar increase in potassium permeability that prevents a depolarized membrane from remaining depolarized until the much slower pumping action can restore the resting potential.

4. The increase in permeability to sodium and potassium does not occur at the same rate. This is the critical factor that accounts for the rapid rise and subsequent fall of the membrane potential, and thus the spike-like shape of the waveform.

It is important to point out that not all of the questions we have raised regarding the ionic basis of the spike action potential can be fully answered at present, although there are fairly sound, internally consistent answers available for some of them. It is not known, however, why the threshold of regenerative depolarization is located at the precise level observed for each nerve membrane. It is generally agreed, however, that at the threshold a rapid change in permeability of the membrane to sodium and potassium ions is triggered, and that the time function of change for the two ions is quite different.

The Hodgkin-Huxley model of the spike is not universally accepted, although it is certainly the most complete and comprehensive of current theories. Its main premise is that a dynamic change in membrane permeability results in the passive flow of sodium and potassium ions into and out of the cell. But some theorists disagree. For example, Landowne (1973), on the basis of his low-temperature experiments on ionic transport, believes that the simple permeability model cannot work. He has suggested an alternative model in which the transported sodium is actually stored within the membrane, and the quantity of that ion that can enter is limited to only that stored amount. His model, therefore, involves active metabolic processes for the production of the spike, unlike the purely passive model championed by Hodgkin and Huxley that has such wide currency today.

The following discussion, nevertheless, presents what is now the conventional and more complete Hodgkin and Huxley approach, but the reader is forewarned that no molecular or ionic theory fully encompasses all the experimental data accumulated in the past 20 or 30 years.

Regenerative depolarization. Figure 7.7 shows the time course of the permeabilities for sodium and potassium ions, respectively, after the threshold of depolarization is crossed. It can be seen that there is an immediate increase in the permeability of the membrane to sodium, which quite rapidly (in about 1 millisecond) returns to its original level.* The amplitude of this increased permeability for sodium during the rising edge of the spike has been estimated to be about 500 times that found at the level of the resting potential. Recalling that the sodium pump had developed a substantial deficiency of sodium in the cell's interior, this increase in permeability of the membrane to sodium allows a substantial amount of sodium to rush back into the cell, passively driven both by chemical concentration forces and by the electrical gradients. This influx of sodium tends to decrease

* Very careful measurements of the electrical currents at the outset of the increase in sodium permeability show that there is another and smaller ionic current preceding it. This early signal has been called the *gating current.* Gating currents are believed by some investigators (Armstrong & Bezanilla, 1973; Bezanilla & Armstrong, 1974; Keynes & Rojas, 1973) to be produced as the configurations of molecules that constitute the sodium channels are altered to produce the increased sodium permeability.

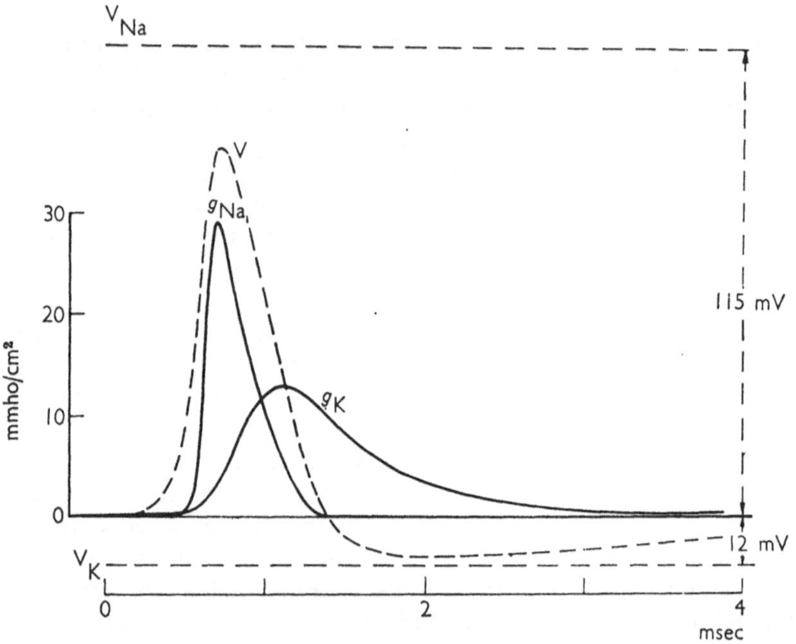

FIG. 7.7. The time course of the membrane permeability changes to sodium (g_{Na}) and potassium ions (g_K) during a spike action potential (V). The difference in the time course of the two ionic permeability changes accounts (according to Hodgkin and Huxley) for the rapid rise and fall of the spike action potential. V_{Na} and V_K indicate the sodium and potassium equilibrium potentials, respectively. (Drawing from Hodgkin & Huxley, 1952b.)

the transmembrane negative potential, adding to the tendency of the membrane potential to continue depolarizing. This further decrease in polarization increases the permeability of the membrane to sodium, adding to the continuing depolarization of the cell membrane. A positive feedback loop is thus established: Decreased potential leads to increased permeability, which in turn leads to decreased potential, resulting in regenerative depolarization of the membrane.

This positive feedback loop provides an explanation of the regenerative and explosive depolarization of the nerve cell membrane during the rising portion of the spike action potential. Every force in action pushes the ionic equilibrium toward a state in which the amounts of sodium will be equal both inside and outside this local membrane region.

It should be emphasized that although the use of the words "rush," "explosive," and "regenerative" might seem to imply great influxes of sodium, this is true only in relation to the amounts that *move about* under resting potential conditions. Relative to the *total* amount of sodium in the cell interior, there occurs only a slight, localized shift in ion concentrations. This is apparently all that is necessary.

The spike does not continue in its positive-going direction indefinitely, of course. The main driving forces behind the sodium influx are the electrochemical potential for sodium, and the diffusion forces resulting from the difference in sodium concentration inside and outside the cell, which were originally generated by the sodium pump. But very quickly, in .25 millisecond or so (a time determined by the rate at which sodium pours into the cell), the membrane potential in that region reaches a new equilibrium point at which concentration inside and outside are equal, and the sodium influx ceases.

Before discussing the restorative processes that return the membrane potential to its prespike resting level, another important feature of the spike—its peak amplitude—must be considered. In 1902 Bernstein proposed a relatively simple model to explain the change in membrane potential during production of the spike. He supposed that the spike was produced by a relatively direct, complete (although transient) increase in the permeability of the cell membrane, allowing a large flux of all the ions involved. On the basis of this theory, he predicted that the spike should vary from the resting potential to zero millivolts at the peak of the spike. As noted in Chapter 2, subsequent research (Curtis & Cole, 1942; Hodgkin & Huxley, 1939) showed that the potential does not return merely to the zero potential level, but actually goes from its negative resting level to one considerably more positive than zero—as much as 40 millivolts in some cases. This fact seemed to be compelling evidence that some sort of active force must be driving the leading edge of the spike. But this fact was not correct!

After Hodgkin and Huxley established the electrochemical processes involved in sodium and potassium transport and it was realized that the membrane equilibrium potential of sodium itself is quite positive, the currently accepted notion of a totally passive, sodium-specific, permeability collapse was reestablished. The active pumping processes were once again exclusively reassigned to the generation and long-term maintenance of the resting potential, and excluded from participation in the rapid spiking process. It should be specifically emphasized that the pump is now thought to be responsible only for the long-term restoration and maintenance of the resting potential. The rapid restoration at the trailing edge of the spike is due to quite another process, the dynamics of potassium permeability. Whether or not this general, passive explanation for the genesis of the spike will be maintained in the light of future research is hard to say, for various combinations of several active and passive processes could all produce the same spike-like shape.

Restorative repolarization. We shall now consider the contemporary explanation of the restorative forces that drive the membrane potential back to its resting level after the depolarization that produced the leading edge of the spike is complete.

Two distinct phenomena have been invoked to explain the action of the membrane potential during this recovery period. The first involves the time course of sodium permeability. We have seen that the supernormal sodium flux begins to be reduced at the peak of the spike. It is then that sodium permeability begins to decrease and the sodium driving force is minimal, since local concentration differences are then at their smallest. This decrease in sodium permeability occurs at a point at which the total transmembrane potential is equal to a considerable positive voltage, but remember that the potentials associated with each ion species exist independently of each other.

At this point, whatever process opened the flood gates to a flow of sodium (a process still not well-understood) now reverses, and the permeability of the membrane to sodium rapidly returns to its very nearly normal low level. Given a sufficient amount of time, this change in the permeability of the membrane would result in the reestablishment of the resting potential. In fact, however, the return occurs much more rapidly than could be accounted for by such a passive process. Therefore, other forces must be invoked to explain the restoration of resting polarization levels. One suggested possibility is the sodium-potassium pump itself. But that active mechanism is also known not to perform sufficiently rapidly to accomplish the necessary restoration.

If we recall that the time course of the change in potassium permeability is slower than that of sodium and look once again at Fig. 7.7, it can be seen that at about the same time sodium permeability is being "turned off," potassium permeability is nearly at its peak value. Potassium is therefor rushing out of the cell in large amounts, relative to the normal resting level influx. It is as if the membrane permeability gates, which had opened and then closed to sodium, were now open to potassium. The term "gates" refers to the channels for sodium and potassium transport. It is not known certainly whether these channels are the same for both sodium and potassium ions, or whether they are separate mechanisms, although their distinctive susceptibility to various pharmacological agents suggests that they are different.

Consider the effect of this increase in potassium permeability. The potassium equilibrium potential across the resting membrane was about —90 millivolts, due to the excess in internal potassium ion concentrations compared with the outside of the axon at resting potential levels. The excess potassium inside the cell now begins to rush out, as membrane permeability to it increases. Once again, it should be remembered that these amounts are large only in relation to the normal influx or efflux. Nevertheless, the absolute amounts of ion flow across the local region of a membrane that is undergoing spike activity are electrochemically significant, and as they change the membrane potential also changes.

Potassium ions continue to flow out of the cell under simple passive diffusion and electrochemical forces, and the total transmembrane potential begins to repolarize (that is, the inside goes more negative) as the inside of the neuron is depleted of these positive ions. This increase in polarization continues until the membrane approaches its original resting potential level. Since sodium permeability is so much a function of the membrane potential, this increasing negativity further curtails the passive flow of sodium into the cell, by reducing the permeability of the membrane channels to that ion.

Although the potassium-mediated restorative process brings the membrane potential back to nearly the original resting potential level, not all conditions are exactly as they were. There is an excess of sodium and a deficiency of potassium inside the cell, due to the movement of these ions during the regenerative and restorative periods of the spike. Now the slower-acting sodium-potassium pump takes over and gradually reestablishes normal conditions. How long this takes is apparently reflected in the duration of the after-potentials.

The entire spike generation process is described by Table 7.1, which tabulates the various ion flows during the four major periods of the spike

TABLE 7.1
Summary of Ionic Actions during the Course of a Spike Action Potential

State of membrane	Inside potential	Membrane permeabilities[a]	Inside ion concentration
1. Normal resting potential	–	In accord with sodium pump P_K normal, P_{Na} normal	Mainly K
2. Regeneration-rising (depolarizing), leading edge of spike	Going +	P_K normal $P_{Na} >$ P_{Na} normal	Mainly K but some small amount of Na enters
3. Restoration-falling (repolarizing), trailing edge of spike	Going –	P_{Na} turned off $P_K >$ P_K normal	Mainly K but some small amount of K leaves
4. Long-term recovery	–	P_K and P_{Na} slightly different than normal	Mainly K some Na pumped out some K pumped in

[a]P = permeability

action potential. We should also elaborate on this summary by reemphasizing some of the most important features of this remarkable process. First, a spike results from the sudden change in membrane permeabilities to two different, positive ions. The brevity of the action seems to preclude the mediation of an active process in this permeability change. For this and other compelling reasons, it is now thought that the ion flows that produce the spike are merely passive, in the sense defined in Chapter 5. Second, the difference in latencies of the permeability increase for sodium and potassium ions is critical to the entire process. Although sodium latency is shorter, the peak of the potassium efflux is reached exactly at the instant the sodium influx begins to decrease. But these swift changes leave a residue of abnormal ion distribution. The return of normal concentration conditions is delayed many milliseconds, until the slower sodium-potassium pump can return concentration levels to their normal resting levels. This is without doubt one of the most remarkable evolutionary developments known in biology, and as we shall see, these differential latencies specify almost optimum response times and information-carrying capacity on the part of spike action potentials.

D. SOME EXPERIMENTAL METHODS

In discussing the ionic mechanisms underlying both the resting and acting potentials, some answers to important questions have been presented without an explanation of the experimental methods electrophysiologists use to explore these issues. Two especially important electrochemical techniques, *manipulation of ion concentrations* and *voltage clamping,* have been especially significant in membrane potential research. These two methods have been applied in investigations seeking answers to two essential questions concerning membranes:

1. Which specific ions are important in membrane potential processes?
2. What is the functional relationship between the membrane permeability to specific ions and the membrane potential?

1. Experimental Manipulation of Ion Concentrations

We have seen how the semipermeable nerve membrane demarcates two regions of differing ion concentration. The mixture of ions in the two solutions, one external to the cell and the other internal, is quite complex; traces of many different ions are found when the solutions are chemically analyzed. However, as has been repeatedly pointed out, only two or three ions are apparently responsible for almost all the membrane potential effects observed in electrophysiological transmission. Substantial changes in the concentrations of most of the other ion species found in the external solution bathing the neuron fail to produce any significant changes in membrane action or resting potentials. However, only slight alterations in the concentration of two particular ions, sodium and potassium, have massive effects on the membrane resting and action potential.

The experimental procedure of external ion substitution is relatively straightforward. A portion of a neuron is placed in a tank containing a solution with the necessary salts (like sodium and potassium), in the same proportions normally found in the animal's body. The external solution may then be replaced by perfusing in one with a different ionic constituency. The main requirement is that the replacing solution be one that is isotonic (that is, exhibits the same osmotic pressure) to the solution being replaced, in order to avoid creating differences in the internal pressure of the neuron. Complete internal substitutions of such substances as isotonic potassium sulfate or potassium isetheionate for potassium chloride can be made, and the neuron may still continue to function quite normally for many hours. Sodium chloride can also be substituted for potassium chloride, or vice versa, or sucrose solution substituted for sodium chloride. In these cases, the effects on the action and resting potentials can be quite profound.

Potassium. It was observed in the early 1930's that alterations in potassium ion concentration specifically affect the resting potential. Several research groups established the fact that increases in the external potassium ion concentration make the resting potential less negative, reducing it to what may often be a zero value. Similarly, Baker, Hodgkin, and Shaw (1962b) showed that if the internal solution of a neuron is altered by substitution of isotonic sodium chloride for potassium chloride, thereby depleting the internal neuron fluids of potassium ions, the resting potential will also be eliminated.

Since the interior of the cell normally has a superabundance of potassium ions, and since potassium permeability is relatively large compared to sodium permeability, it is not surprising that increasing or decreasing the potassium ion concentration mainly affects the resting potential.

If high enough concentrations of potassium ions are applied to the external environment of the neuron, depolarization of the resting potential may be sufficiently intense to evoke action potentials. This is probably the mechanism behind the strong excitatory effects of topical applications of potassium chloride solution to the cortex. Paradoxically, this sort of excitation leads to the phenomenon known as *spreading depression* (Leão 1947), in which a wave of depressed neural activity moves slowly (25 mm per minute) away from the excited region to the surrounding cortical tissue. (Spreading depression can be caused by other forms of excessive stimulation, in addition to potassium chloride application.)

One of the most important features of this technique is that it is possible to temporarily "turn off" the brain by altering the external concentration of a given ion. This is essentially a "reversible lesion" of cortical tissue and is a most exciting research tool for determining the function of various parts of the brain. Butcher and Fox (1968) have described an ingenious method in which manipulations of neuronal ion concentrations can be made with the use of ion-exchange resin beads loaded with the specific ion whose concentration is to be altered.

Sodium. Changes in sodium ion concentration in the solution bathing a neuron turns out to have a relatively minor effect on the resting potential. Changes in external sodium concentration, however, have a significant effect on the spike amplitude, since it is the sodium membrane equilibrium potential of about 60 millivolts to which the spike peak is rising during its regenerative phase. The greater the external sodium ion concentration relative to internal concentration, the greater the potential passively driving the depolorization and the larger the amplitude of the spike. Similarly, if external sodium concentration is depleted (see, for example, Nathan & Sears, 1962), the passive driving force is less strong and the spike action potential is reduced in amplitude.

Iontophoretic injection. In recent years a useful device for intracellular ion substitution has been developed. Earlier, it was possible in a few unusual cases (such as the giant axon of the squid) to expell the intracellular cytoplasm by rolling the axon, like a tube of toothpaste, and then substitute other solutions (Baker, Hodgkin, & Shaw, 1962a). This technique has been very useful in identifying the specific effects of internal manipulation of certain ions. However, attempts to substitute ions in the interior of smaller neurons are fraught with much greater difficulties.

Nevertheless, this feat has been partially accomplished by means of a technique known as *iontophoretic injection* of solutions, although complete ion substitution has never been achieved. A micropipette containing the desired solution can be inserted into a neuron, just as a recording microelectrode can be. If a voltage is then applied to the electrode, there is a tendency for ions to be driven into or pulled out of the cell interior by electrostatic repulsion or attraction, depending on the polarity of the voltage. In this manner specific substitutions can be made in the contents of some local regions inside the cell. This technique has the added advantage of allowing direct measurements of the number of ions injected, since both the charge-carrying capacity of each ion and the magnitude of the ionic current (usually measured in nanoamperes) produced by the applied voltage are known. Armed with the additional information about the duration of the driving voltage, the investigator can then directly determine the total number of ions transferred during the iontophoretic injection period.

2. The Voltage Clamp

The fact that in the plasma membrane a sudden increase in permeability to sodium occurs when a nerve cell fires a spike action potential is one piece of evidence that suggests that some functional relationship exists between membrane potential and ionic current. Under ordinary conditions it is very difficult to determine the details of this relationship because of the complexity and rapidity of the chemical dynamics involved. An experimental alternative that locks the parameters of a membrane at a constant value and allows their static measurement has been developed by Cole (1949). His technique for electrically forcing the membrane potential to a predetermined value and holding it there is now known as a *voltage clamp.*

Voltage clamping of a neuronal membrane is implemented by a special type of feedback amplifier. The device is designed to sense the momentary voltage across a membrane and instantaneously apply a "bucking" or counter-current that has the polarity and amplitude required to bring the membrane to any desired constant potential. With proper adjustment, potentials may be maintained at levels quite different from their normal resting potential, and current permeability measurements may be made while the membrane potential is locked at a fixed value. A disadvantage is that rather large electrodes must be used with the voltage-clamping amplifier in order to avoid local variations in membrane voltage, a condition that might lead to ill-defined variability or nonhomogeneity in the response. Such a situation would result in an averaged value that might not reflect the real membrane condition.

Figure 7.8 illustrates a typical voltage-clamping apparatus that has been set up both to detect the membrane potential and to generate the current needed to clamp it at some voltage desired by the experimenter. After the initial transients that are associated with membrane capacitance charge and discharge and the establishment of ionic equilibria have settled down, Ohm's law, $I = E/R_1$, the relation between the current I and the potential E, is a direct measure of membrane resistance R or conductance $(1/R)$.

One typical result of a voltage-clamp experiment is the finding discussed above, that membranes clamped at potentials more positive than the resting potential display very low resistance (that is, high permeability) to currents of the sodium ion charge carrier. The exact function relating permeability and potential can be plotted by varying the clamping voltage from the resting potential level to the peak value of the spike. Figure 7.9 shows such a membrane potential-permeability function for a typical fiber (as a function of the time after the clamp was applied).

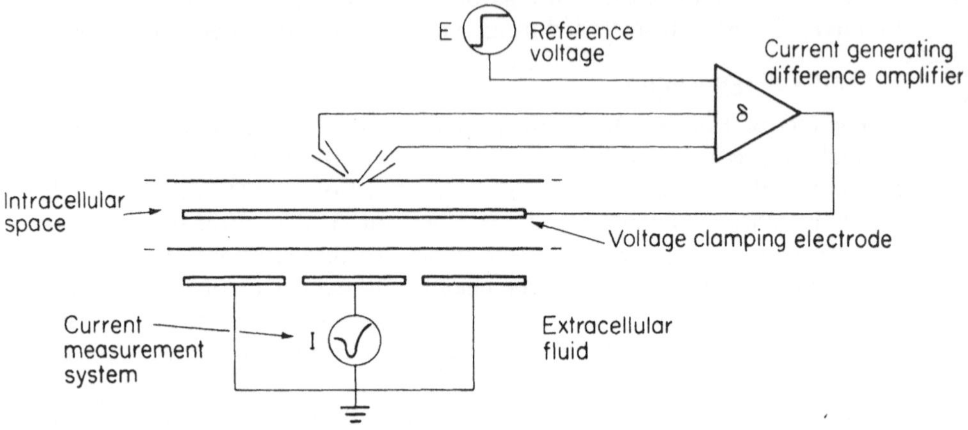

FIG. 7.8. Diagram of a voltage clamping device. Two microelectrodes, one intracellular and one extracellular, measure the membrane potential that is applied to differential inputs of a preamplifier. When a reference voltage is applied to a third input of the preamplifier, the difference between the membrane and reference voltage produces a current that is driven through the large intracellular electrode. This current is sufficient to reduce the difference between the membrane potential and the reference potential to zero by altering or "clamping" the membrane potential. The measurement system can then be used to measure the conductance of the membrane at any given clamped voltage. (Adapted from Cole, 1968.)

The voltage-clamp procedure can be seen as conceptually similar to the freeze-chemistry technique described for the study of visual photochemicals in Chapter 6. It allows measurements to be made in a quasi-static situation, in a manner that provides temporal "snapshots" of processes that are normally changing rapidly. Because there is always a lag between membrane permeability fluctuations and changes in ion flow, this sort of snapshot (like any other) inevitably loses some of the information present in the natural process, but it has been enormously useful in a wide variety of experimental techniques.

Cole (1968), the inventor of the voltage clamp, has pointed out several drawbacks involved in the clamping process, such as its inadequacy in guaranteeing that currents generated are really homogeneous under clamped conditions. Despite these difficulties, the voltage-clamping technique has provided much information that would not otherwise have been available. One of the most important examples of this technique's unique contribution

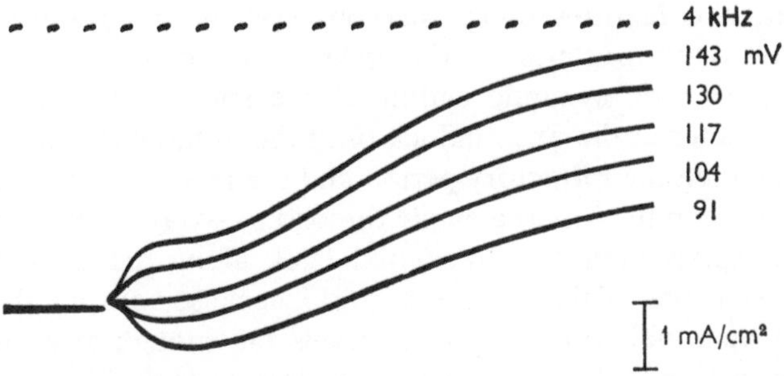

FIG. 7.9. The time course of the membrane currents produced by a sudden change in the clamped voltage as a function of the voltage amplitude. (From Hodgkin, Huxley, & Katz, 1952.)

has been the separation of the normally simultaneous sodium and potassium ionic currents, a feat first accomplished by Hodgkin and Huxley (1952a), and one that played an important part in the development of the model of spike action potentials presented earlier in this chapter.

E. TRANSMISSION PROPERTIES OF SINGLE AXONS

Up to now the spike action potential response has been discussed almost independently of the fact that it has evolved as the medium of communication over the long axons of the nervous system. Let us now concentrate on those properties of the spike that are specifically associated with its transmission function, rather than on the local ionic conditions that power it. It is particularly important to understand the transmission function of the spike action potential because historically it has been considered to be *the* neural code *par excellence.* Patterns of spikes can vary in frequency, in interval, even in amplitude (if the outputs of several neurons are compounded). It is clear that in many parts of any nervous pathway, only the spike is available as a medium for conveying stimulus information.

One of the first considerations in axonal transmission is the matter of directionality of the propagated nerve impulse. The nervous system would explode into an uncontrolled torrent of chaotic nerve impulses if there were not strong constraints imposed on the direction in which nerve impulses can travel from their point of excitation. Nevertheless, it must be understood that the axon itself has no intrinsic directionality, but can convey information equally well in both the antidromic and orthodromic directions, even simultaneously. In fact, the neuron is capable of conveying several different responses, even patterns of response, in different directions down different branches of an axon at the same time.

However, two features of the nervous system do operate to constrain the direction of propagation of the spike action potential: refractory and recovery cycles, and synaptic action. If the spike is initiated at the end of an axon, either at the axon hillock or at the initial postsynaptic or receptor terminal, then the refractory period and the recovery cycle of the resting potential limit conduction to a single direction—away from the initial point of entry. A spike cannot be retrofired back along the axonal path from which it just came. This limitation exists because, during the refractory period, the neuronal spike response travels far enough away from the region it has just activated so that it is no longer capable of eliciting a response in that region.

Directionality is additionally imposed upon the nervous system by the action of synapses. Synapses are strongly directional and therefore act as rectifiers. Synaptic action is discussed in detail later in Chapter 9, and we shall defer further discussion of this matter until then.

A second important property of axonal transmission is its velocity. Conduction velocity is intrinsically interesting because it is one property of the neuron that is a known, if not a directly derivable, function of neuron structure. The fact that different neurons conduct at different speeds was discovered rather early (Gasser & Grundfest, 1939), and it was obvious from these early neurophysiological experiments that conduction velocities are closely associated with fiber diameter.

How extraordinarily precise the relationship is for the large *A* fibers of the somatic system actually was demonstrated by Hursh (1939), who showed that there was a strikingly linear function relating the diameter of a fiber to conduction velocities in mammalian nerves (Fig. 7.10). The upper limit observed by Hursh was about 140 meters per second. This value, of course, is substantially below electronic conduction speed.

The degree of myelinization of an axon is another important factor in determining its conduction velocity. It was demonstrated in Chapter 3 that many kinds of neurons are typically encased in a fatty sheath of myelin, produced by the unusual "wrapping" action of the Schwann satellite cells. It was pointed out that on many of these cells, in particular the long peripheral axons, the myelin is interrupted at intervals by nodes of Ranvier. At these nodes the myelin sheath is completely absent (Bunge, 1968), and its electrochemical resistance is also very much reduced. These nodes, therefore, literally provide short-circuits for the propagating ionic currents that sequentially trigger the spike action potential along successive portions of the axon, allowing the spike to "skip" at a high velocity. Spike action potentials in an unsheathed cell, however, must sequentially activate all locations on the membrane if they are to propagate themselves. Because

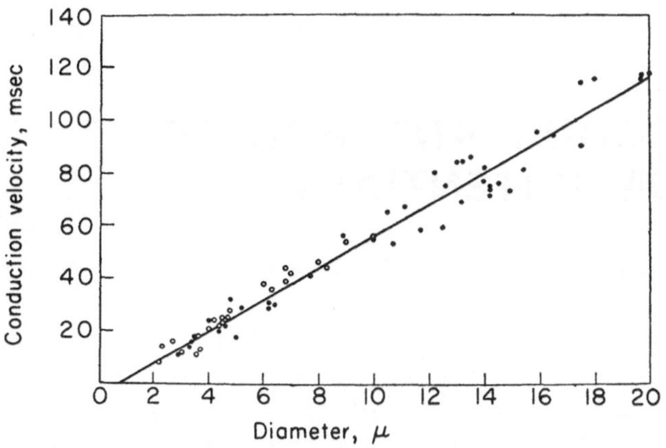

FIG. 7.10. Plot of the conduction velocity of axons as a function of their diameter. Note the highly linear correlation. (From Patton, 1965, after Hursh from Gasser.)

the speed of propagation is finite, this requires very much more time, as the wave of activity moves in a continuous series of excitations from one end of the fiber to the other. Unmyelinated fibers are intrinsically slower, therefore, than myelinated ones.

In this chapter we have considered some of the transmission properties of single axons. But no nerve in the body consists of only a single axon, and only under the most unusual conditions will a neuron fire in isolation. It is far more usual for many neurons to fire together. Multiple neuron firing leads to the production of response waveforms that are far more intricate than the classic spike or the simple decremental graded potential. The mixture of responses into pooled or compound action potentials is the subject of the next chapter.

8
COMPOUND AND POOLED ACTION POTENTIALS

Up to this point, our discussion of transmission has dealt exclusively with the responses of single cells. However, this is somewhat of an idealization, for neurons are not found in isolation. Rather, they are bundled together in large numbers in the transmission pathways and ganglia of the nervous system. Many recorded responses are, thus, necessarily the sum of the signals from many individual cells. The discussion in this chapter is, therefore, aimed at understanding the responses that are recorded from groups of neurons.

Earlier chapters also concentrated on experiments that elicited *intracellular* responses by means of a microelectrode, which was actually inserted through the membrane into the cell. Nevertheless, it was noted that individual neural responses can also be recorded with *extracellular* electrodes. In this case the electrode does not enter the cell but picks up some portion of the external field. The exact form of the response recorded with *extracellular* electrodes is hard to specify because it depends upon a number of complicated and interrelated physical variables. The response shape recorded from a single axon depends very much upon whether it is close or nearby, large or small, and in what fluid the entire nerve-electrode complex is immersed.

Indeed any nerve action potential, even one that is found to be monophasic (one-peaked) when measured with an intracellular electrode, must be considered not as a single point potential, but rather as a dipole consisting of a source and a remotely located sink when recorded with electrodes in the external volume conductor. This dipole arrangement, with connecting lines indicating the pathways along which ions are assumed to flow, was shown in Fig. 7.6. The exact way in which the responses of multiple sources add, therefore, can be very complex and depend upon many variables; response shape in any multicellular response may be close to meaningless except in the most trivial way.

Extracellular responses from multiple neurons, however, can be generally classified into two categories. The first is the class of *pooled action potentials,* one sample of which is illustrated in Fig. 8.1. Here we see a form of response in which the action potentials of the single cells are unsynchronized and not abundant enough to interfere or interact with each other. The result is a trace of multiple, but distinct, series of spikes of varying heights, producing a grassy appearance. The second type, usually referred to as *compound action potentials,* occurs when so many cellular responses occur simultaneously that they add to each other to produce an electrical sum. Compound action potentials often occur when impulsive stimuli are used to evoke synchronized responses of many of the cells in a nerve, or in cases where very large numbers of neurons are involved, as in the electroencephalogram. Compound responses do not appear as a "grassy" collection of individual spikes, but as a smooth, cumulated response that varies in amplitude and duration as a result of the specific nature of the stimulated region as well as the number of individual cells responding. An example, a compound evoked response from the brain, produced by an electrical stimulus applied to a peripheral somatosensory nerve, is shown in Fig. 8.2.

The next three sections will discuss pooled responses, compound responses, and a very important experiment that shows that the two may be considered functionally equivalent, when properly analyzed. It is important to understand the nature of pooled and compound action potentials, for their characteristics also represent a significant set of dimensions that

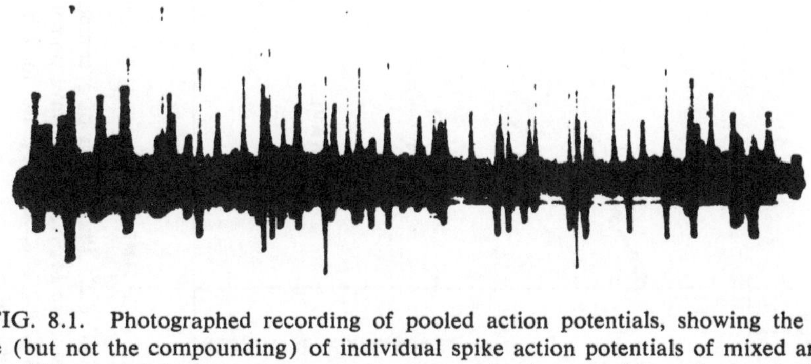

FIG. 8.1. Photographed recording of pooled action potentials, showing the mixture (but not the compounding) of individual spike action potentials of mixed amplitude. This record is from the ventral nerve cord of a crayfish and was made by William R. Uttal and Hedwig Kasprzak.

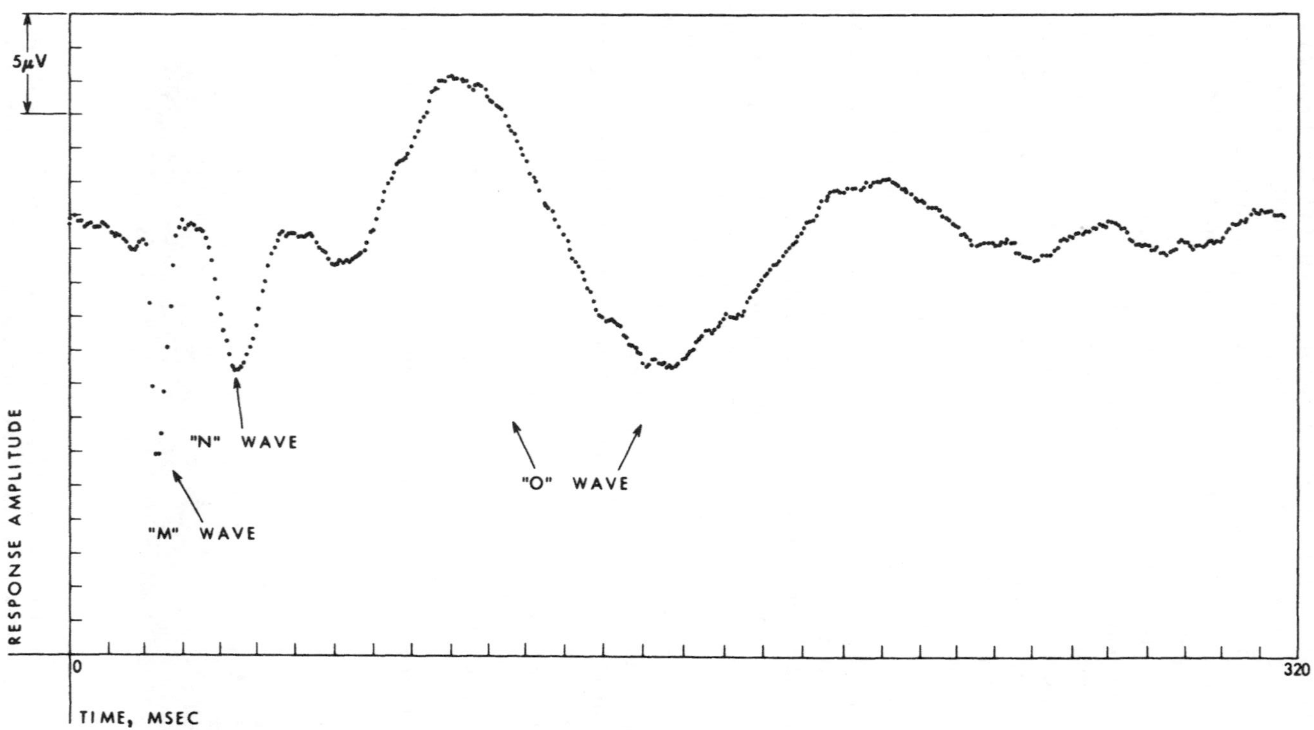

FIG. 8.2. Photograph of a compound potential evoked from the human brain by an electrical shock applied to the ulnar nerve at the wrist. The M, N, and O wave nomenclature is unconventional; more modern workers usually refer to the various portions of the signal as P_{30}, P_{50}, N_{110}, and P_{150}, etc. to indicate their positive or negative polarity and time of occurrence (in milliseconds). (From Uttal & Cook, 1964.)

can be used as transmission codes. There has been considerable interest in compound action potentials, particularly in this regard, because they display roughly the same sort of temporal properties that appear in psychological data while cellular spikes and decremental signals have very different durations and other temporal properties. Major peaks of both evoked potentials and perception times, for example, are often found hundreds of milliseconds following a stimulus. Spikes, on the other hand, last for only a few milliseconds.

Furthermore, since both the behavior of the whole organism ("molar" behavior to the psychologist) and the compound action potentials represent the statistical summary of the action of a large number of individual neurons, a number of workers believe that compound potentials may be better indicators of the relation between the brain and behavior than are the responses of single cells. However, this is still a highly controversial point, for some workers also believe that they may be epiphenomena and their statistical origins might very well be based on entirely different mechanisms, quite unrelated to behavior.

In the following discussion we will consider only the origins and the methods used for recording of pooled and compound action potentials and leave this controversial question unplumbed.

A. POOLED RESPONSES

Figure 8.3 is a diagrammatic plot of a hypothetical combination of a series of four pulse-generating pseudoneurons that could have produced a pooled record similar to that shown in Fig. 8.1. The four pseudoneurons each produce a series of spikes that occur at unique rates and amplitude. All four outputs combine to produce the typical grassy record obtained when the number of neurons is fairly low. Neuron 1 is fairly close and of medium size. Neuron 2 is fairly large but farther away, so there is little difference in its response amplitude in comparison with Neuron 1. Neuron 3 is both small and far away, while Neuron 4 is fairly close and large. The point the reader should note at this point is that it is not possible to determine the distance between the electrode and a given neuron on the basis of its amplitude, since various combinations of the involved factors will all produce the same measured amplitude.

One of the most important analytic problems in neurophysiology is the separation of this sort of pooled responses into their individual neuronal components. This sort of analysis into component signals (which might be compared to a "Fourier analysis" in intent if not in mathematical detail) poses a most refractory challenge. In fact, one particular mathematical analysis (Cox & Smith, 1954) asserts that the only statement it is possible to make is that the interval pattern of any pooled response is essentially

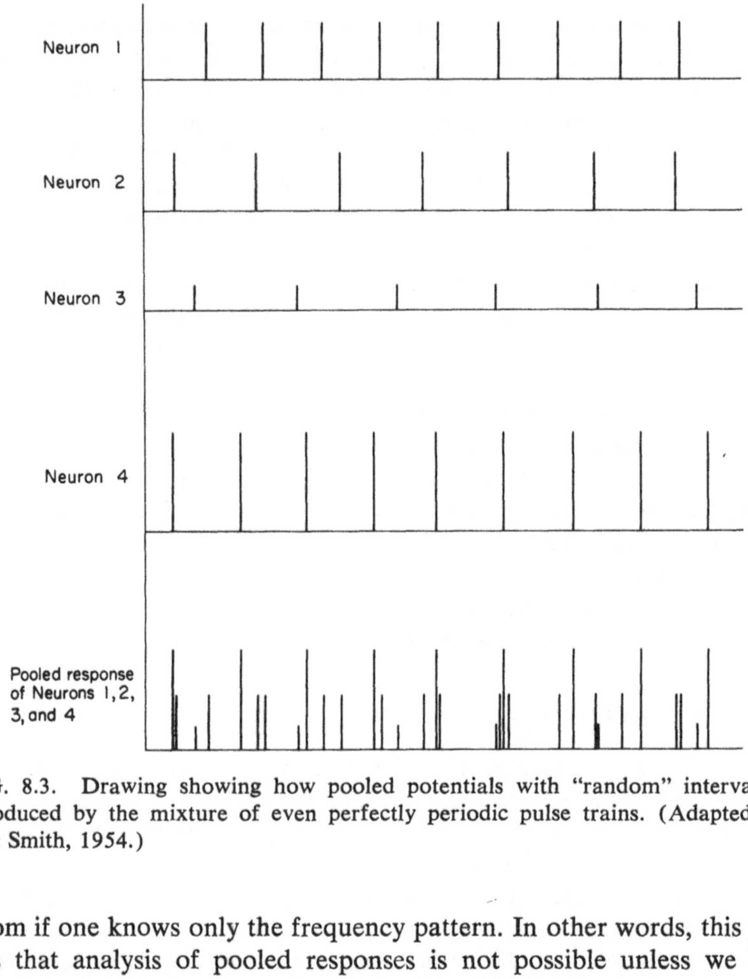

FIG. 8.3. Drawing showing how pooled potentials with "random" intervals can be produced by the mixture of even perfectly periodic pulse trains. (Adapted from Cox & Smith, 1954.)

random if one knows only the frequency pattern. In other words, this proof states that analysis of pooled responses is not possible unless we know more than merely the time of arrival of individual spikes in the pooled train.

In their very important theoretical paper, Cox and Smith discuss the formal limitations on the analysis of *superpositioned processes* exemplified, in this case, by the mixed pool of neuronal action potentials we are presently discussing. In the situation they consider, all amplitude data are considered to be lost. This is a greater restriction than is usually imposed in the neurophysiological case, but their general conclusions are of considerable significance.

Cox and Smith demonstrate that even if the generators contributing signals to a pooled response are perfectly periodic, the intervals between successive responses tend to become more randomly distributed as the number of independent generators increases. This is a major finding, because it says there is no possible technique *using interval alone* that allows the investigator to analyze the pooled signals into their constituents. It also emphasizes the fact that there is no way to tell anything at all about statistical properties of the generators themselves when their number becomes large. However, for smaller numbers of fibers a neurophysiologist can, at least, describe the general statistical nature of the family of generators that contributed to the pooled response. For example, with Cox and Smith's technique, one would be able to distinguish between a set of perfectly periodic generators and a set of randomly firing generators by examining the pooled response produced by each set.

The important conclusion to be drawn from the Cox and Smith paper is that a general analysis of superpositioned or pooled nerve impulses is not possible on the basis of interval alone. Moreover, the conclusion of nonanalyzability may be true, they suspect, even if spike amplitude is added as a criterion for separation of large numbers of pooled neurons. This conclusion must be kept clearly in mind when one reads of promising general-purpose computer neural signal analyzers.

The surgical dissection technique and the multiple microelectrode, therefore, remain powerful tools for the study of complex patterns of neural activity. For less complex mixtures of neural impulses, however, simple statistical manipulations are useful, and in one or two cases (Keehn, 1966; Schmidt & Stromberg, 1969), ingenious computation techniques have taken advantage of specific neurophysiological situations to perform some special analyses, even though a general analysis probably has been forever excluded by Cox and Smith.

B. COMPOUND RESPONSES

If the number of spike action potentials, or for that matter any other action potential, is sufficiently large, then the individual spikes may not be individually discernible on an oscilloscope trace. Rather, the voltages generated by each individual neuron combine to produce a cumulative signal, which is an arithmetic sum of the individual responses, in accord with the usual laws of voltage addition. As the number of individual spike action potentials increases, pooled potentials show increasing numbers of complex spike forms resulting from the superposition of individual spikes. These voltage mixtures can usually be identified by extra bumps and notches on the otherwise smooth rise and fall of the spike; but occasionally deceptively simple, perfectly synchronized responses occur. In the central nervous system almost any electrode not positioned intracellularly, except the most highly selective, will pick up such multiple responses because of the dense neuronal packing.

If the electrode is unselective and has a relatively large field of action, the response will reflect the cumulative response of thousands, or even millions, of both graded and spike action potentials. Under such conditions, no vestige of an individual response is retained, and the potential picked up by the electrode exhibits very different properties than those of individual spikes. For example, it may last for very long periods of time—up to seconds or even longer in some free-running signals such as the electroencephalogram. Second, compound action potentials of this sort always exhibit graded properties as the quantal nature of the contributing signals is obscured by the very large numbers involved.

Compound action potentials represent another important candidate for coding purposes, in addition to pooled responses. They vary in ways that individual spikes do not and thus may be able to represent stimulus information where the individual spikes cannot. They also represent an important possibility as a medium for encoding the collective action or central tendency of large numbers of neurons.

If all the neurons are synchronized by an electrical stimulus, pseudocompound action potentials can be recorded from preparations in which the normal coding process is probably more fairly represented by the pooled type of record. Figure 10.1 shows compound action potentials recorded from the human ulnar nerve by a rather large ($\frac{3}{16}$-inch diameter) electrode, which was placed on the skin and which recorded the signal through it. Though this response looks very much like a single spike action potential, the dependence of its amplitude on stimulus intensity (not shown in Fig. 10.1) indicates that it is not. A single-cell response follows an all-or-none law. This compound signal is clearly the sum of a large number of spikes all occurring at nearly the same time. The duration of the response is almost as brief as an individual spike, indicating a very high degree of synchrony indeed. Very simple, brief compound action potentials of this kind are recorded with a technique originally developed by Dawson and Scott (1949), which allows such signals to be picked up through the skin percutaneously without surgery. As stimulus strength decreases, the signal gradually fades away, until it can no longer be detected. Individual cellular responses, therefore, are not at all detectable with this technique, and it can only be used with the relatively massive compound responses produced by the electrical stimulus.

The general technique of electrical stimulation and compound action potential recording with large electrodes was first developed by Gasser (see, for example, Gasser, 1938) quite a few years ago for use on animals on whom experimental surgery could be performed. When the saphenous nerve of a cat was dissected free and cleared of its outer sheath, simple hook electrodes were able to record a more elaborate response pattern than that observed with the Dawson technique used percutaneously on humans. Erlanger and Gasser (1937) showed that the compound action potential recorded in this situation is composed of several sequential components that conduct at different velocities. A major result of this difference

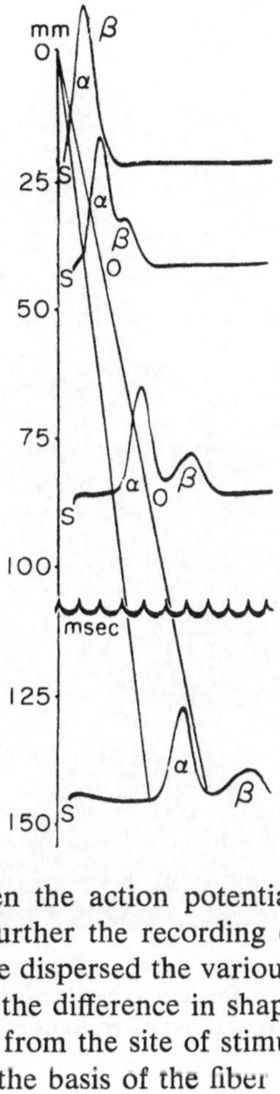

FIG. 8.4. Records showing the dispersion of a single compound action potential into multipeaked waves at various distances as a function of conduction time. The effect is due to the different conduction velocities of the fibers contributing to each peak. S indicates the occurrence of the stimulus. (From Patton, 1965, after Erlanger & Gasser.)

in conduction velocity is observed when the action potential is recorded at increasingly greater distances. The further the recording electrodes are from the stimulating electrodes, the more dispersed the various components of the signal become. Figure 8.4 shows the difference in shape of a typical response recorded at different distances from the site of stimulation. These differences in conduction velocities are the basis of the fiber categorization scheme presented in Chapter 3.

It has been possible to reconstruct the shape of a compound action potential of this sort by algebraic summation of the individual spike action potentials that are assumed to be produced by axons having diameters distributed in accord with anatomical measurements. Figure 8.5a is a micrographic cross section of a typical mixed somatosensory nerve. Clearly the compound action potential in this case is produced by a simple voltage addition of the spike action potentials from the family of individual neurons found in these peripheral nerves.

Figure 8.2 depicted another type of compound action potential. This response is an evoked potential recorded from the surface of the human head by a pulsed electrical stimulus. To be seen at all, this signal must be processed by a computer to extract it from the very severe electrical

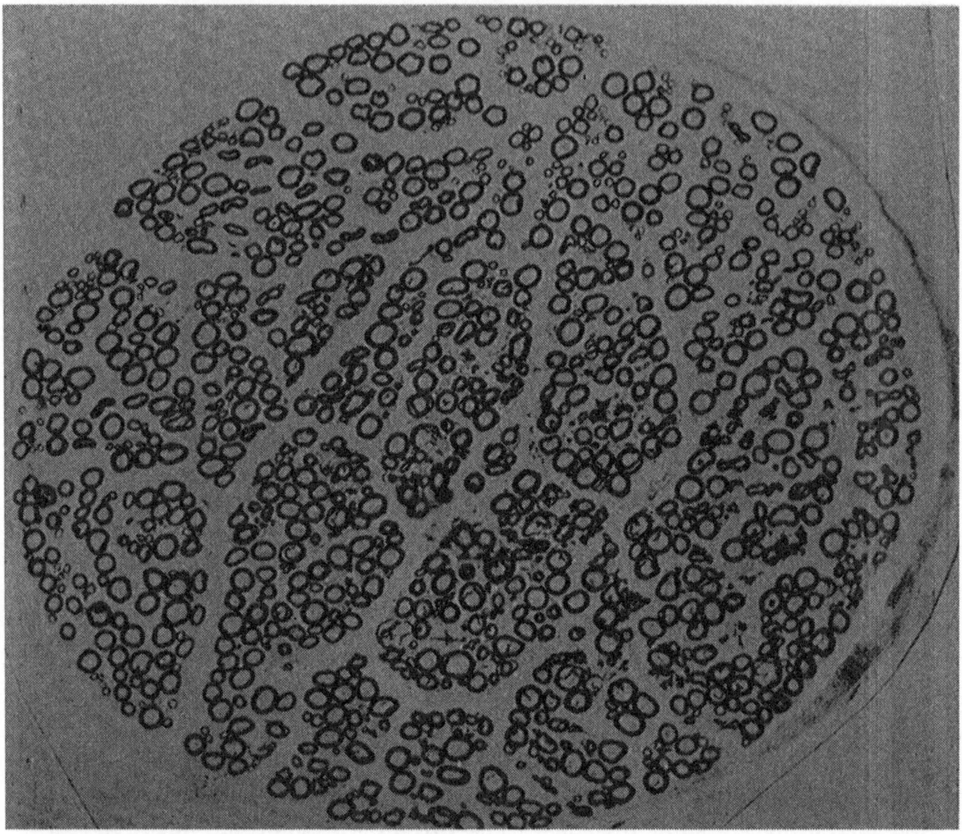

FIG. 8.5. A cross section of a peripheral nerve from a cat, showing the wide variety of fiber diameters typically found in mammalian nerves. (Photograph from Truex & Carpenter, 1964; after Ranson.)

noise conditions produced by interfering electroencephalographic and electromyographic signals. Even so, this tiny evoked response is the cumulation of the potentials of a large number of individual membrane responses. Its properties are not the same as the peripheral compound action potential previously described, for it is obviously much longer in duration and exhibits a much more complex form than that approximation of the classical smooth spike.

1. Detection of Compound Action Potentials

Investigations of the evoked potential are beset with a classic signal-to-noise problem.* The signal is typically so small, particularly when it is

* Much of the following discussion on signal averaging and compound action potentials is adopted from an earlier book by the author (Uttal, 1968).

being recorded through the intact skull and scalp of the animal, that it is hidden in the ongoing noise. The noise consists of the ordinary non-evoked EEG, the electromyograms from widely separated parts of the body, and electronic artifacts introduced by the amplifiers used in the system. One might find, for example, that many important components of the evoked potential from the human somatosensory system are only 15 microvolts in amplitude, whereas the amplitude of the usual electroencephalographic activity is on the order of ½ millivolt. The problem facing the investigator then is to extract the tiny evoked potentials from the other interfering signals. Such a procedure is possible, because of the fact that evoked potentials are synchronized with the impulsive stimulus that generates them. In contrast, the noise is, by definition, activity unrelated to the stimulus. Therefore, at any given delay following a series of stimuli, the amplitude of the unstimulated activity or noise tends to be randomly distributed. When averaged, the randomly fluctuating signal will tend to converge on the value zero. On the other hand, when one averages a signal that is synchronized or timelocked to the stimulus, the average value at each delay following the stimulus will converge toward some representative nonzero value. A mixture of the two types of signal, synchronized and random, will tend, upon repetitive averaging, to reproduce the waveform of the larger, unsynchronized activity.

During World War II repetitive sweep techniques were used to enhance radar displays. The conceptual problem faced in that context was exactly the same as that described for bioelectric potentials, except that the "stimulus" was the original burst of transmitted radio waves and the response was a much smaller radio wave reflected from the target. The various types of receivers capable of receiving the tiny reflected signals all had rather poor signal-to-noise characteristics. In addition, meteorological conditions contributed to irrelevant noise. The method used to extract the significant signals was to send a signal to the target area repeatedly, and to superimpose one response after another on the same display. Thus, very small changes could often be seen with the naked eye.

In 1950, G. D. Dawson suggested that similar superimposition techniques could be used to detect small evoked potentials from the human brain. (He also pointed out, incidentally, that the technique, like the radar solution, was functionally identical to the superimposition idea used by Galton in 1883 to enhance the similarities among a collection of facial portraits.)

Dawson's techniques created an almost unprecedented opportunity to do electrophysiological studies of the intact human brain that has received a great deal of attention in recent years. Although Dawson's first photographic superimposition techniques are primitive in comparison to the sophisticated real-time computer techniques used today, the general nature of the concept is identical. Dawson (1954) also developed what was perhaps the earliest electrical device actually to perform a pseudo-averaging function, a commutator-driven bank of capacitors, each of which charged to a level corresponding to a weighted algebraic sum of the signals at a given time after the stimulus.

Averaging may be formally defined by an equation of the following form:

$$E_t = \frac{1}{N}\sum_{n=1}^{N} E_t^n . \tag{8.1}$$

In this equation E_t is the average value of the signal at a given time, t, following the stimulus; N is the number of sequential responses averaged; and $E_t{}^n$ is the voltage at time t for the nth record. The equation represents but one point on the averaged curve for a particular time t after the stimulus. The variable t must be varied systematically to plot out the entire time function. In practice, the preparation of an average response requires the presentation of a repetitive number (N) of stimuli. For digital computer averaging, each evoked potential is then digitized by means of an analog-to-digital converter, at a rate sufficient to provide an adequate density of sample points. A large number of averages, equivalent to the family of t's is then computed with the use of Eq. (8.1), and the values of the family of averages are plotted or displayed on some appropriate output device to represent the average time course of the response.

In the most sophisticated digital computer averages, a true division by N is made with each successive stimulus presentation. Performing a division, however, is a complicated logical operation that may be avoided by a simple expedient. Thus, many commercial special-purpose averagers do not actually average but are serial accumulators. Each time a stimulus is presented, an analog-to-digital converter samples the data, as described above. These coded values may be summed according to Eq. (8.1), but the division by N is never accomplished. Instead the cumulative value in the register corresponding to $\Sigma_{n=1}^{N} E_t{}^n$, that is corresponding to a single t, increases throughout the entire experiment. In this manner, the values accumulated in each of the registers gradually increase in magnitude up to the maximum capacity of the storage registers.

Since there is as much likelihood that an unsynchronized signal will be above the mean value as below, after any given number of stimuli, once again the contents of each register tend toward some mean value. However, in this case there will be produced an increasingly nonzero mean value as the number of stimulus presentations increases. The division by N, therefore, simply acts as a scaling factor to normalize the amplitude of the signal for any N. It is here that the price is paid for eliminating the arithmetic circuitry necessary to accomplish the division. Since the scale of the oscilloscope display is usually set to display the final reading properly, it is not possible to visualize the unaveraged results of a single sweep during the early states of the averaging process since the summed values are just too small. Only after a considerable number of stimuli have been presented do the characteristics of the signal begin to enlarge sufficiently to become visible to the observer.

2. The Origin of Compound Action Potentials of the Brain

An important general question that arises in dealing with compound action potentials concerns the relation of this waveform to the individual membrane action potentials. For each compound response we would like to know as much as possible about the contributing neuronal population, and in what way their individual responses were summated. The problem of the peripheral nerve compound action potential is relatively straightforward. The calculated sums of the individual spike action potentials have proven to be adequate in predicting the shape of the compound nerve response. However, for such compound action potentials as the evoked brain potential and the electroencephalogram, there is a great deal of controversy among different authorities as to the origin of the potential.

The major controversial issue concerning compound evoked brain potentials is in regard to their cellular origins. Purpura (1959) comprehensively reviewed theories of the genesis of cortical potentials and stated that the early models stressing the contribution of spike action potentials, which were invoked to explain slow brain potentials, could no longer be accepted, principally because spike activity could not be associated experimentally with slow potentials. These models, he further stated, have been replaced by newer ideas in which slower dendritic and synaptic potentials are hypothesized as the basic subunits that are summated to produce the slow electrocortical potentials.

Amassian, Waller, and Macy (1964) expressed this same point of view in one of the first symposia on sensory evoked response in man. For a number of reasons, including the apparent ability of experimenters to disassociate spike and slow activity and mathematical considerations of the volume recording technique, they also concluded that spike action potentials do not contribute directly to the compound evoked brain potential recorded from the surface of the head. They also stated that the slow potentials already known to exist, such as the postsynaptic and the various postspike after-potentials, are more likely added together to produce the waveform.

Much of their evidence supporting that viewpoint rested on the analysis of a small sample of neuronal action potentials. However, modern computing equipment that allows much larger sample sizes to be collected have

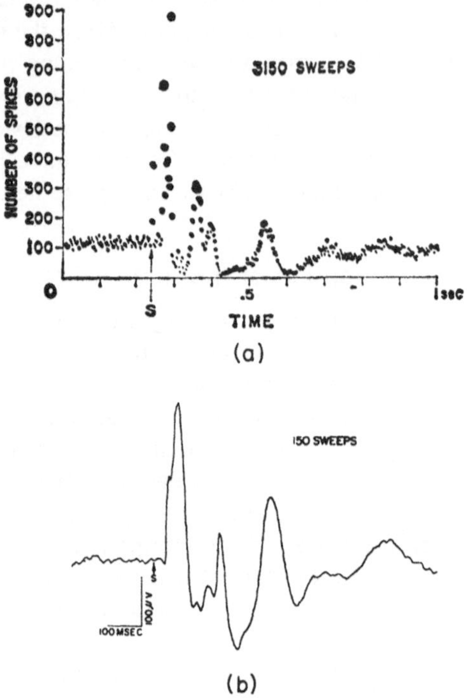

FIG. 8.6 Two records showing the high degree of correspondence between the probability of firing of a single cell (a) and the locally generated evoked brain potentials (b) recorded from the visual cortex of a cat. (From Fox & O'Brien, 1965.)

confirmed that a relation does exist between spike activity and the slow potentials.

Fox and O'Brien (1965) used a combination of computer analysis techniques to demonstrate a very high correlation between the latency histograms of spikes from a single cortical cell, as recorded with a high frequency bandpass preamplifier through an extracellular microelectrode, and the averaged compound evoked brain potential, recorded through the same electrode, but with an amplifier passing lower-frequency signals. The correlation between these two measures is great enough when the sample size is large enough (see Fig. 8.6) to suggest that spike action potentials may have been somewhat prematurely rejected as major contributors to the production of compound potentials. Fox and O'Brien's work suggests that the compound evoked brain potential is a direct correlate of the *probability* of spike firing, and thus they may have demonstrated the missing relation between spike activity and the compound evoked brain potential.

For the time being, it is sufficient to state that the genesis of the compound evoked brain potential is still being vigorously debated; and that while some feel it is a direct reflection of synaptic and dendritic graded activity, others argue strongly for direct spike action potential contributions to the compound evoked brain potential.

Fox and Norman (1968) expanded the earlier notions of the relationship between the spike and the compound evoked brain potential by inventing a new measure they call *congruence*. Congruence is a measure that describes in a quantitative fashion how typical an individual cell's pattern of response is of the total sample of cells contributing to a particular compound evoked potential. The activity of the single cell and of a microelectroencephalogram were recorded in the manner described above, by means of one electrode and two amplifiers with differing bandpass characteristics.

Fox and Norman formed three functions from these recorded data. The first function was the distribution of amplitudes of the microelectroencephalogram recorded with the low bandpass amplifier. A digital-to-analog converter sampled the amplitude of this continuously varying slow potential every millisecond and made an entry into one of 256 amplitude categories each time the amplitude of the signal was within its range. In this way, an amplitude histogram was formed, which showed the number of occurrences of each amplitude value. An example of this type of histogram is seen in Fig. 8.7a.

The second function was based upon the occurrences of the spike action potential from a single cell, as detected with the high bandpass amplifier. Every time the cell fired a spike, it would trigger the computer to determine in which of the 256 amplitude levels the slow potential currently resided. Another computer analysis program then developed a second amplitude histogram, based upon the number of times a spike occurred for each of the 256 amplitude levels. An example of this type of histogram is shown in Fig. 8.7b.

So far, this procedure is a relatively straightforward process in which two simple amplitude histograms are developed. Now consider that some of the amplitudes of the slow potential occur more frequently than others. The extreme values, of course, occur relatively infrequently, and the mid-values occur relatively frequently. The frequency of the spikes shown in the second histogram is, therefore, strongly biased because there are many more mid-value amplitudes of the slow potential than there are extreme value amplitudes. It is, therefore, necessary to normalize these data by dividing the distribution of spike occurrence by the distribution of amplitude occurrence. By carrying out this simple calculation, Fox and Norman controlled for unequal numbers of the slow potential amplitudes. The results of this normalization form a third graph, which plots a function that is essentially equivalent to the number of times a spike occurred at each amplitude value of the slow potential. This typical curve, therefore, is normalized for the number of times each slow potential amplitude value occurred. An example of this third normalized histogram is shown in Fig. 8.7c. If we carefully study the three histograms a most important fact emerges. Even though there is an almost normal and, therefore, at first glance an

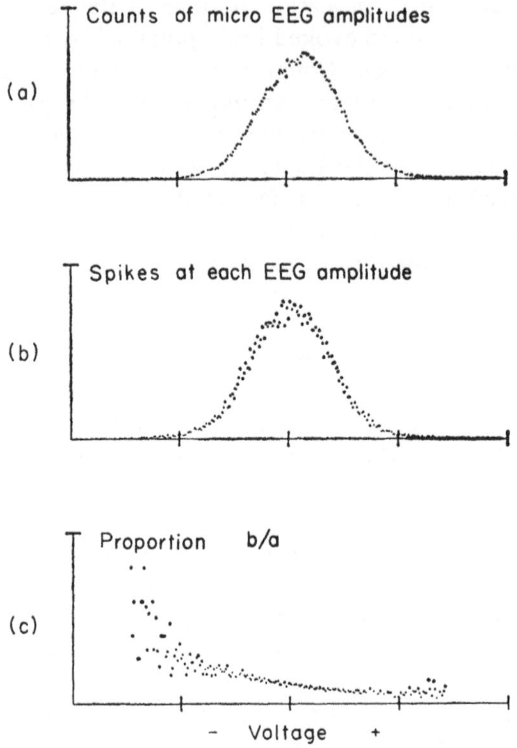

FIG. 8.7. Three distributions that show the close relation between the probability of a spike action potential and the local electroencephalogram (EEG). (a) Histogram of the amplitudes of the micro EEG at various times. (b) A histogram of the number of spikes produced at each EEG amplitude. (c) A new distribution produced by dividing (b) by (a), displaying a high statistical correlation. (From Fox & Norman, 1968.)

almost random distribution of the first two functions, there turns out to be a very high degree of correlation between the microelectroencephalogram amplitudes and spike occurrence. This is indicated by the nearly straight line that is obtained when the data are normalized and plotted.

The specific numerical value of the correlation between the activity of a given single cell and the microelectroencephalogram can be computed and was found by Fox and Norman to vary from experiment to experiment. These investigators feel that this fact in itself is a very significant result, for it confirms the notion that the microelectroencephalogram recorded locally is not a measure of activity of any single cell, but rather represents the action of a large population of cells. Thus, if the single-cell activity varies in its congruence with respect to the microelectroencephalogram, it indicates that other cells in the neighborhood are also contributing to the overall potential. If the single-cell recording had always correlated with the microelectroencephalogram, it would have raised the ever-present specter that the microelectroencephalogram is something quite different from that recorded in a more conventional manner with very large electrodes.

If the notion of congruence is a reasonable one, it should also be expected that when the sensory system into which the electrodes have been placed is driven by an appropriate stimulus, the correlation between spike count and the amplitude of the microelectroencephalogram would increase. This is exactly the result obtained for those cells in the visual cortex of the cat which respond to light flashes. (The visual cortex does include other cells that are not responsive to light, and for these cells the congruence was seen to decrease.)

In summary, Fox and Norman presented a valuable new measure for the estimation of the "coordination" or congruence of a group of cortical cells. This measure has a number of important implications. First, the high degree of correlation with single-cell activity tells us that the microelectroencephalogram, if it is not a direct product of the spikes themselves, is certainly derived from processes that also underlie spike generation. The two are essentially analogs of each other, therefore, and may be considered to be at least informationally equivalent. Second, the microelectroencephalogram gives us a measure that is a cumulative index of the activity of many cells in the recording region of the electrode, rather than a measure of the activity of a single cell. Such a measure may be the only plausible way of investigating the action of ensembles of neurons—and ensembles may be the "name of the game." Third, congruence is a direct estimate of the degree to which the activity of a single cell represents the mean biological activity of all the cells in the region contributing to the slow potential. Congruence should, therefore, prove in future years to be an important measure of brain activity and contribute greatly to an understanding of how it is that the activity of single cells are pooled into compound responses.

One final point with regard to the possible origins of the compound potentials that can be recorded from the brain: As we learn more about glial cells, the supporting units found between neurons, it is becoming clear that they, too, produce bioelectric potentials. Some of these voltages may be produced under the indirect influence of the neurons that are embedded among the glial cells. In some cases (e.g., the Muller cells in the retina), the time course of the glial and neuronal cells' response may be nearly the same. Thus, it is possible that, even though they are not involved in meaningful information processing, glial responses may be correlated with stimuli or psychological processes. It has been further suggested that the direct cause of glial depolarization is the increased potassium concentration in the extracullular space as a result of neuronal activity. This possibility has been considered in detail by Orkand, Nicholls, and Kuffler (1966) and in an extensive review by Kuffler and Nicholls (1966). The fact that such signals do occur raises, at least, the possibility that they may contribute to the production of compound action potentials.

In any event, there is as yet no complete agreement on the origins of electroencephalographic or evoked brain potentials; any or all of the voltage sources mentioned so far may contribute to those compound action potentials. Hopefully, the problem can be resolved in the not-too-distant future.

C. DO HYPERSYNCHRONIZED COMPOUND RESPONSES TRULY REPRESENT NATURAL CODING METHODS?

As we noted earlier, Gasser (1938) and Erlanger and Gasser (1937) originally explored the nature of action potentials in peripheral nerves by a stimulation technique that has become standard in both neurophysiological and psychological laboratories. The stimulus they used was a highly unusual—some would say nonphysiological—event; a brief, but (in terms of normal biological electropotentials) very massive electrical pulse. The description of the peripheral nerve compound evoked potential given in our earlier discussions is primarily based on results from experiments in which stimuli of this sort were used. However, when the activity of a somatosensory nerve trunk is observed by means of more normal stimuli, such as a tap, a brush, or the application of heat or cold, then most observational techniques indicate that the pattern of response is better characterized as being of the pooled type of record. This is entirely reasonable, of course, for a much greater degree of information coding is inherently available in the delicate time relations between the more or less independently responding single-cell spike action potentials.

A major question arises, though, because of the emphasis placed by many experimenters on the electrically-generated compound action potential. Does this artificial stimulation technique impose any spurious or extraneous features on top of the normal coding properties of the nerve? In other words, does the hypersynchronized response in which all neurons respond simultaneously introduce properties of its own that make the recorded potential unrepresentative of a statistical summation of the semi-independent fiber response? To elaborate further by expressing the question in a more formal manner, does this particular statistic—the compound action potential produced by pulsed electrical stimulation—represent in an unbiased way the central tendencies of a family of independently responding axonal potentials?

Clearly, the only way to resolve these questions is to devise some other means of performing the same statistical analysis performed by the nerve itself when it is stimulated with a transient electrical stimulus. Such an analytical technique was developed by Casby, Siminoff, and Houseknecht (1963), who used a specially designed analog computer to cross-correlate the pooled responses evoked by natural stimulation.

Before detailing their study and its reassuring results, let us digress to a brief discussion of the meaning of cross-correlation in this neurophysiological context.

1. Cross-Correlation*

Cross-correlation is a mathematical technique for determining the similarities between two functions. It is defined formally by the following expression when the functions are variables of time:

$$C_c(\tau) = \frac{1}{T} \int_{t=0}^{T} f(t)g(t - \tau)\,dt. \qquad (8.2)$$

$C_c(\tau)$ is defined as the *temporal cross-correlation* between the two functions $f(t)$ and $g(t - \tau)$; τ is a displacement of one function with respect to the other in time—a lag or lead; T is the duration of the signal. It should be noted that $C_c(\tau)$ has a single value for each value of τ. $C_c(\tau)$ is, therefore, not a single number, but a function of the variable delay τ.

What does the function $C_c(\tau)$ mean, and what are its potential applications in the study of bioelectric potentials? Perhaps the best way to illustrate the practical significance of the correlation is to consider the two functions $g(t)$ and $f(t)$ at a single time delay τ. A separate scatter plot can be made between the values of $g(t)$ and the values of $f(t)$ for each time delay τ.

Figure 8.8a is an example of a scatter plot for two sinusoidal oscillations $f(t) = \sin t$ and $g(t) = \sin t$, which are in phase—that is, $\tau = 0$. In this case, each value of the y coordinate of one sinusoid is equal to the corresponding coordinate of the other sinusoid. The scatter plot then becomes a straight line with a positive slope, as illustrated. In one sense such a scatter plot represents a maximum correspondence between the two signals. They can, therefore, be said to be perfectly correlated. By one correlation formula, this would mean a correlation of 1. However, in the generalized correlation, Eq. (8.2), the specific number for perfect correlation is a function of the scaling of the various constants and is consequently unnormalized. Nevertheless, the two sinusoids are maximally related at this phase angle. If we now insert a time delay or phase shift between the two signals, $f(t)$ is no longer equal to $g(t)$, and the perfect relationship between them tends to diminish. Figure 8.8b is a plot of the correspondence between two sinusoids separated by a delay of ¼ wavelength (a phase shift of 90°). The scatter plot now is in the shape of a circle, a plot associated with a relatively low degree of correlation. When formalized with the above expression, it gives a medium value of $C_c(\tau)$. Figure 8.8c is a scatter plot

* This section is adapted from an earlier book (Uttal, 1968) by the author.

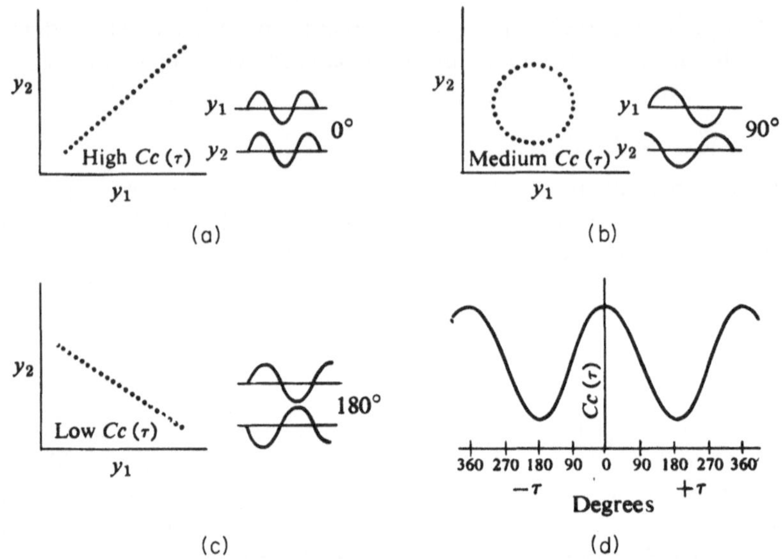

FIG. 8.8. Scatter plots of $y_1 = f(t)$ and $y_2 = g(t)$ at various times. Each of the three figures represents a single delay τ in the total correlation function. (a) The scatter plot for in-phase signals ($\tau = 0$) in which high values of y_1 are matched with high values of y_2. This corresponds to a perfect match and a high correlation and high $C_c(\tau)$. (b) The scatter plot for two signals 90° ($\tau = \pi/2$) out of phase in which values of y_1 are matched sometimes with low values of y_2 and sometimes with high values of y_2. This corresponds to a very low degree of correlation and a medium $C_c(\tau)$. (c) The scatter plot for two signals 180° ($\tau = \pi$) out of phase. In this case, high values of y_1 are associated with low values of y_2 and the signals are therefore inversely correlated to a high degree but with a low $C_c(\tau)$. (d) A plot of $C_c(\tau)$ as a function of τ for two in-phase sinusoids, showing the replication of the original sinusoidal waveform. If they were not in phase, the central peak would be shifted by an amount equal to the phase angle between the two sinusoids. (From Uttal, 1968.)

of the signal for a 180° phase shift that results in a straight line with a negative slope. This represents a perfect inverse correlation and reflects a more negative value of the correlation coefficient described above.

A plot of the entire set of correlations $C_c(\tau)$ for all possible phase shifts also reproduces the sinusoidal waveform, since it reflects the periodic properties of the signal (see Fig. 8.8). In the case of perfect sinusoids, the application of the correlation expression would be trivial, since it only reproduces the sinusoids and indicates the phase angle between them.

Cross-correlation is extremely useful, however, for inspecting two periodic signals whose waveforms are relatively irregular over the total duration of the record. Under these conditions the periodic nature of the signal, if any, tends to emerge and to be displayed in the correlogram. The cross-correlogram can be considered a long-term averaging process that tends to emphasize the periodic similarities between two functions over the aperiodicities. Thus, two sets of data having very long-term periodic trends, or sequences common to both signals, can be reduced to the form of the correlogram. From this point of view, the cross-correlation function can be seen to be closely similar to the Fourier frequency analysis technique. In fact, by making $f(t)$ or $g(t)$ equal to a family of sinusoids it becomes exactly that.

Thus, by actually reducing the amount of information present, both cross-correlation and Fourier analysis help to make the important data more amenable to interpretation.

2. The Casby, Siminoff, and Houseknecht Experiments

Having introduced the general nature of cross-correlation, we can now consider in detail Casby, Siminoff, and Houseknecht's (1963) specific application of this mathematical technique to the study of pooled nerve action potentials.

Figure 8.9 is a sketch of the electrode placement and a block diagram of their recording and analog computing apparatus. The nerve of a cat was dissected free and hung on two well-separated recording electrodes. When the skin receptors of the experimental animal were stimulated with a gentle brushing, the two recording electrodes picked up pooled action potentials from a mixed nerve twig. The temporal patterns recorded by these two electrodes represent the two functions $f(t)$ and $g(t)$ which were to be correlated. In each case these functions were pooled action potentials recorded from the same nerve, but at different locations. This means that

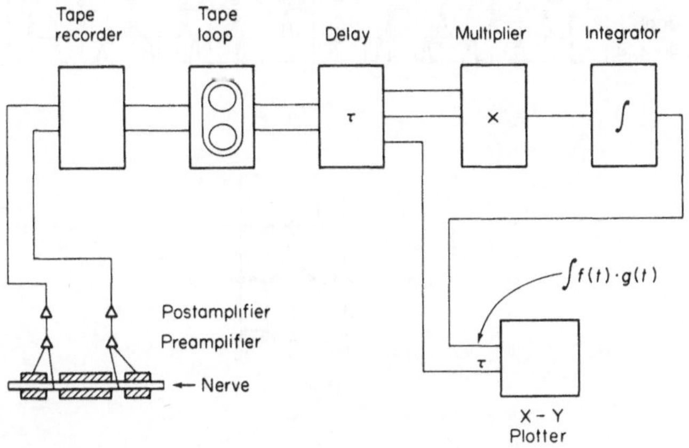

FIG. 8.9. Block diagram of an analog cross-correlator. (Adapted from Casby, Siminoff, & Houseknecht, 1963.)

the two signals should contain roughly the same information, although the two functions should also differ because of the differences in electrode location relative to the individual fibers and because of the time discrepancy produced by the two differing recording locations. They will, however, be similar enough so that a sorting procedure based on conduction time can provide a useful means of replicating the form of the compound action potential that would have been produced if an electrical shock had been used rather than the more normal stimulus of a brush.

Other important parts of the equipment that the reader should especially note are the device that allowed the investigators to adjust the time delay between the two signals and the multiplier that electronically multiplied the two signals at each delay interval. Finally, the integrator that accumulated the sums of the cross-products should be noted. All of these devices are necessary to implement the various substeps of Eq. (8.2).

The effect of cross-correlational procedures that Casby *et al.* performed on these data can best be illustrated by examining Fig. 8.10. This figure shows what would be the result of a cross-correlation of the two signals (Fig. 8.10a) if the delay apparatus was continuously adjusted to sample different delays (τ), and if the signal picked up at the two electrodes happened to be the same spike but arriving at the recording electrodes at different times as it passed down the nerves. The two detections of the single spike would occur at different times at the two recording points for all τ's not equal to the conduction time of the spike between the two electrodes. Multiplying the two functions by each other for all τ's not equal to conduction time, therefore, would result in a relatively low value of

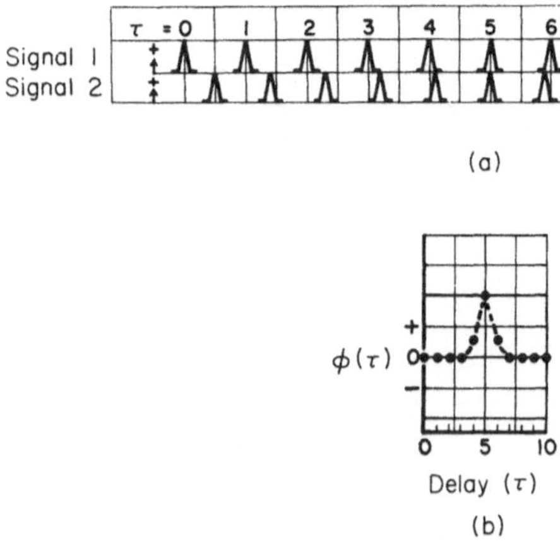

FIG. 8.10. Diagrams show the result of cross-correlating pooled spike action potentials from two different points on the same nerve. (a) the temporal relationship of the two spikes as a function of the delay. (b) The results of the cross-correlation. (Adapted from Casby, Siminoff & Houseknecht, 1963.)

the cross-production. However, in Fig. 8.10b it can be seen that when τ is adjusted to be equal to the condition time for this single spike, the value of the cross-product of the two functions will be higher than at all other τ's, since the two peaks of high value will be multiplied by each other, rather than one peak being multiplied by a lower constant value. As one scans the range of τ's, therefore, a cross-correlation function similar to that shown in Fig. 8.10b would be obtained, with the peak centered at a τ equal to the conduction time.

An understanding of the data presented below depends upon an appreciation of this process. It must be understood that a peak will occur in the cross-product of the two functions when τ is equal to the conduction time of the nerve action potential, recorded at two electrodes that differ in position along the nerve twig. Thus, from one point of view, the device used by Casby *et al.* is a complicated, though accurate, way of measuring the conduction velocity of nerve impulses. However, their interest was in finding an alternative method of examining the statistics of nerve action potential populations in a way that was sensitive to conduction velocity. They set out to reproduce the compound action potential by doing the statistics with their cross-correlator. This would be an independent check on the production of the compound action potential evoked by an electrical shock.

Suppose now that the Casby et al. technique is applied to a pooled response of the type shown in Fig. 8.2. Then the simple single spike used in the calculations in the example shown in Fig. 8.10 would be replaced by a complicated pooled pattern of spike responses. This cross-correlation will produce something quite different than simply the square of the original single spike. It will produce a cross-correlation function that can be seen to be essentially identical in meaning to the compound action potential elicited by an electrical stimulus. This is so even if the stimulus were something as diffuse as a stroke with a camel's hair brush.

The shape of the function elicited by a natural stimulus and processed by the cross-correlater will not be exactly identical to that produced by an electrical shock, however. It is not the simple sum of individual spikes that is obtained with the classic Erlanger and Gasser technique, but a cross-product in which the cross-multiplication of bipolar waves can produce even tripolar waves.

Figure 8.11 is a cross-correlogram of a pooled neural response produced with this technique from an actual preparation. It shows the multiple response peaks corresponding to those produced with electrical shocks and seen in Fig. 8.4. One peak is produced by the most rapidly conducting A_α fibers. A peak occurring slightly later is due to the somewhat less rapidly conducting A_δ fibers. The last peak corresponds to the C fibers responses.

An important point is illustrated when these data are compared to the compound action potentials produced with the Erlanger and Gasser tech-

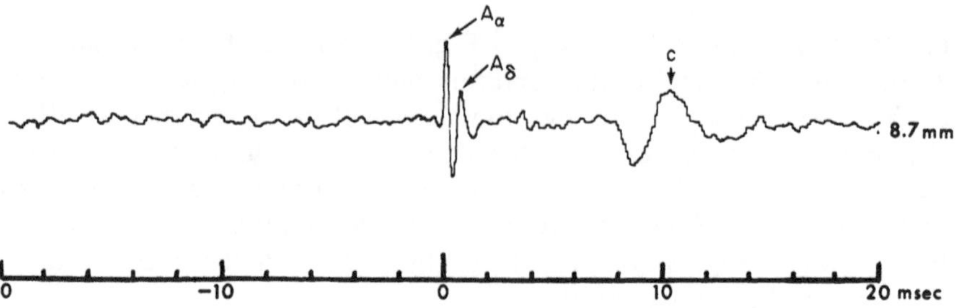

FIG. 8.11. "Compound" action potential waveform produced by cross-correlating the pooled spike action potential patterns produced by a natural stimulus—stroking with a tuft of cotton—recorded at two different points on a cat's peripheral nerve. The various peaks correspond to the waveforms produced when an electrical pulse stimulus is used artificially to synchronize the firing of the individual axons. (From Casby, Siminoff, & Houseknecht, 1963.)

nique and allowances are made for the differences in procedure used to produce each curve. The point is that the functions must be considered to be essentially equivalent; namely, two different ways of producing a statistical summary of the behavior of a population of neurons. In one case, when the electrical stimulus was used the statistical summation was accomplished by the nerve itself. In the other, the statistical summation was produced by an external device—the cross-correlator. The obvious conclusion is that the electrically stimulated response in no way distorts the true pattern of the response. There are no additional or extraneous factors inserted by the nonphysiological synchronization of the responses. If we take into account the constraint that all information regarding the phase and timing of an individual action potential is lost, the compound action potential elicited with an electrical shock is thus seen to be a perfectly good indicator of the performance of individual neurons.

In this chapter we have considered a number of topics that concern the transmission of information in groups of neurons. However, like the previous chapter, it dealt with responses within neurons, in spite of the fact that they were recorded collectively. In the next chapter we turn to a related topic. In this case, however, the emphasis will not be on the details of intracellular transmission, but on communication between cells. The next chapter, therefore, will necessarily concentrate on the action of the synapse, that portion of a neuron especially evolved to signal information from one cell to another.

9
THE SYNAPSE

Chapter 3 emphasized the fact that the contemporary doctrine of neuronal anatomy asserts that neurons transmitting information within the nervous system are not protoplasmically interconnected. Rather, in each sensory modality three to five separate cells are connected "end to end" to sequentially convey afferent signals to the brain. In the brain, neurons communicate through a very large number of interconnecting regions. The regions of communication between neurons, known as synapses, have been the object of intensive study by neurophysiologists because it is only at these points that neural integration and interaction become possible. This interest has been especially great since neural discontinuity was firmly established by advances in electronmicroscopy and staining techniques.

Considering the infinitesimal dimensions of synapses, how is that we know that the neurons on either side of a synapse are separated by their plasma membranes? Eccles (1964) has listed the following specific evidence for accepting the doctrine of anatomical discontinuity between neurons:

1. Axonal degeneration produced by separating the axon from the cell body does not proceed across synaptic boundaries. Thus the metabolism of the axon beyond the synapse is separate and distinct from that preceding the synapse.
2. Most synapses insert a ½ to 1 millisecond delay in the transmission of a neural signal. This is equal to the delay produced by conduction over an inch or two of uninterrupted axon.

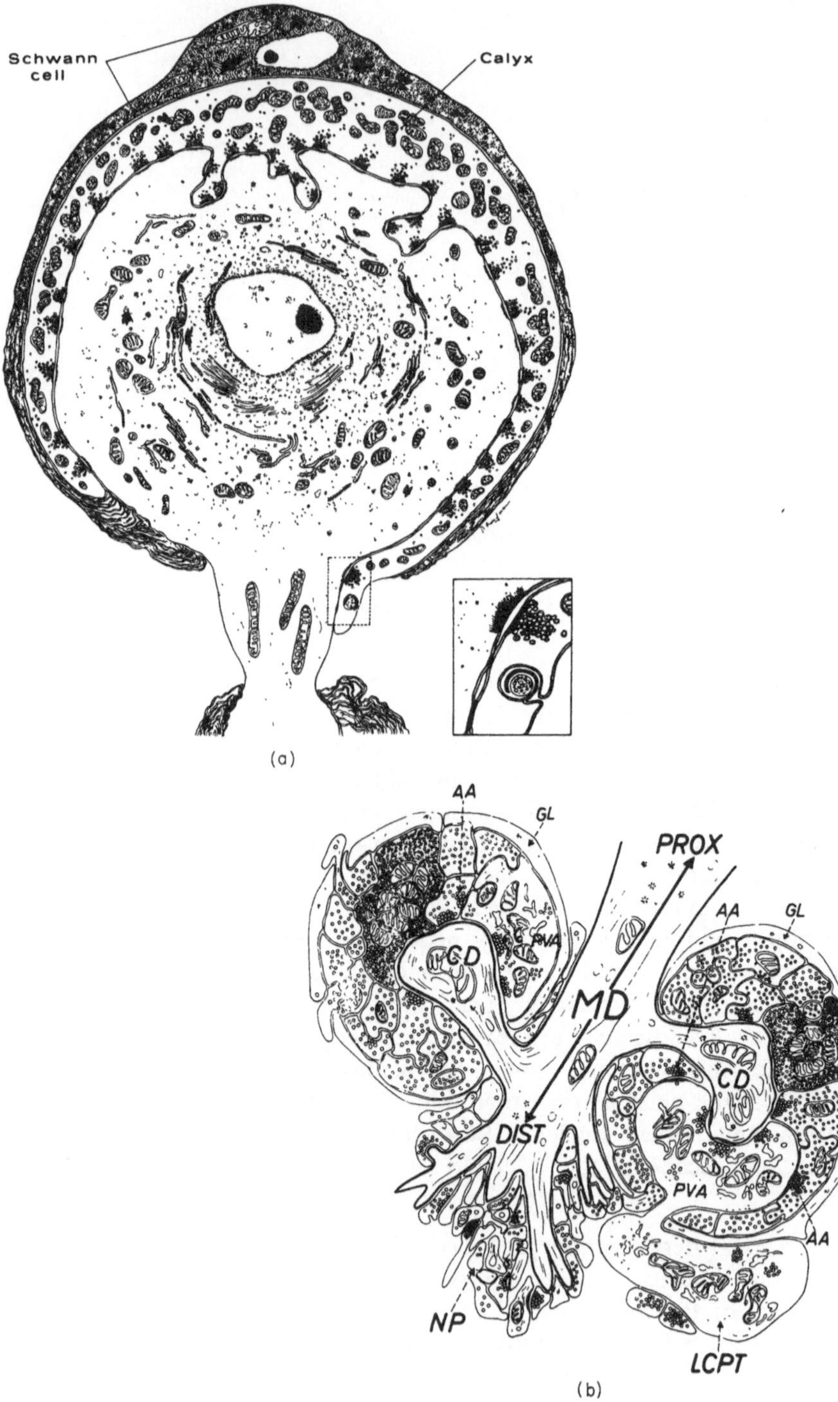
Schwann cell
Calyx
(a)
AA
GL
PROX
AA
GL
CD
PVA
MD
CD
DIST
PVA
AA
NP
LCPT
(b)

FIG. 9.1a, b

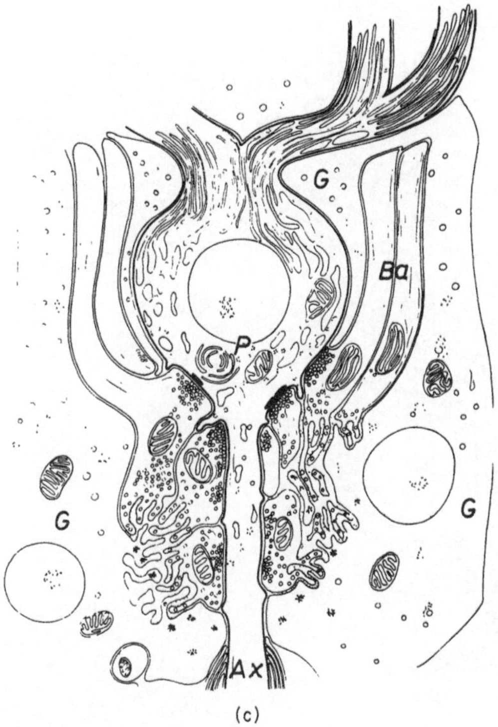

FIG. 9.1c

FIG. 9.1. In addition to the various synaptic contacts shown in Chapter 3, a number of other examples, from McLennan's (1970) authoritative book on *Synaptic Transmission* are reproduced here. (a) Synapse (shown in the inset) of a ciliary ganglion cell from a chick (from de Lorenzo, 1960); (b) a synaptic complex from the mammalian pulvinar. The main dendrite (MD) branches to form secondary structures (CD) that have large numbers of synaptic contacts with presynaptic cells containing groups of synaptic vesicles. Some of the synapses are indicated as AA (from Majorossy, Réthelyi, & Szentágothai, 1964); (c) synaptic contacts at the base of a Purkinje cell(like that shown in Fig. 3.22f). P = the Purkinje cell body; Ba = basket cell axons; G = glia; Ax = axon (from Hámori & Szentágothai, 1965).

3. Special staining techniques, such as the Cajal silver stain, act selectively on a single cell, and the stain is often sharply demarcated at the synapse.
4. Electron micrographs show directly that the cell membranes of the two cells are separate from each other in all except some electrical synapses and that the cell membrane is continuous around the end of the axon.

Proceeding from the premise that synapses do not represent regions of protoplasmic continuity between neurons, we must next inquire into the mechanisms that account for the transmission of information across this cleft. The purpose of this chapter is to consider this question at a molecular level comparable to that used to explain the dynamics of resting and action potentials.

A. ELECTRICAL AND CHEMICAL SYNAPSES

Figure 9.1 displays several different kinds of synaptic connections. Anatomically these junctions look quite different from each other. However, in spite of this apparent variety of form, only two different mechanisms

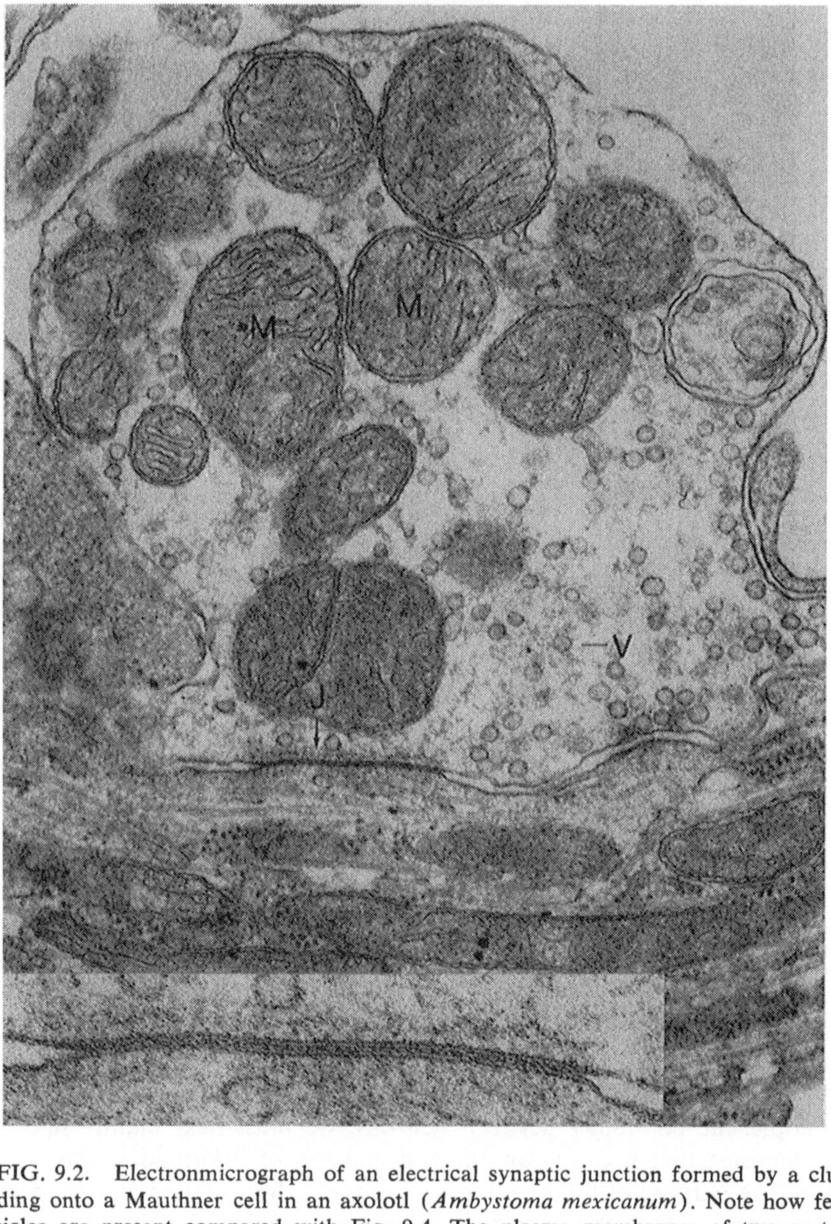

FIG. 9.2. Electronmicrograph of an electrical synaptic junction formed by a club ending onto a Mauthner cell in an axolotl (*Ambystoma mexicanum*). Note how few vesicles are present compared with Fig. 9.4. The plasma membranes of two motor neurons are seen to be nearly touching (a tight junction), allowing synaptic interaction by means of close electrical coupling. M = mitochondria; J = junction; V = vesicles of unknown function. Magnification is 56,000×. Inset shows the junction at a magnification of 154,200×. (Photograph courtesy of Dr. George D. Pappas and J. Keefer of the Albert Einstein College of Medicine.)

of synaptic action are known at present. The first mode of synaptic action is based upon direct electrical activation, similar to the way in which a nerve impulse is propagated along the axon itself. The second, and apparently much more common, mode of synaptic action is mediated by means of highly specialized transmitter chemicals released from the terminal or *presynaptic region* of the cell from which the information is coming. Information is transmitted by the diffusion of transmitter substance across the space between two neurons, producing an alteration in the membrane potential of the second or *postsynaptic cell.*

Electrical synapses between two neurons are less frequently found in vertebrate preparations than are the much more common chemical synapses, and only very rarely in mammalian tissue (Sotelo & Palay, 1970). Electrical synapses require a much higher degree of electrical coupling to function than do chemical synapses and, therefore, the two neurons are generally more closely spaced. How close the spacing may be in such an electrical synapse is evidenced by the electronmicrographic work of Bennett, Aljure, Nakajima, and Pappas (1963) and Pappas (1966). One example, a Mauthner neuron from an axolotl, an amphibian that retains its gills into adulthood, is shown in Fig. 9.2. The important feature of this photograph is that the spacing between the two neurons is very close—only 20 Å—and there is no evidence of any chemically active vesicles. In some other cells there is a suggestion that the two membranes may actually be fused. Pappas believes that this closely spaced region is the morphological structure responsible for electrical synaptic conductivity.

It is important for the reader to understand that the distinction between an electrical and a chemical synapse does not mean that one is exclusively chemical or the other exclusively electrical. The voltages and currents that are the primary excitants in electrical synapses do result from the same sort of ionic flows described in our previous discussion of the resting and action potential. The chemical synapse, while it does seem to be relatively insensitive to electricity as a primary excitant (some neurophysiologists feel that it is completely inexcitable electrically), does exhibit the same sort of electropotentials as any ionic system. (See the sections on pre- and postsynaptic potentials later in this chapter.) It must be remembered that our technology for observing the magnitude and time course of even these chemical events is almost exclusively an electrical one. Chemical events can be only indirectly observed with the usual sort of electrophysiological electrodes and amplifiers. The key difference between chemical and electrical synapses is the nature of the primary excitant that best is able to alter the resting potential of the postsynaptic region.

In the epilog to his distinguished book, (Eccles, 1964) also lists some more specific distinguishing features of the chemical and electrical synapse which have been paraphrased as follows:

1. A much smaller synaptic delay occurs across an electrical synapse than across a chemically mediated synapse.
2. The passive (electrotonic) spread of the produced potentials extends much further in an electrical than in a chemical synapse. An electrode placed in a presynaptic neuron will pick up much more of the activity in the postsynaptic neuron if the synapse is electrically mediated. In other words, electrical synapses are more closely electrically coupled than chemical ones.
3. The size of the synaptic cleft between two neurons is much larger in a chemically mediated synapse than in an electrical synapse, where the plasma membranes of the cells may be actually fused. This reduced spacing in an electrical synapse may partly explain the high level of electrical coupling described in (2) above.
4. Two neurons that electrically interact at a synapse are usually of roughly the same size. If the synaptic element of the first cell is very much smaller than that of the second cell, chemical transmission is almost obligatory, since direct electrical action between small presynaptic and large postsynaptic neurons would require some sort of an electrical energy amplification process. No such process is known to exist.

Because of the relative rarity of electrical synapses, and because there is a well-accepted theory of chemical mediation of synaptic action, we will concentrate primarily in this chapter on current views of chemically mediated synapses.

The main points of the chemical synaptic action introduced in this preliminary discussion are illustrated in Fig. 9.3, which is adapted from a diagram originally drawn by Grundfest (1957). This figure emphasizes

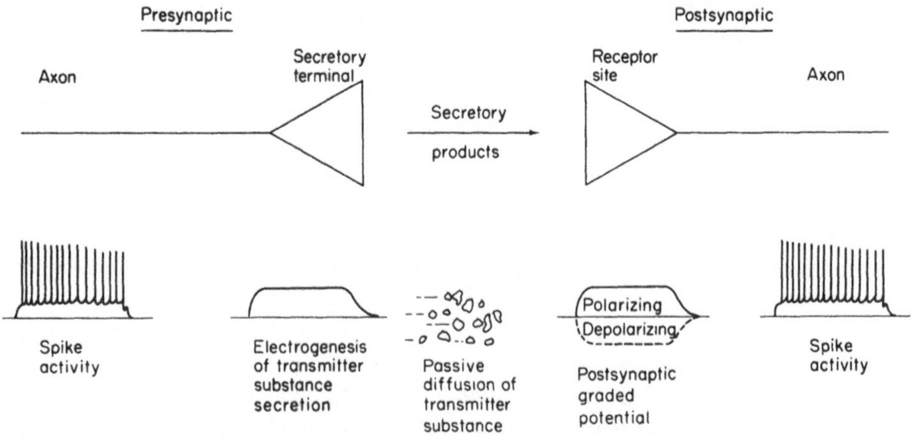

FIG. 9.3. Schematic sketch of the various stages of neuroelectric and chemical coding at a chemical synapse. (Adapted from Grundfest, 1957.)

several important concepts. Note that even within the microscopic synaptic region—only a few hundred angstrom units in width—several different levels or modes of information coding can exist. Transmission in the presynaptic axon up to the region of the synapse takes place by means of the frequency-coded spike action potential discussed earlier. Presynaptic tissues then convert this pulse-coded information into a graded potential, which in some way not yet completely understood brings about the release of the transmitter chemical that serves as a chemical code for the transmitted information. The transmitter chemical itself then diffuses in what appears to be a purely passive manner (i.e., no active force drives the transmitter molecules) across the synaptic cleft. The molecules of transmitter substance are subsequently captured by certain specialized receptor sites on the postsynaptic tissue, whose resting potential is altered to produce another graded postsynaptic potential.

The postsynaptic potential may have either an inhibitory or excitatory effect on the postsynaptic neuron. If it is excitatory, it increases the amount of spike action potential activity in the postsynaptic axon. If it is inhibitory, postsynaptic activity is reduced. Now that we have established a general qualitative picture of the two kinds of synaptic action, let us consider the chemical process in greater detail.

B. CHEMICALLY MEDIATED SYNAPTIC TRANSMISSION

The major puzzle in the chemical synaptic transmission process is: How is presynaptic graded activity able to evoke postsynaptic membrane depolarization? This section will attempt to fill in the details of the currently accepted model of chemical transmission that was introduced in the previous section.

The earliest evidence indicating the presence of some sort of chemical mediation of the synaptic transmission process was quite indirect. Investigators working the first half of the 20th century encountered a number of different chemicals that produced exceptionally strong neural effects. Loewi (1921), for example, demonstrated the very strong inhibitory effects of acetylcholine on heart rate. Cannon and his colleagues (Cannon & Bacq, 1931) found that epinephrine and some of its chemical relatives (specifically norepinephrine) could accelerate heart rate. In each case the action was produced by a chemical that was normally emitted by stimulated portions of the parasympathetic and sympathetic nervous systems, respectively. The notion that parts of the nervous system secrete chemicals thus began to seem highly plausible.

On the basis of a number of pieces of evidence it is now agreed by most workers studying chemical synapses that the chemical transmitter substances are organized in a quantal fashion. That is, many molecules of the chemical transmitter substance appear to be clustered into groups or packets of almost constant size that are released as units. One of the earliest and most compelling arguments for the packet concept was some work that dealt with very careful recordings of the microstructure of the postsynaptic potentials. Del Castillo and Katz (1954) observed in motor end plates and Eccles, Eccles, Iggo, and Lundberg (1961) and Nishi and Koketsu (1960) observed in central neurons that when a microelectrode was inserted into the postsynaptic tissue, spontaneous depolarizations occurred with rather irregular periodicity. These spontaneous depolarizations were very small, with an amplitude of only about .5 mV. Because of their very small size, they have come to be called *miniature postsynaptic potentials*. It has been determined that the miniature postsynaptic potential results from the action of a relatively large and constant number of molecules (typically several thousand), since electrophoretic experiments in which only a few hundred ions are injected by a driving electrical field do not produce equivalent depolarizations.

Since the miniature potential so frequently occurs spontaneously, and since it does seem to be the result of the action of a substantial number of molecules, the notion that large numbers of molecules of the transmitter substance form constant size aggregates or packets has gained wide currency. Further direct support for this concept has come from electronmicrographs of the sort shown in Fig. 9.4. (The first observation of the vesicles now believed to be the packets of transmitter substance is usually attributed to De Robertis & Bennet, 1954.) Clusters of these structured vesicles can be seen in the presynaptic terminal in close proximity to the synaptic cleft. It is currently thought that the molecules separate either upon or shortly after their release and that they then travel independently to the postsynaptic receptor sites. Supporting the notion that the molecules of the transmitter substance diffuse passively across the synaptic cleft are correspondence between the cleft size, the calculated diffusion time, and synaptic delays observed electrophysiologically.

A wide variety of biochemicals are now believed to act as synaptic transmitter substances in the nervous systems of various animals. For example, acetylcholine has been shown to be an effective transmitter substance for several synapses. While this substance was long thought to be solely involved in neuroeffector junctions and parasympathetic action in vertebrates, "cholinergic" synapses are now thought to be more generally distributed throughout the entire phylogenetic tree. Similarly, norepinephrine was classically suggested as the transmitter substance in the sympathetic portion of the autonomic nervous system. Other possible excitatory transmitter substances in mammalian brains include 5-hydroxytryptamine, glutamic and aspartic acids, and dopamine. Glutamates also seem to be excitatory transmitters in the squid. Serotonin is now also known to be a powerful transmitter in some parts of the vertebrate brain.

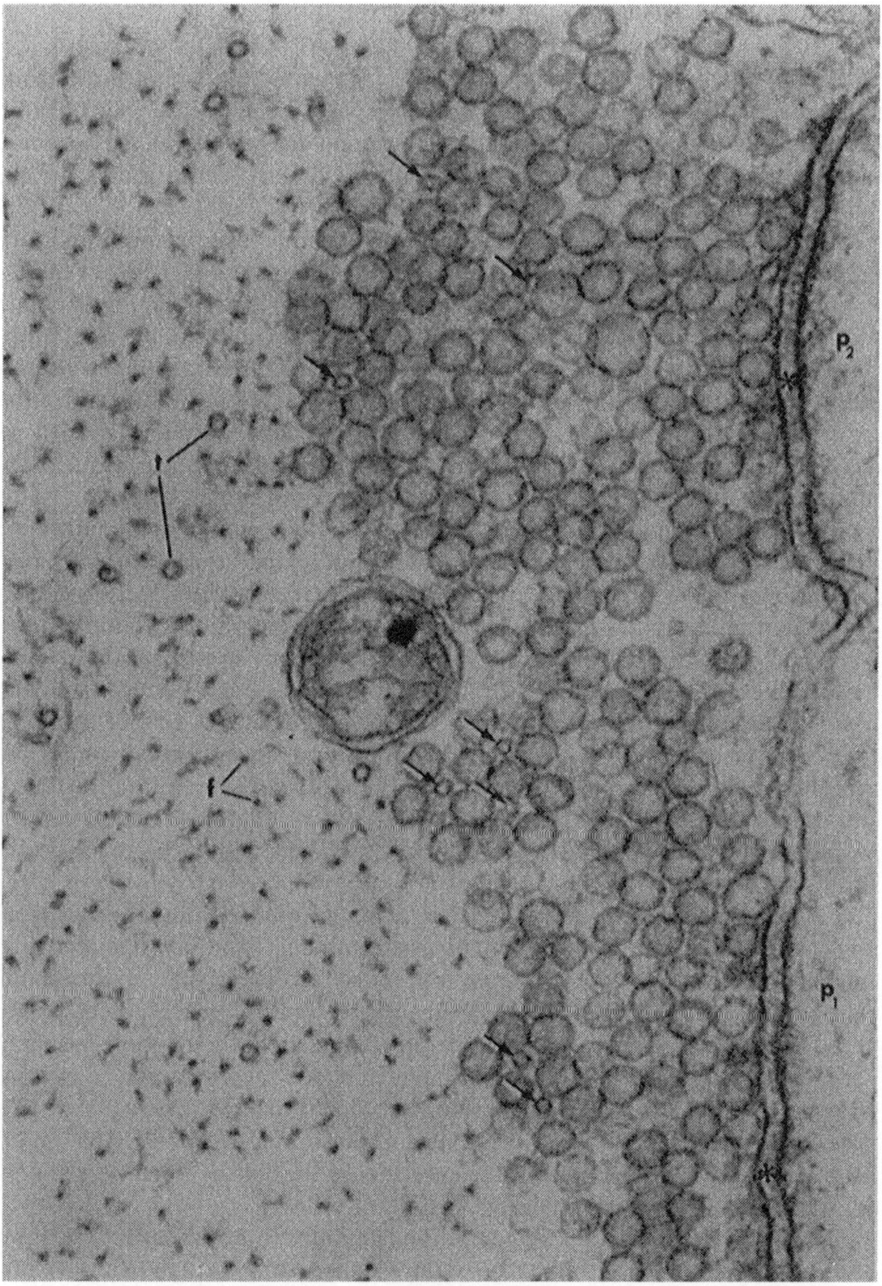

FIG. 9.4. Electronmicrograph of a synaptic region of a Muller cell from a lamprey. Note the synaptic vesicles clustered on the presynaptic side of the cell. A = presynaptic cell; P_1 and P_2 = two postsynaptic cells; t and f = microtubules and microfilaments, respectively. Arrows point to regular arrangement of vesicles around microtubules. Magnification is about 130,000×. (From Smith, Järlfors, & Beránek, 1970.)

Recent work (see Chapters 10 and 11) indicates that some of the substances that had been thought to be exclusively excitatory may in fact be either excitatory or inhibitory, depending upon the nature of the receptor site. Acetylcholine, for example, is now known to serve either role in invertebrates like *Aplysia.*

Substances that have classically been considered to be solely inhibitory in mammalian brains include the amino acids glycine and GABA (gamma-aminobutyric acid). On the other hand, a substance like strychnine, while not strictly a natural transmitter substance, is known to inhibit inhibitory action and thus can produce a pseudo-excitatory effect. Strychnine's general effect is, therefore, to elicit neural activity, and in large doses this activation may be so strong as to result in severe convulsions. Chemicals such as strychnine appear to decrease the sensitivity of postsynaptic region to the inhibitory substances that normally stabilize the neural net.

The general concept of the secondary inhibition of inhibitory effects or deactivation of excitatory effects is exceedingly important. Furthermore, some chemical means of deactivating transmitter substances must exist in the postsynaptic tissue in order to prevent a perpetuation of the inhibitory or excitatory effects of the transmitter. In cholinergic systems, postsynaptic tissues seem to be rich in AChE (acetylcholine esterase), an enzyme that quickly breaks down acetylcholine. The synaptic effects of acetylcholine therefore last for only a few milliseconds before the transmitter molecules are destroyed and thus deactivated. This is particularly important in speeding up the action of neurons and motor units, since perseveration of a response would reduce the speed at which changes could be made in the state of a neuron.

Similarly it has been suggested that some general anesthetic drugs work by interfering with chemical synaptic transmission. The postsynaptic sites, chemoreceptors that they are, are especially likely to bind a number of chemicals, but not all that are absorbed excite the neuron. Specifically, the suggestion has been made that blocking occurs when the anesthetic drug occupies many of the postsynaptic receptor sites without producing the effects that would be required for nervous transmission. Since the sites are already occupied, any of the usual transmitter substance that might arrive would be ineffective. Other anesthetics, particularly local ones, must directly act to reduce axonal activity.

Although several major stages of the synaptic transmission process have been distinguished, not all are yet fully understood. It is not known, for example, if spike action potentials produce an intermediate graded potential in presynaptic tissue, or if the potentials detected there are merely degraded spikes. A question of equal importance concerns the action of the specific release mechanism. How does a graded potential release or trigger the release of very large numbers of transmitter molecules? In other words, what ionic mechanism suddenly allows the membrane to pass large numbers of transmitter molecules (the vesicles presumably having burst on the presynaptic side of the synapse), when previously only infrequent spontaneous passages leading to miniature postsynaptic potentials occurred. These and some related issues represent the frontier of current research in the field. A full and excellent discussion of synaptic biochemistry can be found in Albers, Siegel, Katzman, and Agranoff (1972).

C. EXCITATION AND INHIBITION: DEPOLARIZATION AND HYPERPOLARIZATION

Another major issue concerning synaptic transmission is the relation between the polarity of the postsynaptic potential and the excitation or inhibition of the neuron. As indicated earlier, synaptic activity may be either excitatory or inhibitory. *Excitatory* synaptic action results in the stimulation of the postsynaptic membrane in a way that increases the response of the cell. Synaptic action may also be *inhibitory* in a way that blocks or reduces the response.

The graded postsynaptic potentials that can be measured during excitation are defined in terms of the direction of the effect and not the polarity of the membrane potential. Thus the potential measured during excitation is referred to as the *excitatory postsynaptic potential* (commonly abbreviated e.p.s.p.), while the potential recorded during synaptic inhibition is referred to as the *inhibitory postsynaptic potential* (i.p.s.p.).

It must be very clearly remembered that these definitions are based upon the actions of excitation and inhibition and not the polarity of the measured voltage. The question of the relation between the polarity and the direction of the response is another matter.

It turns out that e.p.s.p's are always depolarizations and therefore the term "e.p.s.p." has often acquired the additional meaning of a depolarizing synaptic potential. But to be very precise this is not exactly correct. This becomes clear when the situation for i.p.s.p's is considered. Here there appears to be no simple relation between hyperpolarization and inhibition.

Achieving an understanding of the relationship between inhibition and the polarization of the membrane is particularly important because there is considerable controversy among researchers concerning some critical points. Direct association between inhibition and hyperpolarization might at first also be assumed. Thus, the reasoning goes, if the postsynaptic membrane is hyperpolarized, excitation would require greater depolarization than would be required with a normal resting potential. In other words, inhibition might be thought of as having the sole effect of simply shifting the threshold for excitation and thus adding to the magnitude of the e.p.s.p.

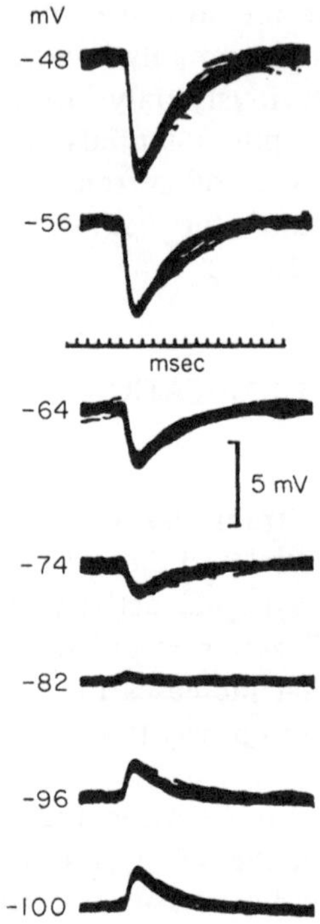

FIG. 9.5. Records indicating the effect on the polarity of an inhibitory postsynaptic potential of variations in the membrane resting potential produced by a voltage clamp. (From Coombs, Eccles, & Fatt, 1955.)

that must be generated to produce the necessary depolarization for activation. However, a number of experiments (e.g., Coombs, Eccles, & Fatt, 1955) have shown that *inhibitory action may be either hyperpolarizing or depolarizing,* depending upon the clamped resting potential of the postsynaptic tissue at the time the inhibitory transmitter substance arrives. One recording of such an experiment is shown in Fig. 9.5. An inhibitory depolarization is seen to occur when the membrane resting potential has been locked at a hyperpolarized level by means of a voltage clamping apparatus. At other clamped levels, the inhibitory potential is clearly hyperpolarizing.

The dependence of the polarity of the inhibitory postsynaptic potential on the level of original membrane potential conclusively rules out any notion of a direct and simple association between inhibition and hyperpolarization. A more accurate statement might be that the i.p.s.p. polarization shift tends toward a membrane potential that is intermediate in amplitude between the potassium concentration potential and the chloride equilibrium potential.

A new ionic mechanism has been hypothesized to explain inhibitory membrane potential shifts that are either hyper- or depolarizing. This theory has been proposed by Eccles (1964). He believes that the fundamental inhibitory action is not hyperpolarization per se but is a change in the normal membrane permeabilities to potassium and chloride ions.

According to this model, the membrane changes its permeability in a complex manner. The arrival of a packet of an inhibitory chemical transmitter seems to increase drastically the permeability of the membrane to the flow of potassium and chloride ions. This change in permeability is the fundamental reason for the polarity changes observed in inhibitory postsynaptic potentials.

A major, related question concerns whether or not one neuron can exert both excitatory and inhibitory effects. It was generally assumed until very recently that inhibitory and excitatory postsynaptic potentials were produced by separate cells manufacturing specific chemical products that compose the transmitter substance. A given cell was presumed to produce only one substance that was exclusively excitatory while another might solely produce a different substance that was exclusively inhibitory. If the first cell was connected to a neuron, the postsynaptic cell was always excited; if another that secreted an inhibitory substance was connected, the postsynaptic cell was inhibited.

However, Wachtel and Kandel (1967) found a very significant exception to this presumption. They discovered that one of the neurons in the well-mapped abdominal ganglion of the sea hare (*Aplysia*)* has both inhibitory and excitatory effects on the cells with which it makes synaptic contact. Even more surprising, it seemed that the cell could also have either an excitatory or an inhibitory effect at the same synapse! The difference in action in each case was related to the rate at which the presynaptic fiber was electrically stimulated. Wachtel and Kandel inserted a microelectrode into the presynaptic cell and, by passing a pulsed current through the microelectrode, they were able to produce a single spike action potential each time the current was pulsed. When the neuron was stimulated at low and moderate rates, another microelectrode in the postsynaptic neuron picked up an e.p.s.p. sufficiently large for it usually to produce a spike action potential each time the presynaptic fiber fired. However, when the presynaptic fiber was stimulated at a faster rate, the response of the postsynaptic fiber changed drastically. The depolarizing e.p.s.p. changed to a hyperpolarizing i.p.s.p. and the cell ceased responding.

Even though this is a most unique preparation and the effect occurs in an animal very far down on the phylogenetic tree, it raises the possibility that vertebrate synapses also may be either inhibitory or excitatory, depending on something as seemingly irrelevant as the frequency of the incoming nerve impulses. If so, the possible complexity of interactions among neurons in a simple nerve net increases drastically.

A supplemental experiment was carried out by Wachtel and Kandel to identify the transmitter substance. Curare, which is known to have a specific blocking action on acetylcholine, was infused into the solution bathing

* A complete discussion of the anatomy and physiology of this particularly useful model animal will be presented in Chapter 11.

the *Aplysia* ganglion preparation. The effect of this infusion was to inhibit *both* excitatory depolarization and inhibitory hyperpolarization. They concluded that both effects are therefore due to the action of one transmitter substance, acetylcholine. Thus, Wachtel and Kandel's experiment demolishes two previously held hypotheses. First, they establish that a single neuron can have both inhibitory and excitatory outputs. This is even more striking in light of their finding that the antagonistic effects can occur at the same synapse if not at the same receptor site. Second, they establish that the same transmitter substance can have either excitatory or inhibitory effects. This second finding is, perhaps, even more important, for it strikes at a very fundamental biochemical assumption, namely, that there is something specific about the chemistry of the substance that makes it either an excitatory or inhibitory neural transmitter.

We must now ask: How is it possible for such a two-neuron system, which remains constant in all properties except presynaptic spike frequency, to produce such drastically different postsynaptic effects? Wachtel and Kandel speculate that the presynaptic terminal sends the transmitter molecules to postsynaptic tissue that possesses two kinds of receptor sites. One type produces an excitatory depolarization when stimulated by acetylcholine, while the other produces an inhibitory hyperpolarization when stimulated with this same chemical. They further hypothesize that the time constants of the two receptor sites are quite different. The excitatory receptor site adapts very quickly and ceases operation at high stimulus rates, while the inhibitory process is very resistant to this sort of accommodation and continues to function at high frequencies.

This important study leads to a key question regarding chemical synaptic action: What ionic mechanisms account for the polarization changes on postsynaptic tissues after they receive the transmitter substance? This question unfortunately brings us to the limits of current knowledge. The exact structure of receptor sites on the postsynaptic membrane is not known, nor is their mode of action well-understood.

Upon limited evidence, however, some authorities have hypothesized differences between the ionic effects involved in excitatory and inhibitory polarization changes. It has been suggested that the depolarizing effects of an excitatory transmitter substance are specifically due to an increase in the permeability of the membrane to both sodium and potassium ions, in somewhat the same way as the Hodgkin and Huxley model of the spike action potential operates (see Chapter 7). These ions, less constrained by high membrane impermeability, begin to pass through the membrane under the influence of the usual passive diffusion and potential forces. This ionic flux thus always tends towards a depolarization of the membrane potential.

On the other hand, we have already considered Eccles' model of how either hyperpolarization or depolarization could result as the membrane permeability to potassium and chloride ions is increased. Passive forces are also all that need to be invoked once the membrane permeability is altered to produce either polarity change.

A possible mechanism explaining how a permeability change capable of producing such shifts in the postsynaptic membrane potential may occur has been described by De Robertis (1971). He reports the extraction of a substance he believes to be the postsynaptic receptor material upon which the action of the transmitter substance is exerted. Like most cell membrane components, this material is a combination of a protein and a lipid, but it is unique because it is highly sensitive to one or another of the possible transmitter substances. De Robertis suggests that these substances are organized in macromolecules that are roughly as long as the nerve membrane is thick. He further suggests that four of these macromolecules are organized in a circular arrangement to form a gatable pore through the nerve membrane. When acetylcholine (or one of the other specific transmitter substances) passes from the presynaptic tissue to the receptor molecules, the effect is to open the "gate" and allow a flow of sodium ions. This is essentially the cause of the membrane permeability and the resulting depolarization. The "gate" is closed when the acetylcholine esterase located on nearby molecules deactivates the acetylcholine, thereby returning the pore to its original state of low permeability. This hypothetical mechanism is diagrammed in Fig. 9.6.

Another view of the problem of the generation of the graded postsynaptic potentials suggests that some intermediate chemicals are formed, and they produce the changes in potentials. For example, it has been suggested by Greengard (see Marx, 1972) that a substance known as cyclic AMP (3′,5′-adenosine monophosphate), known to be involved in nerve action potential effects, may also be specifically involved in postsynaptic inhibition. Cyclic AMP is thought to be produced by the action of a transmitter substance like norepinephrine or dopamine, and it leads to the hyperpolarization of the cell and thus a reduced excitability of the neuron. Though this idea does not change any of our views about the primary processes involved in inhibition, namely the change in membrane permeability to potassium and chloride ions, it does suggest an alternative way in which the permeability changes may be effected. It also reflects a change in the direction of much neurophysiological research from the electrophysiological to the biochemical. Some of the mechanisms we had been speaking of only qualitatively have now begun to be understood in terms of the ultramicroscopic chemical reactions involved.

The final stage in the postsynaptic process is, of course, the conversion of the e.p.s.p. into a spike action potential. This appears to be performed in exactly the same manner described earlier for the conversion of the axon local potential into a propagated spike action potential. However, no direct

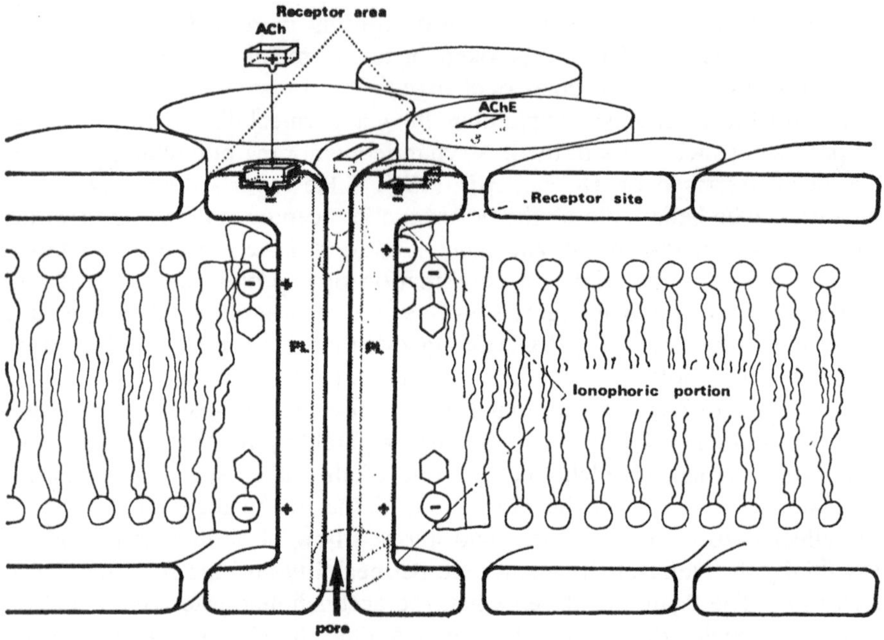

FIG. 9.6. Diagram illustrating De Robertis' hypothetical mechanism possibly accounting for the variable permeability of the postsynaptic membrane to ions when a transmitter substance impinges upon it. PL = four proteolipid molecules; ACh = acetylcholine; AChE = acetylcholine esterase. (From De Robertis, 1971.)

test of this hypothesis has yet been accomplished, and it does take some extreme values of the constants involved in the cable model to enable the relevant equations to substantiate this hypothesis.

The interaction of multiple e.p.s.p.'s and i.p.s.p.'s at the synaptic junction is one of the most important biological concepts ever postulated. Transmission in other parts of the neuron occurs pretty much on a one-for-one basis. There is little integration of information from multiple sources. It is the responses of synapses, and only of the synapses, that possess the capability to interact and to produce an output response that is representative of an integrated or cumulative effect of several inputs. Thus, it is only at the synapse that the capacity for adaptive behavior, response plasticity, creative thought, and learning can be manifest.

Synapses also perform the very important function of adding directionality to an otherwise nondirectional neural communication system. In doing so, they rectify the conduction patterns of the naturally bidirectional axons. Among other useful results of this rectification action is the stabilization of the nervous system. Without the rectifying properties of the synapse and other stabilizing forces, any single signal, once started, could result in an explosive spread of activity throughout the system. Clearly, much of what we are, at the molar level, is a function of the ultramicroscopic integrative synaptic region between neurons.

D. PRESYNAPTIC AND POSTSYNAPTIC INHIBITION

As we have seen, the interactions that occur between inhibitory and excitatory postsynaptic potentials can determine the course of the neural action produced in the postsynaptic fiber. We have discussed a form of inhibition that probably acts at some molecular level to modulate the effects of the excitatory substance, by altering the permeability of the neural membrane to the ions that produce the excitatory state. The locus of this form of inhibitory action is clearly the postsynaptic tissue, and it is thus referred to as *postsynaptic inhibition.* Figure 9.7a depicts a simple circuit arrangement that could allow this sort of postsynaptic inhibition to occur.

It has also been demonstrated, however (see, for example, the work of Frank & Fuortes, 1957), that there is another form of inhibition, mediated by a different sort of synaptic interconnection. This alternative form of interaction, which could be mediated by the circuit pictured in Fig. 9.7b, is characterized by the fact that the inhibitory signal-carrying neuron itself forms a synapse with the presynaptic portion of the exciting synapse and modulates the amount of transmitter substance it is able to release. Because of the locus of the effect in this case, this process is called *presynaptic inhibition.*

Presynaptic inhibition seems to be the direct result of a reduction in size of the presynaptic membrane potential. Apparently the amount of the transmitter substance released is nonlinearly related to the magnitude of

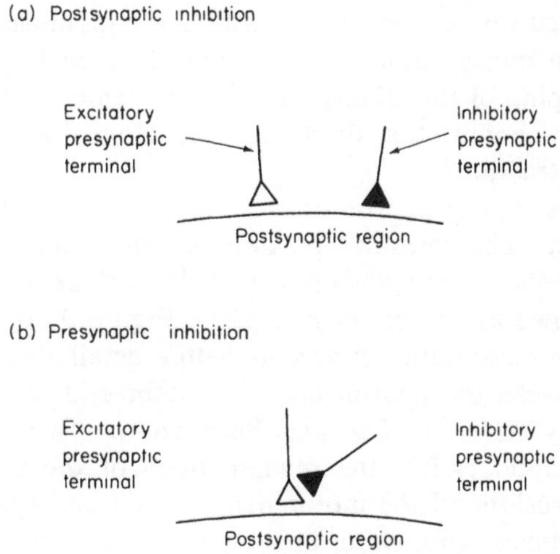

FIG. 9.7. Diagrams distinguishing between (a) postsynaptic and (b) presynaptic inhibition.

membrane depolarization, and a reduction of only a few percent in the presynaptic membrane potential can reduce the amount of transmitter substance released by more than 50%.

E. A SPECIAL FORM OF SYNAPSE: THE NEUROMUSCULAR JUNCTION

The junction between an efferent neuron and a motor unit of the musculature exhibits so many common features with the true synapse (defined as the functional junction between two neurons), that it is essential to discuss it at this point. The final act of communication along the chain of neurons from receptor to effector is the converse of transduction. In this final step, neural electrochemical energy is converted into physical energy of quite a different sort than that which coursed through the neuron. The counter-transducer, as we might consider it, is the motor unit. A full description of how muscle cells carry out this conversion of electrochemical voltages to mechanical forces is not relevant to this book, but the final stage of communication at the junction between the motor neuron and the effector, the neuromuscular or myoneural junction, is important to our discussion.

Armed with the information already presented about synaptic action, it is quite easy to understand the basic concepts of neuromuscular transmission. All the presynaptic functions are virtually identical, and the reader has only to learn a slightly different vocabulary specific to muscle cell response to understand the key features of the process.

Anatomically, the main difference between the synapse and the neuromuscular junction is that the postsynaptic element in the junction is not a neuron but one of several kinds of muscle cell. The contractile action of a muscle unit is now believed to be largely due to the variable overlapping of the sliding subcellular filaments shown in Fig. 9.8. The question pertinent to this discussion is: How does the neural action trigger this contraction?

A typical neuromuscular junction, or motor end plate, is shown in Fig. 9.8a. The terminal portions of the neuron are seen to rest in shallow trenches or troughs on a specialized thickening of the motor cell membrane known as the *motor end plate*. Figure 9.8b, a drawing made from an electronmicrograph, shows in better detail the fine structure of the junction between the neuron and the motor end plate. The reader should particularly note the fine gap between the two structures (reminiscent of the synaptic cleft), the invaginations of the muscle cell's end plate by tiny extrusions of the motor neuron, and perhaps most important, dark stained particles reminiscent of the packets of transmitter substance in the synapse.

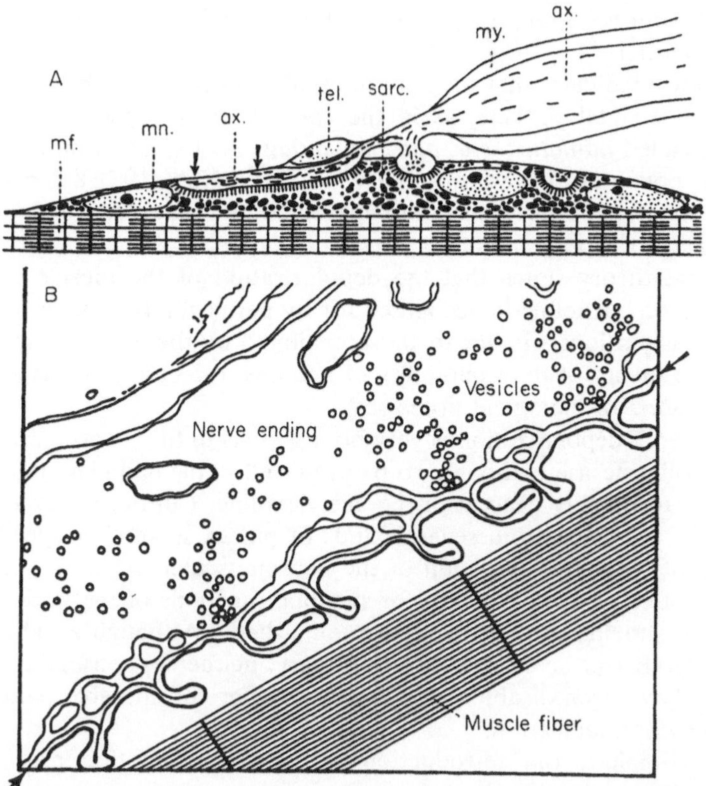

FIG. 9.8. The motor end plate shown at two levels of magnification. Ax = axon; my = myelin; sarc = sarcoplasm; tel = terminal Schwann cell; mn = muscle cell nucleus; mf = myofibrils. In the more highly magnified drawing, the end plate can be seen to invaginate the muscle cell. (From drawings prepared by Eccles, 1964; from micrographs produced by Couteaux, 1958; Birks, Huxley, & Katz, 1960.)

The details from this point on are familiar. The generally accepted notion of neuromuscular transmission is that a spike action potential conveyed along the motor neuron causes the release of substantial amounts of some chemical transmitter substance. It is well established that the transmitter substance for the neuromuscular junction is usually acetylcholine. Acetylcholine has been shown to be released at the ends of motor neurons (Dale & Feldberg, 1934; Dale, Feldberg, & Vogt, 1936), and has been demonstrated to produce a muscular response when injected into the end plate (Dale *et al.*, 1936; Nastuk, 1959).

How acetylcholine acts to cause the muscular contraction is still not completely clear, but like the postsynaptic potential, there occurs a graded potential response upon the arrival of the transmitter substance. This is the *end plate potential,* a signal exhibiting many properties of the postsynaptic potential.

Evidence indicates that the neuromuscular transmitter substance is transmitted in quantal packets, as in the case of the synapse. Tiny electrical signals, called *miniature end plate potentials,* analogous to miniature postsynaptic potentials, have been observed (Fatt & Katz, 1952) to occur more or less spontaneously. These are believed to be due to the random release of packets of acetylcholine.

Current theory states that the depolarization of the muscle cell membrane by acetylcholine is accounted for by permeability changes to sodium and potassium ions, similar to those exhibited by the neuronal membrane. The ionic theory of the excitability of neurons, therefore, also is applicable to other sorts of biological processes.

In general, depolarization is very brief and, as in the synapse, the enzyme acetylcholinesterase is thought to be produced in the end plate in quantities sufficient to quickly deactivate the acetylcholine. Curare, a strychnine-like substance extracted from several kinds of plants, is another specific agent capable of blocking the action of the neuromuscular junction without discernible effects on other neural or synaptic functions of the body. For this reason, it should never be used as an anesthetic. Although an animal may be paralyzed and lie quietly during surgery, neither its sensory nor central neurons have been disabled, and it might suffer considerable pain without any behavioral indication.

This completes our introduction to the communication processes between neurons themselves and between neurons and motor units. So far the emphasis has been on the action and interaction of individual cells. To understand fully the action of the nervous system, we must now turn to some of the properties that underlie interactions between *networks* of these cellular components.

10
NEURONAL INTEGRATION

A. INTRODUCTION

With the exception of Chapter 8, previous chapters have dealt exclusively with the physiology and anatomy of single neurons. This material is of great interest to researchers and absolutely necessary to the further unraveling of neurophysiological mysteries. But in a larger sense, the study of single-cell biology would be merely an exercise in esoterica were it not for the fact that neurons interact with each other. Make no mistake—the neurophysiology of the single cell is fascinating but, in the opinion of this author, the enormous investment in single-cell neurophysiological research should be directed at broader goals than those defined solely at the level of the single neuron. The real importance of research in the area of single-cell neurophysiology lies in the fact that these cells are the unit building blocks that make up the complex interacting network of the nervous system. It is the system and its function in relation to the behavior of the total organism that makes neurophysiological research of such great significance.

One distinguished neurophysiologist said, *sotto voce* to the author following a lecture, "although most neurophysiologists record single-cell action potentials and report the input–output relations, I am pretty well convinced that most of them think they have found the key to the explanation of the soul." Although most neuroscientists would disclaim this accusation, many would agree with Horridge's (1969) statement: "The aim of neurobiology is the explanation of behavior [p. 1]." Single cells explain virtually nothing important about the molar or total behavior of an organism. Molar behavior can be explained only by invoking processes that involve the interactions of large numbers of neurons. Any attempt to do so solely with the premises of single-cell physiology quickly runs into difficulty. Not only is there a persistent logical problem in explaining why the activity of any one cell or group of cells should be more important than any of their neighbors, but the actual time dimensions of single-cell responses are also different from the temporal properties of behavioral responses. Where percepts last for tenths of a second, neural responses last for milliseconds. Where neurons can respond to separate stimuli a millisecond apart, the human observer is unable to separate two events that occur less than ten or twenty milliseconds apart (Uttal, 1959).

Behavior, throughout the animal kingdom, is characterized by discrepancies between the input stimulus and the output response. This fact has already been alluded to as a critical, and at first glance, an incompatible, difference between the communicative functions (in which signal features must be maintained) and integrative functions (in which signal features must be modified) of the nervous system.

Adaptive behavior must be mediated by equally flexible neuronal patterns. This means that an input signal pattern must be modified to form or select some new neural pattern by interacting with other sources of neural information. However, if this modification is to serve an adaptive function, the output pattern must remain related in some way to the input. Otherwise, stimulus inputs would simply trigger random activity. To maintain this functional relationship while still maximizing opportunities for flexibility and adaptation, the input stimulus pattern must undergo a process of selection of and interaction with mixed, multiple inputs. This process of adaptation through mixing and selection has come to be known as *neural integration.*

Perhaps the best formal definition of integrative activity in nervous nets is that proposed by Bullock and Horridge (1965):

> . . . [Neural] integration is a process or set of processes that result in an output not identical to the input but still some function of the input [p. 257].

This definition emphasizes the fact that the change in input pattern must retain relevance to the environmental needs of the organism, yet still be different enough so that the response is not simply a direct and inevitable result triggered by the input. This definition of integration does not speak strongly enough to another aspect of the integrative process, however: the fact that the inputs to an integrative neural net must be multiple. The word "integration" includes a connotation of the accumulation or mixture of *multiple inputs.* It is from this mixing that behavioral flexibility must be assumed to originate through deterministic, not mystical, processes. To assume otherwise would require the invocation of processes that imply the rejection of most of the best of modern science. However complicated these integrative mechanisms are, and however poorly we understand the transformations of input patterns into useful responses, it is important to remember that the output is the functional derivative of the input. Even if part of that input comes from information stored earlier in the animal's genetic history and experience, the fact remains that any output is no more or no less than some function of the inputs.

The notion of integration is most often linked to the functions of *interneurons* (neurons that are neither receptors nor motor neurons, or as Purpura, 1969, has so succinctly defined them—"neurons that connect neurons to neurons"), but this linkage is not necessarily valid. It is quite clear that even the highly specialized receptors are capable of some integrative behavior. Figures 3.8 and 3.10 displayed evidence of both efferent and afferent synaptic connections in these most peripheral receptor neurons.

This fact is very important, for it means that even the output of a receptor is very likely to be the result of a mixture of multiple inputs, rather than the transmission of a stimulus involving only a simple energy transduction.

It would have been necessary to hypothesize integrative processes at the neuronal level on the basis of molar behavior, even if they had not been directly observed. Man, particularly, exhibits such variety and flexibility of behavior, and such plasticity in modifying responses as a result of previous experience (that ubiquitous process called "learning"), that some sort of multiple-input mixing process seems obvious.

Humans rarely display what appear to be pure "trigger responses." Even the simplest reflexes of newborn infants appear to depend upon the conditions of multiple stimulus input. Both previous experience and the direction of "attention" strongly modulate the results of even the simplest stimulus. An important related fact is that stimuli that trigger complex behavior in people can become altered in significance so that they initiate completely different responses from one moment to the next. For example, a click may initially produce some sort of a "reflexive alerting or orienting response." Subsequently it may produce no response at all; and later it may recall a complex mental image of a scene verbally described. Obviously, the relation between the inputs and the output in this case are multiple; equally obviously, the nature of the neural interconnection pattern underlying the molar behavior is *plastic, labile,* or *modifiable,* all three terms connoting the same concept. The speed with which these alterations in stimulus significance occur prohibits consideration of neural growth or "rewiring" as an explanation of adaptive behavior. Rather, the system in these cases must be thought to be functioning in some sort of a switching mode, in which multiple inputs are sorted and selected by processes that rapidly route and reroute signals along selected, preexisting pathways. On the other hand, changes that last for longer times, such as the memories of a lifetime, must certainly have some more permanent structural or biochemical changes associated with them.

Understanding how neuronal integration serves as the necessary and sufficient mechanism for mental and behavioral flexibility must be considered among the major intellectual challenges of our time. Yet our past success in acquiring an understanding of the function of individual neurons is somewhat misleading in any evaluation of just how far we have progressed towards meeting that challenge for interacting groups of neurons. For vertebrates, we know surprisingly little about the collective behavior of neurons beyond simple reflex mechanisms. However, through the judicious choice of "model" nervous systems on which to focus research energies, some progress has been made in understanding how information signals are integrated by neuronal nets.

B. MODEL BIOLOGICAL SYSTEMS

A neuronal net that can produce behavior of human complexity is most certainly not a simple one. There are probably something like 10 billion neurons in the central nervous system of a human, and the multiple interconnections among them are very complex. Figure 3.22d showed that even a single neuron can have a very large number of synaptic connections. This means that the possible number of combinations of neurons is vastly larger than even the large number of cells involved. No realistic estimate of the significant number of neuronal combinations in the vertebrate brain is possible because no one knows how many of the myriad synapses that can be observed under a microscope are simply redundant. Furthermore, in most vertebrate systems the cells are so numerous, so small, and apparently so inconsistent in location and evolved function that the recording of responses is pretty much a blind sampling. Thus, no replication of the manipulation of specific neurons or of precisely defined input-output relations is possible.

In a situation so complicated, the experimenter is constantly on the lookout for some less complex version of the problem. He would like to work with some lesser part of the overly complex total problem, one that can be isolated or abstracted to a level of workable simplicity—a neural subset that still maintains the basic properties of the more general neuronal net under investigation.

There are two ways to achieve this goal. One way is to look to a part of the vertebrate nervous system that is simple and well ordered in structure. The plexi of the receptor organs meet this qualification in part, and the research on interactive processes in the retina and the cochlea has been constructive and extremely influential. We now know much about the neurophysiology of the perceptual processes that are mediated in these simple peripheral nets. However, there is a problem in focusing on a simpler network in this manner. The retina is not the seat of the most interesting behavior exhibited by vertebrates or any other organism, and thus is not a suitable model for most kinds of neuronal integration. In "going to the periphery," we may have indeed thrown out the baby with the bathwater!

The second way of using a simple model to attack the general problem is to use another, less complicated species. Following this alternative, the experimenter attempts to find an organism that possesses a simpler nervous system, but also a full range of interestingly complex behavior. Unfortunately, nowhere in the vertebrate phyla does such an ideal creature exist; thus, in recent years investigators have increasingly used invertebrates as research models to study the action and interaction of nerve nets.

Some invertebrate nervous systems are immeasurably simpler than vertebrate central nervous systems. Whereas the number of cells presumably involved in the simplest vertebrate behavior must be in the tens of millions at the very least, the number of cells in an invertebrate ganglion controlling some useful, adaptive molar behavior may involve only a few hundred neurons. Furthermore, these invertebrate neurons are often very large and, perhaps even more important, their structure and function tend to be quite similar from animal to animal. Also the same cells are routinely found in pretty much the same place, performing pretty much the same function. Another advantage (although of somewhat less importance in these days of high-speed electronic recording equipment) is that many invertebrate preparations operate on what seems to be a slower time scale than comparable processes in the vertebrate. For example, in the crayfish the latencies of responses induced in the caudal photoreceptor by a flash of light are of the order of 5 seconds (Uttal & Kasprzak, 1962). Comparable latencies in vertebrates may be as short as a few milliseconds. Other processes in invertebrates also extend over many seconds, often presenting slow-motion opportunities to investigate the fine temporal structure of neural integration processes.

The classic example of the influence that an invertebrate model system could have on understanding more complex processes was the work carried out on the network of neural interconnections in the compound eye of the horseshoe crab by Hartline, Wagner, and Ratliff (1956). Many important concepts of inhibitory interaction between adjacent arthropod receptors, processes that are now known also to exist in vertebrate receptors, were first directly investigated in this preparation. Indeed, much of our current understanding of diverse perceptual processes, including such illusions as contour intensification, has arisen out of the neurological studies carried out on this simple animal. The orderliness of the neural structure led to the clear demonstration of certain findings that in turn led directly and compellingly to some very important concepts of great generality in perceptual phenomena. The implications have been far reaching, extending to art and even such handicrafts as textile weaving. An excellent and complete discussion of this interplay between those neurophysiological findings and various aspects of human perception can be found in Ratliff (1965).

Other invertebrate preparations have also played important roles in various types of neurophysiological investigations where vertebrate tissue would have been too complex. Wiersma (1958), Kennedy (1971), and Kuffler (1954) have worked extensively with the crayfish, mapping out many of the major neurons as well as their interconnections and defining their functions. Nicholls and Kuffler (1965) have concentrated on the leech's simple nervous system, and Maynard (1965) and his colleagues have worked extensively with the spiny lobster.

In the next sections of this chapter we consider in general terms some of the processes considered to be examples of neural interaction at both cellular and network levels. The next chapter will then be devoted to clarifying and making these concepts more concrete by presenting specific instances of nervous integration using a single marine mollusk, *Aplysia,* as our model.

C. BASIC MECHANISMS

The compound and pooled action potentials that were discussed in Chapter 8 are processes produced by a mixture of nervous signals in a way that does not, by any stretch of the definition, involve true neural integration. These types of responses are merely the physical superimpositions of individual electropotentials produced by the various parts of individual neurons. The composite signals are defined solely by the physical and geometrical relationships of the neurons and the fluids that constitute their environments.

In general, even neurons that are closely bundled together do not interact functionally except in the most extraordinary situations. Nerves passing by each other are so well insulated that the electropotentials induced in each one do not often affect their neighbors. Although there had once been some interest in *ephaptic interaction* (information transmission between adjacent axons) this phenomenon appears to remain mainly a laboratory curiosity. In the somatosensory system of man, despite the long parallel runs of many neurons, there is no opportunity for integrative interaction until the axon reaches the spinal cord, for the simple reason that no synapses exist prior to the spinal centers.

This last statement emphasizes a major point in our discussion of integration. The word *integration,* as it is used by the majority of contemporary neurophysiologists, *always involves synaptic action.* It is now generally accepted that the convergence at synapses of neural information patterns encoded by summatable, graded action potentials constitutes the best opportunity for neurons to interact. As we shall see, most of what are now called "integrative processes" are reflections of such interactions between graded potentials involving more than two neurons.

This restriction does not mean, however, that there are no intracellular processes that can modulate the fidelity of an information pattern. Such processes certainly do exist, and they do modify the signal patterns within neurons. While these mechanisms should not be considered truly "integrative," their role is important enough to be considered at this point. Trans-synaptic integrative mechanisms will be discussed in the next section.

1. Intracellular Mechanisms That Modulate the Fidelity of Information Transmission

(a) Refractory periods, excitability, and the recovery cycle. If an axon is stimulated within a few milliseconds after a previous response, it will be incapable of responding to the same extent it originally did. If the second stimulus follows the first within a millisecond or so, there will usually be no response at all, no matter how strong the stimulus. This period of time is referred to as the *absolute refractory phase* of the neuronal action potential. For a few milliseconds after the absolute refractory period the neuron requires a very much larger-than-normal stimulus to be excited; this period is called the *relative refractory phase.* During the relative refractory phase the response amplitude of the individual spike may be somewhat altered, but the main effect seems to be an elevation of the threshold for evoking a response. In some preparations a brief, superexcitable period may also follow the absolute refractory phase.

The time course of the periods of altered excitability following a stimulus may be established by use of the two-pulse experimental paradigm. Two stimulus pulses separated by a variable interval are applied to the neuron. Some measure of the responsiveness of the neuron or nerve to the second of the two stimulus pulses is then plotted against values of the interstimulus interval. A time plot of the recovery function is then produced, which describes a number of important properties of the neural membrane mechanism. One of the most important is that the refractory periods define the maximum possible information transmission rate of a neuron.

In Fig. 10.1 the recovery pattern of a compound action potential recorded from a peripheral nerve in the human arm is displayed. (See Chapter 8 for a discussion of the technique used to obtain these records.) The recovery of the second response can be clearly seen as the interval between the two stimuli increases. In this particular case the variation in amplitude primarily reflects the variation in the number of cells firing at any given

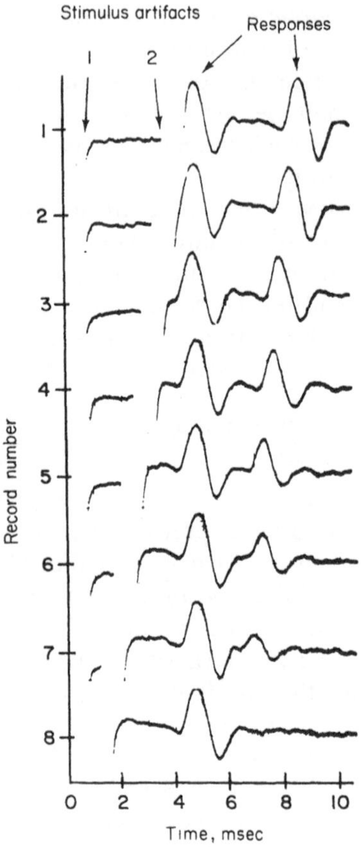

FIG. 10.1. Photograph of compound action potentials from the human ulnar nerve, produced by a two-pulse electrical stimulus applied to the nerve. The refractory period is demonstrated by the réduction in the amplitude of the second response as the interval between the two stimuli is reduced. (From Uttal, 1959.)

time, rather than changes in the amplitude of each individual axonal response. However, the time course of the refractory period is quite typical of that of individual neurons, in which case the stimulus amplitude required to produce a response would better be used as the criterion of excitability, rather than response amplitude.

Obviously the effect of the refractory periods will be to alter the pattern of the transmitted signal from the input one. The neuron will simply be unable to pass all incident information, and presynaptic trains of stimuli containing intervals that are less than the refractory periods will undoubtedly have some spike action potentials skipped. The neuron in this case would be acting as a frequency demultiplier. In general, there is a substantial reduction in the average response rate of individual neurons as one ascends the sensory pathways. Thus, frequency demultiplication appears to be an important and ubiquitous neurophysiological phenomenon.

(*b*) *Neural adaptation or accommodation.* Neural adaptation is a process that is probably related to the two-stimulus recovery cycle described above. The experimental paradigm used to measure accommodation is quite different, however, than that used to measure the recovery cycle, for the effect is an accumulated one that lasts over a prolonged period involving multiple interresponse intervals. The classic neural adaptation experiment involves the application of a continuous, long-lasting stimulus to a normal receptor channel, in contrast to the two pulses of the refractory period experiment. The results of such a stimulus on the response of a nerve fiber are diagrammatically shown in Fig. 10.2. In this case the spike action potential activity of a single cell has been recorded with a microelectrode. A gradual decrease in the frequency of responses is observed even though the stimulus remains equally intense and constant throughout. This diminution of response to a constant amplitude stimulus is behaviorally useful, since it tends to reduce incoming information from these parts of a stimulus that are not changing, and which thus contain mostly redundant information. Such a process tends especially to emphasize turn-on and turn-off times or spatial regions in which the stimulus amplitude is changing. It thus selects the most important information from more abundant but irrelevant stimuli.

(*c*) *Oscillation of the compound action potential.* A phenomenon, possibly related to neural adaptation, has been observed in a somewhat different experimental paradigm (Uttal, 1966). When a long train of pulse stimuli is presented to a compound nerve, oscillations in the amplitude of the sequential responses occur, and the amplitude of individual responses is seen to wax and wane in a very regular fashion. This phenomenon is shown in Fig. 10.3. The period of this oscillation appears to be unrelated to the frequency of stimulation.

It is not yet certain exactly what causes this oscillatory phenomenon. Some mathematical models suggest that it could represent the "hunting"

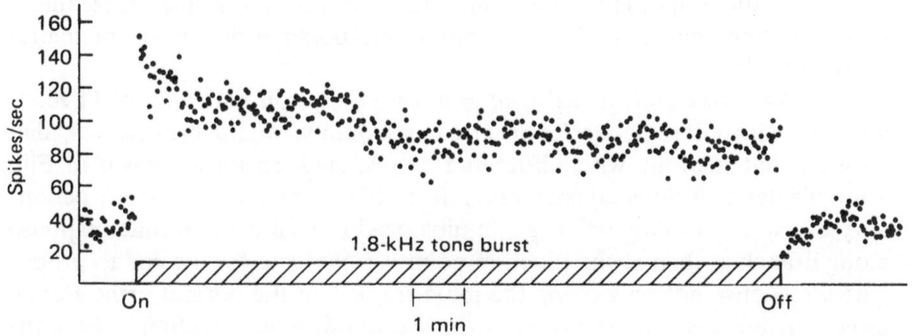

FIG. 10.2. Plot of the reduction in the rate of firing (adaptation) during a maintained acoustic stimulus in the auditory nerve fiber from the cat. (From Kiang, 1965.)

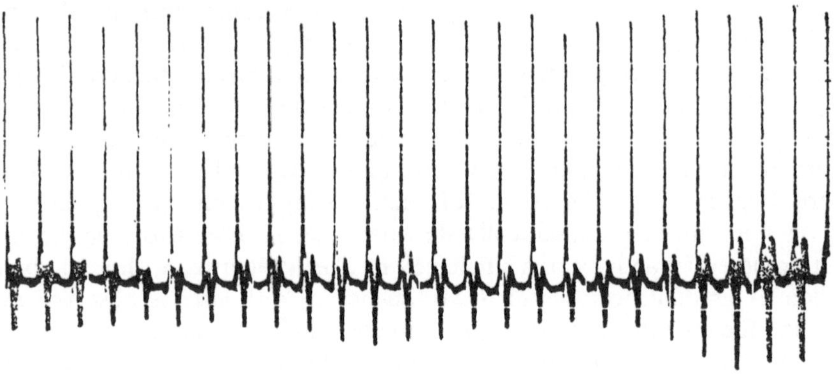

FIG. 10.3. Oscillations in the amplitude of the compound action potential evoked in human peripheral nerves by repetitive electrical stimuli. Downward-going larger spikes are stimulus artifacts. (From Uttal, 1966.)

of the membrane potential for a new stable equilibrium level when a prolonged train of stimuli changes the average membrane potential level. However, the insensitivity of the phenomenon to stimulus frequency suggests that it may be related to some constant dimension of the nervous system, such as the length of the involved axons.

This set of circumstances, along with the fact that similar oscillations have been observed by Larrabee and Bronk (1947) in a preparation involving a presynaptic stimulator and a postsynaptic recording site, suggest that the phenomenon may, in fact, be mediated by feedback from some reflexive or synaptic mechanism in the spinal cord, even though the stimulus and recording electrodes in this case were not separated by a synaptic junction. Whatever the cause of these oscillations, it is clear that the signal transmitted along the nerve is very much different than the input signal, at least in the amplitude dimension, and the result is that the initial information pattern has been distorted along one possible dimension of neural encoding.

(d) The axon and dorsal root reflexes. Fullerton and Gilliatt (1965), working with microelectrodes inserted into human peripheral nerves, discovered that neurons with bifurcated axons, such as those shown in Fig. 10.4, display a curious nonsynaptic reflex, due to the intrinsic nondirectionality of axonal conduction. If a stimulus produces an orthodromic response along branch *a*, then at the junction point it is possible for the cell to propagate a response not only along the main trunk *b* in the normal orthodromic direction, but also along branch *c* in the antidromic direction. The antidromic response propagated along the normal sensory or afferent fiber can collide with an orthodromic response moving along the same branch, extinguishing both. In this manner a considerable amount of information can

FIG. 10.4. Diagram illustrating the axon reflex.

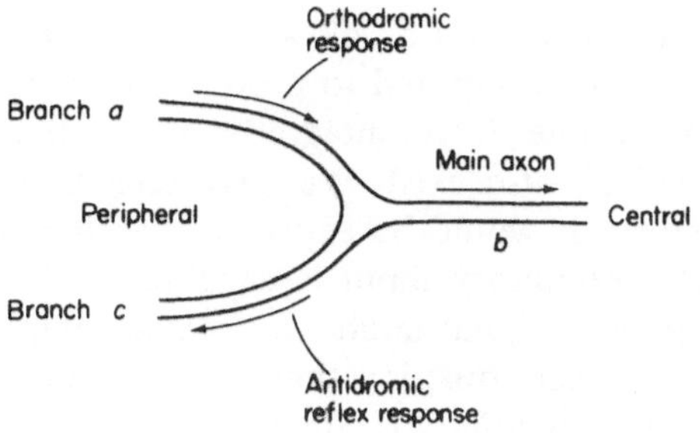

be lost as orthodromically conducted spikes along neuron twig *c* are cancelled.

A similar reflexive action was described by Toennies (1938). He called the phenomenon the *dorsal root reflex,* since it seemed that some synaptic action in the dorsal root ganglion was reflecting an antidromic spike action potential down the same sensory fiber that had just carried an orthodromic spike into the ganglion.

Both these reflexive actions must be distinguished from the better-known *sensory motor reflexes,* to be described later. In the case of sensory motor reflexes, signals are reflected in the normal orthodromic direction along motor fibers as a result of synaptic connections between sensory and motor fibers in the central nervous system. In the axon and dorsal root reflexes, information is sent back by way of the same axon along which the afferent activity was conducted. This leads directly to the possibility of a collision blockade with subsequent action potentials, a situation not usual in the normal reflex situation, where afferent and efferent signals travel along separate lines.

2. Transsynaptic Integrative Mechanisms

(*a*) *Spatial summation of graded synaptic inhibitory and excitatory inputs.* In the discussion of the action of synapses in Chapter 9, it was made clear that all synaptic action is mediated by graded potential codes or representations of the impinging information. The quantal nature of the transmitter substance has also been discussed, evidenced by the presence of miniature postsynaptic potentials and the manner in which these miniature potentials can cumulate into a summed, but still graded, postsynaptic potential. The cumulative characteristics of this coding mechanism make it a highly likely candidate as a mechanism for the integration of multiple inputs. In the simplest case of spatial summation, two separate synapses that terminate in close proximity to each other might jointly produce a postsynaptic potential sufficiently strong to exceed the threshold for the elicitation of a spike, where either synapse alone could not. This would be an example of an excitatory spatial summation that actually represented what logicians would call a logical "and" operator, in which both inputs *A* and *B* were required to produce the output signal *C*.

On the other hand, integrative interactions more complicated than simple summations also exist. We have seen that some synaptic contacts can be inhibitory. It would be expected therefore that is some situations in which a single excitatory input is capable of evoking an output, a simultaneous but opposite signal might operate to inhibit an output. This circuit implements a logical function that is called "and not."

A third possibility is somewhat more complex. Given what we now know about synaptic action, it might be expected that some sort of a "majority" logical unit could occur, in which the output was determined by the relative weighting of a large number of simultaneous inhibitory and excitatory inputs. Thus a postsynaptic neuron might fire only if the sum of the incident excitatory inputs exceeded the sum of the incident inhibitory inputs.

A fourth possibility also exists. If any one of the inputs into a postsynaptic neuron is capable of eliciting a response, then the circuit functions very much like an "or" logical unit.*

All four of these forms of spatial summation provide alternate means of processing input information in a systematic and meaningful manner. Integrative logical processes of this sort build on even more fundamental organizational properties of the neuronal net, which permit these forms of spatial summation as well as other integrative functions such as transformation from temporal to spatial dimensions, and vice versa.

Neural *divergence* as shown in Fig. 10.5b is one such fundamental organizational property, which has the effect of drastically changing the pattern of the transmitted information. In a divergent system information moving along a single axon may be replicated in such a way that it emerges along several different lines. This is essentially an amplifying process that acts to involve many cells when only a few were initially activated.

A similar organizational property is the neural *convergence* depicted in Fig. 10.5a. In a convergent system, information moving along several input links is pooled into the activity of a single neuron. Such systems allow spatial patterns of information to be re-encoded by temporal patterns and provide the basis for the spatial summation described above. Spatial codes

* It is interesting to note that in computer theory only an "and" and an "or," or either an "and" or an "or" together with an inverting circuit (capable of changing an asserted to an unasserted signal, or vice versa), is required to form the "simplest set." A simplest set refers to the smallest set of logical units capable of implementing any logical process, no matter how complicated. Though use of the simplest set is uneconomical in terms of the total number of units that must be used to produce any particular logical circuit when one is dealing with computers, in a system like the brain in which units are extremely abundant it may be that this analogy is plausible.

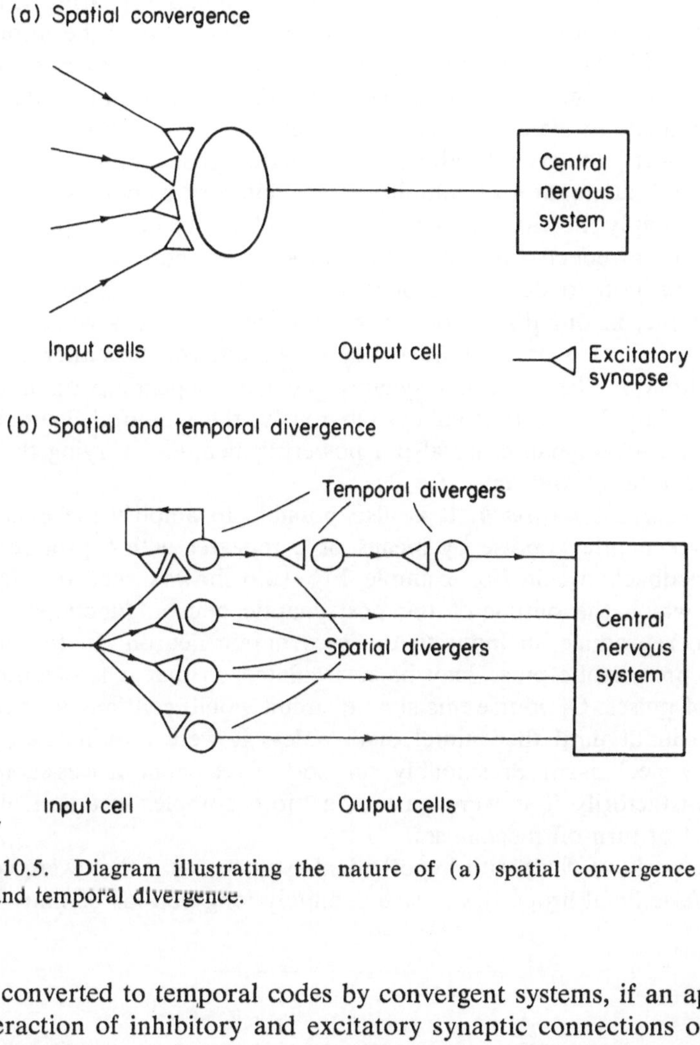

FIG. 10.5. Diagram illustrating the nature of (a) spatial convergence and (b) spatial and temporal divergence.

can be converted to temporal codes by convergent systems, if an appropriate interaction of inhibitory and excitatory synaptic connections occurs at the synapses between the many presynaptic cells and the single postsynaptic cell.

Neural divergence and convergence and the notions of logical interactions between graded inputs may ultimately provide the basis of a theory fully adequate to explain integration of neuronal signals of virtually any level of complexity.

(*b*) *Temporal summation.* The four logical circuits described earlier illustrate integrative mechanisms that operate to spatially summate disparate information. It was implicitly assumed in that discussion that all responses were functionally simultaneous. It is possible, however, that in addition to spatial integration of this sort a sequence of signals appearing at a single neuron might interact temporally with each other to alter the information pattern. In this case we would refer to the process as *temporal summation* (or *inhibition*). This term usually connotes the ability of an excitatory signal to cumulate its effects in such a way that a series of subthreshold signals can, in concert, achieve something that a single signal cannot.

Temporal summation of inhibitory or excitatory responses of this sort does not simply increase or decrease the probability of a response. It can have profound effects on what information is passed and act in a highly selective fashion to define the decoding capabilities of a given synapse. For example, in our discussion of Wachtel and Kandel's work we noted that the effect of a presynaptic volley of stimuli can be either excitatory or inhibitory in the same postsynaptic neuron, depending upon the frequency of impulses in that volley. Obviously, the temporal characteristics of an integrative synapse are also a powerful means of varying the nature of the information transmitted.

(*c*) *Recurrent feedback.* It is also possible to amplify the effects of a single presynaptic impulse by means of a repeater action produced by a simple feedback circuit. For example, Fig. 10.6 shows a recurrent feedback loop in which the output of the postsynaptic neuron functions to delay and then reproduce an input to the presynaptic neuron. In this manner, a single presynaptic pulse may be recirculated so that it is extended into a train of pulses. Of course this simple circuit would continue to recirculate a simple input until the animal died, unless it were terminated by some inhibitory mechanism. Presumably, this sort of recurrent process could only work satisfactorily if it were part of a more complex net that also had some sort of turn-off mechanism.

So far we have dealt with hypothetical synapses as if they were constant in their functional properties over indefinitely long periods of time. It must

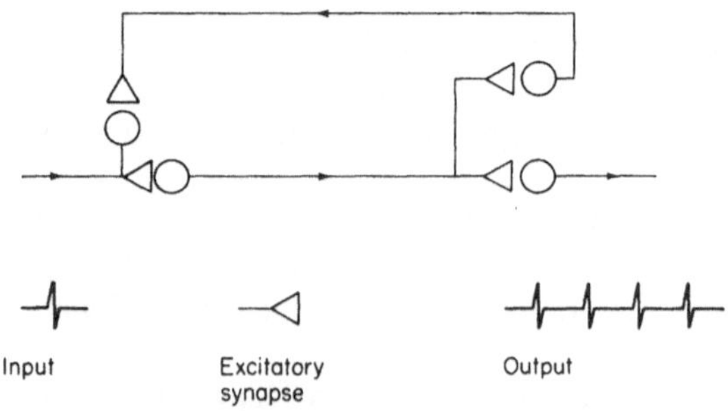

FIG. 10.6 Diagram illustrating the nature of reverberation produced by positive feedback in neural nets.

be presumed that this is an unrealistic assumption. The observed variability and plasticity of molar behavior demands that an equivalently adaptable mechanism be inherent in the neural network. Once again it is the synapse to which we look as the most likely source of neuronal plasticity.

The following discussion of transsynaptic effects includes phenomena of synaptic variability. It has often been suggested that these variations in synaptic action might underlie the learning of analogous behavioral processes. However, it should be made very clear that it is not known exactly what the relation is between learning and synaptic action, if indeed there is one. In the absence of hard data, we must keep open the possibility that learning may be mediated by other, functionally equivalent mechanisms not yet identified. The reader may wish to refer to E. Roy John's (1967) book on learning mechanisms for some of these other possibilities.

(*d*) *Post-tetanic potentiation.* Assume the following experimental preparation: A stimulating electrode is inserted into a presynaptic axon, and a recording electrode is placed in the postsynaptic cell. The postsynaptic electrode may be recording either postsynaptic potentials in the region of the synaptic interface between the two cells, or spike activity in the postsynaptic axon. In either case, its purpose is to measure the activity of the postsynaptic cell. Now let us stimulate the presynaptic cell with a train of pulses. Such a pattern of stimulation is said to be *tetanic* because of the assumption that a similar stream of nerve impulses is responsible for the tonic contraction of a muscle, a phenomenon known as a *tetanus*. The effect of a single tetanic stimulus burst on many types of synapses is to produce a prolonged hyper-responsiveness, or *potentiation* to a single test stimulus pulse. Post-tetanic (or *postactivation* as it is sometimes called) potentiation may be evidenced by a greatly enlarged postsynaptic potential or by an increase in the number of postsynaptic spike action potentials produced by the single presynaptic test stimulus. Figure 10.7 shows a typical set of data illustrating the very long duration of potentiation obtainable. Not all post-tetanic potentiations last as long or produce as large a change in relative response magnitude, but this record is probably not atypical, and it illustrates the very elongated durations that can be obtained with this procedure.

Post-tetanic potentiation is of particular interest because it is one of the few single-cell plastic changes in vertebrates with durations comparable to those of behavioral phenomena. Eccles (1956) specifically has speculated that this type of increase in long-term cellular responsiveness may be one of the underlying mechanisms of human short-term memory. However, since it is reversible it could not possibly explain long-term memorial storage.

Post-tetanic potentiation has been found frequently in vertebrate cerebral neurons, as well as in peripheral cells; thus it is apparently a ubiquitous

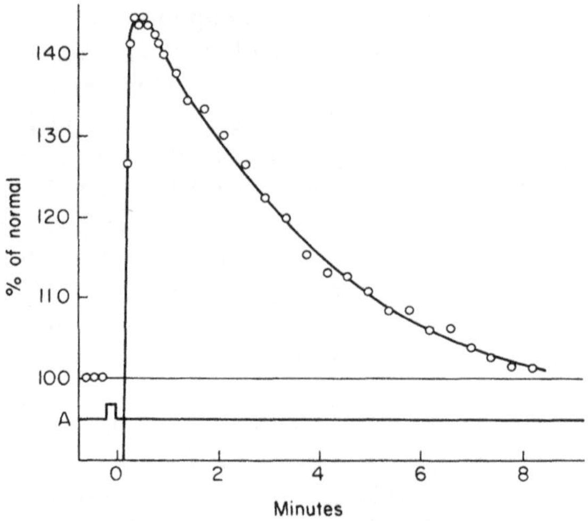

FIG. 10.7. Plot of the time course of post-tetanic potentiation as reflected in the amplitude of a postsynaptic graded potential. The pulse on line A indicates the occurrence of the tetanic burst. (From Eccles, 1956; after Liley & North.)

phenomenon that may well have an important bearing on neural theories of learning. It is also important to point out that some post-tetanic effects are depressive, rather than potentiating.

(*e*) *Long-term synaptic changes.* The potentiating effects following tetanic stimulation, while they may possibly play a role in short-term memory, cannot relate to the very prolonged phenomena observed in studies of long-term memory. A great deal of attention, therefore, has been paid to synaptic changes of a more permanent nature. A number of investigators have speculated about synaptic effects that *might* be associated with long-term memory, but it must be again emphasized that much of this is speculative and that there is yet no soundly based neurophysiological model of the learning process.

Synapses stimulate the speculations of learning theorists because they share, with learning, certain analogous organizational and plastic properties. Both molar learning and synaptic action are flexible and can generalize to similar stimuli along common dimensions. The correspondence is far from exact, however. The time scales of changes in synaptic properties, as they have been measured so far, do not correspond with the lifetime retention time scales of long-term learning; the synaptic effect is much too brief. Nevertheless, to many, the synapse still seems the most likely site of the neural processes corresponding to learning and memory.

Some investigators (Hyden & Eghazi, 1962; McConnell, 1962) have speculated that the information contained in the structural organization of large molecules is, in fact, the critical means of long-term retention. However, no plausible mechanism has been postulated for "reading out" macromolecular memories, and here, too, the time scale is incompatible with the learning process. The molecular processes simply take too long to exert their effects. The decoding of information stored in macromolecules is a process that can take generations to be fully interpreted, rather than the fractional seconds characteristic of memory and recall behavior. These writers, among others, have also suggested specifically that RNA (ribonucleic acid) and DNA (deoxyribonucleic acid) are involved in the storage process. There is somewhat better evidence for this, but the nature of the involvement may be something very much different from the molecular memories proposed by some adherents of this theory. The involvement, for example, may be in terms of synaptic growth processes controlled by RNA or DNA, rather than in terms of specific information storage in those macromolecules.

Other chemicals have also been involved in some of the biochemical changes that occur in the brain of an animal as a result of experience. Rose (1967) has studied the changes in the amount of lysine incorporated in particular fractions of the brains of rats and found that the amount of the material incorporated was increased and then decreased in sequence in the period immediately following exposure to light stimuli. Similarly, Horn, Rose, and Bateson (1973) report changes in the incorporation of this same material as a function of imprinting experience in chicks.

Whatever the details, it is clear that experience is reflected in a large number of biochemical, anatomical, and electrophysiological ways in the brain. Because of the observed changes occurring in synapses and their constituent biochemicals, it is to the synapse that most workers turn for plausible memory models at the neuronal level. The following list (summarized from Eccles, 1956, and John, 1967) describes some of the mechanisms that have been postulated as possible mechanisms for long-term alterations of synaptic transmission characteristics.

1. The terminal boutons or knobs of the presynaptic dendrite have been observed to swell (Eccles, 1956) when the neuron is repeatedly stimulated. This is thought to be due mainly to the infusion of some water, but is considered by Eccles to be semi-permanent. Eccles has speculated that this swelling leads to higher synaptic conductance, in some way. This increased conductance might be caused by a reduction in the size of the synaptic cleft, thus shortening the distance the chemical transmitter substance must travel.

2. It has also been observed that frequently used dendrites in some cases seem to grow completely new synaptic boutons, in addition to altering the size of the old ones. This sort of growth might be the way in which

the macromolecules RNA and/or DNA are involved in learning. These molecules are known to be involved in growth processes, and they may mediate their neural effects by stimulating synaptic growth, rather than serving as storage mechanisms themselves.

3. Decreases in synaptic resistance and increased effectiveness of synaptic conduction might also occur if synaptic exercise gave rise to an increased supply of available transmitter substance, resulting in a larger e.p.s.p. for a given presynaptic stimulus strength. Similarly, growth or changes in the quality of receptor sites might also account for this increased synaptic effectiveness.

4. Alterations in synaptic conductivity as a result of changes in the electrical properties of material surrounding the synapse have also been suggested as a possible explanation for increased synaptic conductivity. However, current theory in which chemical transmission is emphasized has made such ideas less influential.

5. Another possibility that is receiving increased attention is that only a few of a very large number of pre-existing synapses may be selected as a result of the deactivation of the others. This process suggests that learning consists of a selection of neural pathways from a system that initially, or after some period of early development, consisted of an almost totally interconnected plexus. This sort of plastic change by elimination would be comparable to the punching of codes on materials that originally were fully intact—for example an unpunched paper card.

But, to restate what hopefully is obvious by now, possible mechanisms that might account for changes in synaptic conductivity and, thus, for plastic behavior in neuronal nets, are almost entirely speculative. We simply do not have any direct evidence that these changes, even if they actually occur, are truly related to learning. They are the sorts of changes that might occur and that seem plausible explanations of such phenomena as post-tetanic potentiation, which, itself might or might not be related to learning changes in molar behavior. The chain is certainly not yet well established.

D. NETWORK PROCESSES

Now that we have considered some of the cellular mechanisms that might underlie integrative behavior at the level of the neuron, it is appropriate to discuss relevant network processes that might contribute to higher level integration. The phrase "network processes" refers to the functions of groups of neurons that are the result of the cellular mechanisms just described. These network processes are also not equatable with molar responses, but they do represent functions of a level of complexity that begin to exhibit comparable behavior.

1. Inhibition

Although the synaptic and cellular aspects of inhibition have already been discussed, the prime importance of this process must be emphasized once again at the network level. The overriding importance of a capability for inhibition can best be emphasized by reconsidering the fact that if a network did not possess some intrinsic inhibitory mechanism, it would continue to respond to even the simplest input until physiologically exhausted. The behavioral ramifications of an inhibition-free network would be disastrous. Suppose that some organism mutated suddenly so that inhibitory processes were no longer part of its network properties. The organism would then exist in a constant state of uncontrolled activity; no coordinated or integrative activity of any kind would be possible. Obviously such a situation is absurd; no creature could live long with such a handicap, and certainly no species could evolve under the enormous burden of the absence of inhibition. Thus, all neural networks must possess some sort of general inhibitory capability.

Specific inhibitory mechanisms have been observed in different parts of many nervous systems. Lateral inhibitory interaction among the receptor units of vertebrate and invertebrate receptors and other sensory neurons is now a widely accepted concept. Hartline and his co-workers were the first to establish the general notion of reciprocal interactions between neighboring receptors of the compound eye of the horseshoe crab. In recent years the concept of antagonistic inhibition and excitation has been extended to the vertebrate eye by workers such as Kuffler (1953) and Barlow and Levick (1965). It is now widely accepted that lateral interactions in the periphery are fundamental to such important perceptual processes as the intensification of contours.

In the horseshoe crab eye, Hartline and his collaborators (e.g., Hartline, Wagner, & Ratliff, 1956) have demonstrated that the receptor cells are interconnected in a purely inhibitory fashion. Activity in one cell exerts an inhibitory influence on activity in a neighboring cell as a function of the distance between the two cells and the amount of activity in the inhibitor. The action is reciprocal in the sense that each usually inhibits the other, and thus simultaneous stimulation of both leads to a much smaller amount of activity in each than when they are separately stimulated. Figure 10.8 shows the results of one of these experiments.

A somewhat different interactive organization appears to characterize the vertebrate eye. Concentric and antagonistic arrangements of receptive fields that may interact in excitatory as well as inhibitory modes seem to be fairly common, not only in the periphery, but at higher levels of the nervous systems of many animals.

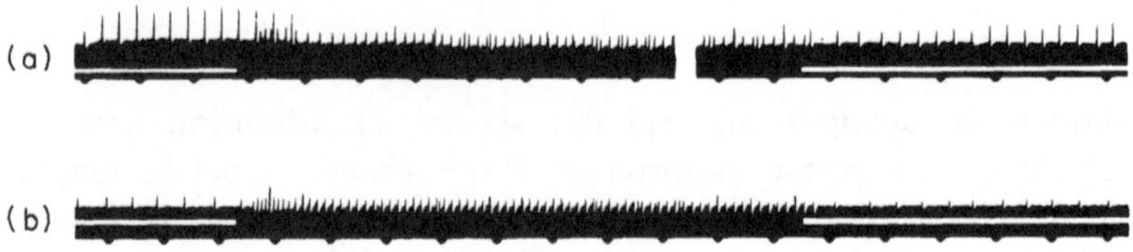

FIG. 10.8. Sample records showing the reduction in activity in one neuron (a) as a result of lateral inhibitory interaction from a second one (b). (From Hartline, Wagner, & Ratliff, 1956.)

Thus, within one of these concentric receptive fields, some receptor cells appear to excite each other mutually while others inhibit each other in a manner very much like that observed in the horseshoe crab eye. Receptive fields of this sort, with antagonistic inhibitory and excitatory regions, can give rise to rather specific spatiotemporal sensitivities on the part of the various neurons. For example, a light moving at a particular speed in a particular direction over a restricted range may be the unique stimulus for a given central neuron.

Inhibition between receptors is not the only kind of inhibition to be found within complex organisms. The coordinated action of muscles is critically dependent upon an antagonistic inhibitory interaction between opposing muscles. To move, an organism must not only be able to contract a muscle but simultaneously to inhibit or relax the action of an antagonistic muscle.

An important attribute of inhibition is that it is very easy to imitate the action of an excitatory unit, by a process that has become known as *disinhibition.* Consider Fig. 10.9. If unit *A* is inhibited by unit *B* and unit *B* by unit *C,* then activation of unit *C* will have the effect of decreasing the activity of *B,* thus reducing its inhibitory activity on *A*. The net result of activating *C,* therefore, will be to increase the activity of *A*, suggesting an excitatory interaction between *A* and *C,* but in fact illustrating a purely inhibitory relationship between all three parts. Such a disinhibitory process is of theoretical importance because it mimics an excitatory process with one that is, in fact, based purely on inhibitory relationships among the constitutent neurons. It is sometimes difficult, therefore, to determine in a neural net whether the observed effects of some stimulus are due to excitatory or inhibitory interconnections.

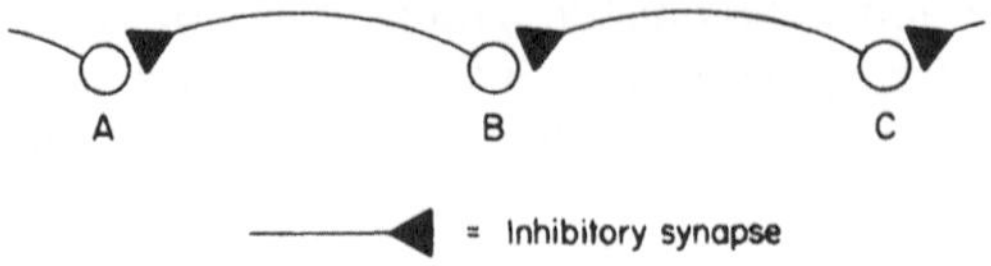

FIG. 10.9. Diagram illustrating the nature of disinhibition—the increase in activity in neuron A because of the inhibition by C of the inhibition exerted by B on A.

2. The Reflex

Inhibition, it should be appreciated by now, is a necessity for the successful operation of a neuronal net at any level of integration, but nowhere is this need as clear-cut as in reflexive actions. Reflexes are defined as the more or less automatic, rapid motor responses of an effector to some stimulus. In vertebrates the locus of reflexive action has classically been considered to be the spinal cord. In invertebrates a reflexive analysis of behavior may be equivalent to a full statement of the animals' response repertoire, and it may indeed be that the dichotomy of reflex and central integration perishes at that level of the animal kingdom.

3. Rhythmic Activity

Inhibitory and reflexive mechanisms give rise to a number of other, more complex response processes as they occur in concert. One of the most interesting derived processes is the elicitation of rhythmic or cyclic activity, a network process frequently observed at many different levels of behavior.

Examples of rhythmic behavior are common throughout the animal kingdom. Some single-cell animals pulsate in their entirety or rhythmically wave flagella in their environments. The rhythmic action of human locomotion or the locust wing (Wilson, 1967) also can be seen to depend on a combination of reflexive and inhibitory interactions, even though the basic rhythm is controlled by efferent signals from the central nervous system.

4. Cellular Learning and Neuronal Plasticity

The literature concerned with human learning at the molar level is so abundant that it is often taken for granted that the ability to learn is a characteristic common to all living organisms. Over the last few decades researchers have studied learning in many animals far down the phylogenetic tree. *Paramecia, Planaria, Aplysia,* cockroaches with and without heads, and many other invertebrates, as well as goldfish, rats, and a host of other mammals have been the subjects of psychological research. In recent years, however, some investigators have raised the question of whether or not the concept of learning, seemingly so highly developed in vertebrates, is actually so common in invertebrates. Many studies of learning in lower animals have been highly equivocal, and there is considerable controversy over whether or not "learning" in lower animals is the same type phenomena that occurs in higher animals. Some have gone so far as to suggest that learning is almost uniquely a human trait, that it has been overemphasized, and that it should be used only with the most careful consideration in describing the behavior of infrahuman species. At least one investigator has seriously suggested that animal learning is an artifact of the laboratory, and that it is not a process commonly utilized by most animals in their natural environments.

This viewpoint may be too radical, but its expression does raise a number of important issues. The most significant one in the context of this discussion is whether the variations in performance observed at the cellular, network, and molar levels are identifiable with each other in such a way that one level is "explained" by the one below it. In our discussion of the *Aplysia* experiments in the chapter to follow, certain behavioral changes we describe seem to possess all of the qualifications for true learning. However, the analogies that have been drawn between learning in humans and *Aplysia* have been severely criticized. It has been suggested that the "learning" demonstrated in these primitive animals may more correctly be characterized as pseudoconditioning—a change in performance due to stimulus repetition rather than a true adaptive response change. Whatever the outcome of this controversy, it is clear that some changes do occur as a result of prior experience in the performance of synapses, cells, networks, and of both simple and complex whole organisms. Whether or not these processes are identifiable with each other will have to be decided by future research.

5. Feature Filtering

Another general aspect of neural net function is the process that has come to be called *feature filtering.* Feature filtering refers to the general process by which only selected attributes of a complex stimulus pattern are passed through the neuronal transmission system. A classic example is the accentuation of contours produced by lateral inhibitory mechanisms, already mentioned.

In the early 1960's a major revolution occurred in neurophysiological research. Until then almost all attention had been directed to neural responses produced by impulsive electrical stimuli, flashes of light, clicks of sound, or taps to the integument. Some investigators (Hubel & Wiesel, 1959; Lettvin *et al.,* 1959) were concerned by the fact that these did not represent stimuli to which most animals are evolved to respond. Quite to the contrary, most animals deal with stimuli that are subtle variations in the arrangement of their natural environment. The essence of their innovation was the realization that, for example, a moving spot that might indicate a fly to a frog, would be of greater biological significance than the artificial impulsive stimuli that had been popular in the laboratory. Therefore, both groups of investigators began experimenting with spatially and temporally patterned stimuli, and the results were truly astounding. Whereas unpatterned visual stimuli, even very bright ones, had evoked only the most feeble neurophysiological responses, a much less energetic stimulus with some particular movement or shape could evoke a very strong response.

A particularly strong example of such neural feature specificity is the orientation and moving line-detection sensitivities of central visual neurons of the cat and the monkey, as described by Hubel and Wiesel (1961, 1968); but Table 10.1 lists a wide variety of other specific spatiotemporal features that can selectively trigger feature-filtering neurons in various species.

The notion that these feature-filtering single neurons are correlates or codes for comparable sensitivities at the behavioral level has wide currency in today's psychobiological literature. The general idea is that if certain brain cells are more or less uniquely tuned to particular features, then activation of these neurons by the appropriate stimuli constitutes all that is necessary for the correlated perception to occur.

A relation between specific sensitivities of neurons to straight lines oriented in a particular direction and the adaptive effects observed in humans as a result of viewing grid patterns has been suggested by a number of authors. In psychophysical experiments that test for grid adaptation, a subject looks at a grid composed of lines aligned in a particular direction, as a preadaptation stimulus. An example of this type of pattern is shown in Fig. 10.10a. Following the adaptation period, the detectability of another grid (shown in Fig. 10.10b) is measured as a function of its orientation. The general effect is that test grids aligned along axes close to that of the preadaptation grid lines are seen less well than those that are oriented differently.

One problem with this hypothesis had been that few experiments that were entirely comparable had been carried out in the neurophysiological realm. Those that were most often cited as examples of possible neurophysiological correlates usually had very different time constants. Where the psychological phenomena occurred over periods of several seconds, the neurophysiological adaptive processes occurred much too briefly. However, a new report (Maffei, Fiorentini, & Bisti, 1973) may possibly fill in the missing link in this case. These workers have found in the visual cortex of the cat neurons that respond to exposure to gratings with a prolonged inhibition of response comparable to that of the psychophysical experiment. In their experiment the test of the adaptive effect is performed by using a test grating that is of a lower contrast than the preadaptation grating. Both the human psychophysical phenomenon and the neurophysiological findings in the cat show a reduced response following preadaptation. Figure 10.10c displays the results of the response of a neuron to a test stimulus prior to and following adaptation and after recovery from the 30-second-long inhibited period.

This kind of psychophysical phenomenon illustrates situations that are presumed by many investigators to be direct molar concomitants of the selective filtering characteristics of single neurons in the nervous system.

TABLE 10.1
Trigger Features of Single Neurons at the Indicated Level of the Visual System of Various Animals[a]

Anatomical location	Trigger feature	Anatomical location	Trigger feature
	Goldfish		*Cat*
Retina	Local redness or greenness	Retina	Local brightening and dimming
	Directed movement	Lateral geniculate	Local brightening and dimming
	Frog		
Retina	Convex edge	Visual cortex area 17	Simple cells: Moving, slits, bars, edges with specific orientation
	Sustained edge		Complex cells: Combinations of simple cell outputs of same orientation.
	Changing contrast		
	Dimming		
	Dark		
Optic tectum	Newness		
	Sameness	Visual cortex area 18	Hypercomplex I cells: Ends of lines
	Binocularity		Hypercomplex II cells: Line segments and corners.
	Pigeon		
Retina	Directed movement		
	Oriented edges		

Ground squirrel	*Ground squirrel*
Retina	Local brightening or dimming
	Local blueness or greenness
	Directed movement
Lateral geniculate body	Color coded units
Optic tectum	Directional units
	Oriented slits or bars
	Complex units
	Rabbit
Retina	Local brightening or dimming
	Directed movement
	Fast or slow movement
	Edge detectors
	Oriented slits or bars
	Uniformity detectors
Lateral geniculate	Greater directional selectivity
Tectum	Habituating units

	Infrequent types in cat
Retina	Directed movement
	Uniformity detectors
Lateral geniculate	Local blueness or greenness
	Binocular, directional, and orientational units
Optic tectum	Directed movement
	Complex units
	Monkey
Retina	Local brightening or dimming
	Local redness, greenness, or blueness
Lateral geniculate	Various forms of color coding
Cortex	Similar to cat; some color coded
Inferotemporal cortex	Very complex; possible hand detector

[a] Adapted from Barlow, Narasimhan, and Rosenfeld (1972).

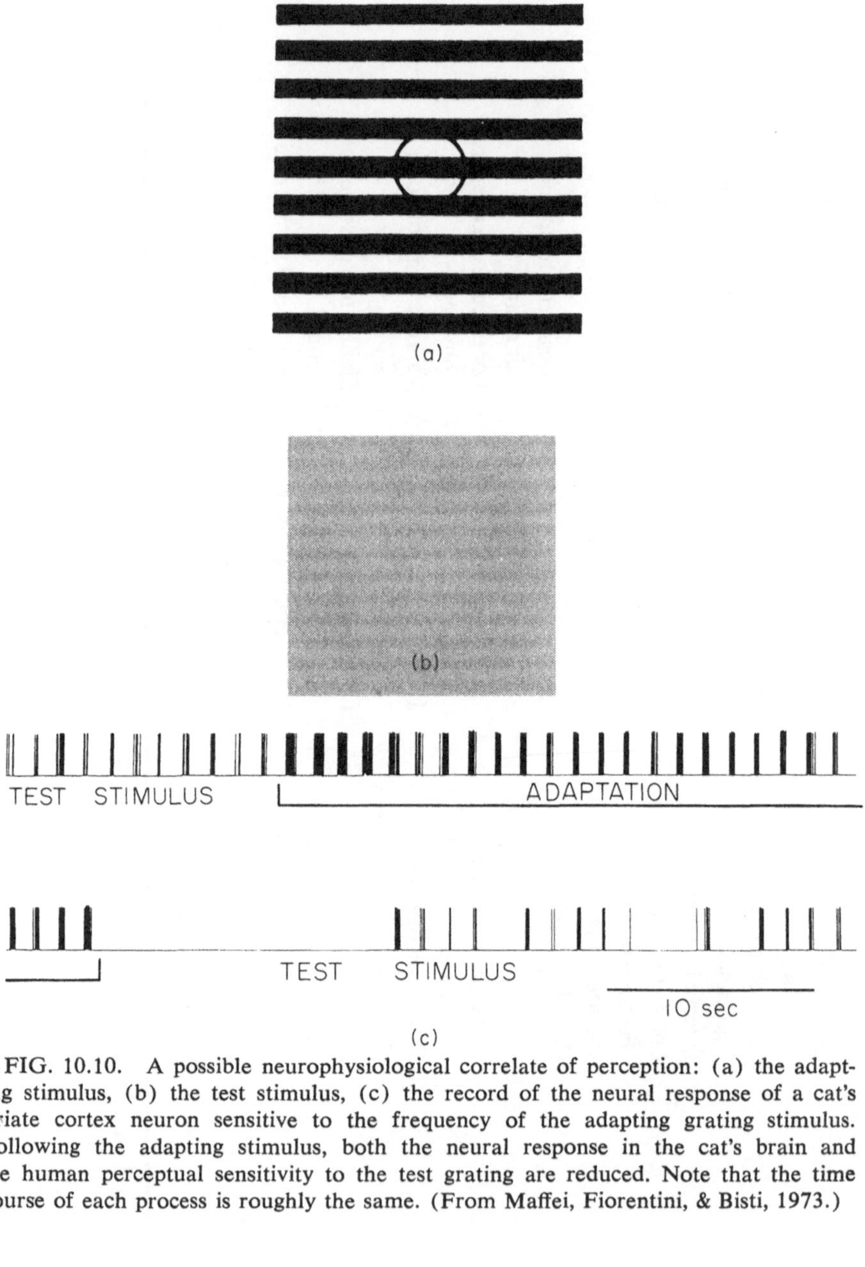

FIG. 10.10. A possible neurophysiological correlate of perception: (a) the adapting stimulus, (b) the test stimulus, (c) the record of the neural response of a cat's striate cortex neuron sensitive to the frequency of the adapting grating stimulus. Following the adapting stimulus, both the neural response in the cat's brain and the human perceptual sensitivity to the test grating are reduced. Note that the time course of each process is roughly the same. (From Maffei, Fiorentini, & Bisti, 1973.)

However, some opposition to this general notion has appeared and a number of authors have raised philosophical and technical arguments against direct relations between the actions of single cells and the psychophysical response (Weisstein, 1969; Uttal, 1971). Needless to say, the question is of fundamental importance in understanding the relation between brain and behavior.

6. Centrifugal Processes and Central Feedback

The next set of network processes that should be mentioned as sources of variation in afferent information control are classified under the general term of *centrifugal signals*. Centrifugal signals are those that are generated by some action of the central nervous system and then, like effector signals, are transmitted outwards to the periphery. The definition of centrifugal signals is specifically constrained to include only those efferent signals that affect sensory inputs. Some, as we saw in Chapter 3, may actually modulate the sensitivity of receptors, while others may interfere in one way or another with signals generated by receptors at some other point along the afferent pathway. The interplay between afferent sensory and centrifugal signals represents one of the more complex forms of interaction of the sensory processes with the rest of the nervous system.

Centrifugal feedback processes appear to exert their influence in a number of ways. They may directly influence the activity of the receptors, or they may act on second-order cells. They may also work by interacting with incoming signals at the level of the brain stem. In some cases, it is even reasonable to include interaction among different regions of the brain in the category of centrifugal interaction. In fact, any effect on a lower level of the nervous system that results from stimulation at a higher level is generally considered to be a centrifugal feedback process.

Most of the experiments concerning centrifugal feedback were conducted with the use of an experimental paradigm in which the higher levels of the nervous system were stimulated artificially and the differential effects on a response produced by a second, often natural stimulus, were observed in a more peripheral portion of the system. As Livingston (1959) so eloquently points out, several other kinds of experimental paradigms exhibit the same sort of centrifugal effect on peripheral structures. The distinguishing characteristics of these other experiments is that the more central nervous action is triggered not by an artificial stimulus, but by some natural means.

Recent experiments in peripheral habituation, which have attracted a great deal of attention (see Thompson, 1967, for a review of this material), are closely related to centrifugal effects. In habituation experiments, the ability of the sensory pathways to transmit information changes as a direct result of increasing experience with the stimulus pattern. These experiments must necessarily include adequate controls to ensure that the effect is due not to fatigue in the peripheral mechanisms, but to some almost perceptual recognition by a central mechanism of the decreasing biological importance of the signal. Presumably, when this recognition occurs the central mechanism transmits some sort of inhibitory signal back down to the peripheral units, reducing their sensitivity. This is, of course, an incredibly complex process that can hardly be described, much less explained, at the present time. But the enormous adaptive utility of such a mechanism is obvious even at this stage of our ignorance.

Even some of the most "obvious" experimental observations regarding vertebrate habituation seem to be quite ambiguous. Worden (1966), using the cochlear nucleus of the auditory system as a model system, has written a most provocative review of the matter of habituative effects on the sensory pathway. He considers some of the methodological issues involved in the modulation of sensory inputs. Worden was particularly concerned about some reports by Hernández-Peón (e.g., 1960) that ascribed a reduction in the evoked potential from the cochlear nucleus to inhibitory feedback signals from the central nervous system. These signals were supposedly associated with a reduction in attentiveness to stimuli on the part of the experimental animal—a cat. Worden demonstrated that much of what was thought by Hernández-Peón to be habituation and centrifugal effects was actually accounted for by head-turning behavior on the part of the cat, when it was not attending to the stimulus. Thus the stimulus intensity was actually reduced.

There are other ways in which feedback effects may be produced indirectly—that is, without a complete neural circuit. For example, the well-known gamma motor neuron system shown in Fig. 10.11 is a feedback circuit in which muscle spindle sensory fibers vary their activity on the basis of feedback signals, which affect the muscle fibers served by the sensory fibers. Surprisingly, the most effective feedback motor signals are not the direct ones activated by the muscle spindle afferents, but the output of cutaneous receptors that are indirectly activated by the muscular contractions.

7. Coding and Decoding

Finally, in this catalog of network processes, we come to a very general topic that actually transcends the specific functions of a network, yet originates from and utilizes the properties of the network for its operation. We, thus, introduce the topic of *coding*—the means by which messages communicated about the nervous system are represented at any particular point. A code may be defined as a set of dimensions and the rules for using these dimensions in the representation of a message. The topic is extraordinarily broad and is the subject of an extensive discussion by the

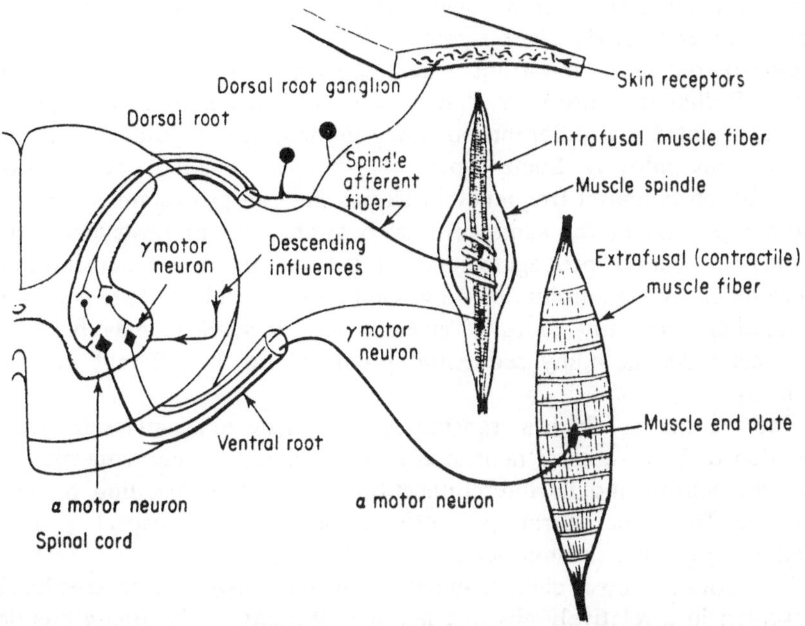

FIG. 10.11. The gamma motor neuron system, illustrating a feedback loop that involves direct neural links as well as sensory-motor mediated nonneural links. (From Thompson, 1967.)

author elsewhere (Uttal, 1973). Here we shall deal only with some of the major issues concerning neural coding.

The representation of information patterns or messages by patterns of neural signals in the nervous system is a multifaceted process. Conventional neurophysiological texts usually describe coding only in terms of the relationship between the intensity of a stimulus and the frequency of the induced nerve action potential patterns. However, each dimension of every message in the nervous system must be represented by some equivalent kind of activity in the nervous system. Thus, for example, visual stimulus intensity, ultimately interpreted as brightness, might be represented by spike action potential frequency at some point in the nervous system. But color (the stimulus quality) and the spatial and temporal aspects of the stimulus must also be represented. Not only are there many dimensions to a message, but there are also numerous neural dimensions or codes that may be used for the representation process. These possible neural dimensions or candidate codes, as they are known, include not only the frequency of the spike action potential, but also the dimensions of the graded potential, such as ampliitude and temporal pattern. Such derived statistical measures as the standard deviation of the interspike interval may also be a separate code, distinct from even such a closely related dimension as the mean frequency of the same signal.

One of the most important general concepts of modern neural coding theory is that it is probable that a single dimension of a message might be represented by different encoding dimensions at different levels of the nervous pathway. Some aspect of a message might be represented by spike action potential frequency in axons that can propagate spikes; but in another portion of the same cell (such as the pre- or postsynaptic membrane) that cannot propagate spikes, this same message dimension may be represented by the amplitude of a graded potential. Thus, there is no unique code; there are only codes at given levels. A message may be encoded and reencoded at many sequential points along the afferent or efferent pathways.

The manner in which a transmitted code is detected and selectively responded to by a receiver neuron is closely related to the properties of the receiver neuron and its interconnections. Thus, the decoding of nervous messages from one neural code into another is another aspect of the integrative action of neuronal nets.

This brief survey of cellular mechanisms and network processes has been presented in a relatively abstract manner. We have only briefly mentioned some of the experiments that make these notions concrete and have passed over some very basic concepts in a way that must certainly seem cryptic to the reader. There are two reasons for this rapid scan. One is that the complexity of the problem is awesome. The function of nerve nets in their role as integrative mechanisms is so involved that it is obviously impossible to describe them adequately, or to expect to understand their functions completely if all of the details were thrust upon the reader at once. Second, the enormous abundance of published material dealing with those aspects of integrative function that are understood have been obtained with a wide variety of biological preparations. Not only are the results extremely variable, but the techniques and preparations that have been used would have to be explained over and over again for each new experiment to make this material fully intelligible.

One method of making such complex material intelligible is to pick one model preparation, in this case the mollusk *Aplysia;* define its relatively simple neuroanatomy and the appropriate experimental techniques precisely; and then present examples from the research carried out only on this animal to illustrate more concretely the general network processes we have described. In the following chapter we proceed to do just that.

11

THE *APLYSIA:* ANATOMY AND PHYSIOLOGY OF A MODEL PREPARATION

A. TAXONOMY AND ANATOMY

The various species of *Aplysia* (*californica, depilans, fasciata, juliana, limacina, rosea, etc.*) are gastropod mollusks of the subclass Opisthobranchia. The full taxonomic classification of this genus is shown in Table 11.1. This animal is fairly common in tropic and subtropic waters and may grow to 8 inches or more in length. Figure 11.1a is a drawing of an intact specimen displaying the obvious reason for its common name, the sea hare: Two antennae-like structures, very much resembling a pair of rabbit ears, are common to all species of the class. Figure 11.1b shows the animal in a restraining device typical of one that might be used in an experimental study.

Aplysia have shells like all other mollusks but, like many members of this class, its shell is internal and vestigal, and the animal's body is enclosed only by a soft integument. In motion the animal is especially attractive, for it swims by a graceful undulation of the mantle with rhythmic, almost sinusoidal, waves of motion moving down its "wings." Some *Aplysia* secrete a purplish dye, a feature shared in common with many other gastropods such as squid, octopus, and the Tyrian snail, source of the true royal purple.

Figure 11.2 is a diagrammatic sketch of the *Aplysia*'s nervous system, which can be seen to be roughly bilaterally symmetrical, with the exception of distinct asymmetries at the anterior and posterior ends of the animal. Although the animal has no true "brain," there are three pairs of ganglia

TABLE 11.1
Taxonomic Classification of Aplysia

Phylum:	Mollusca
Class:	Gastropoda
Subclass:	Opisthobranchia
Order:	Anaspidea or Aplysiacea
Genus:	*Aplysiidea*

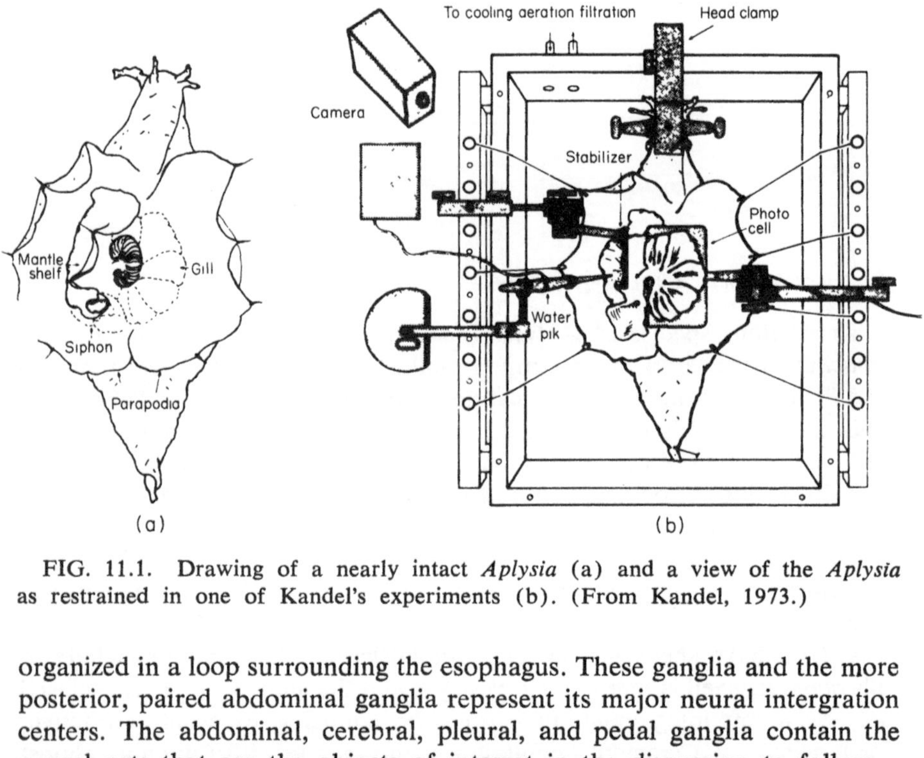

FIG. 11.1. Drawing of a nearly intact *Aplysia* (a) and a view of the *Aplysia* as restrained in one of Kandel's experiments (b). (From Kandel, 1973.)

organized in a loop surrounding the esophagus. These ganglia and the more posterior, paired abdominal ganglia represent its major neural intergration centers. The abdominal, cerebral, pleural, and pedal ganglia contain the neural nets that are the objects of interest in the discussion to follow.

The major circumesophageal ganglia are more clearly diagrammed in Fig. 11.3 which shows better the structure of the connectives between the six ganglia. It can also be seen that the two abdominal ganglia are actually fused into a single structure, covered by a common sheath. An additional important point to note is that the cell bodies of some neurons are located in the cerebral ganglion, but these same neurons have fibers extending through other ganglia all the way to the abdominal ganglion. One such cerebral ganglion cell whose fibers extend through the pleural ganglion and then to the abdominal ganglion can be seen.

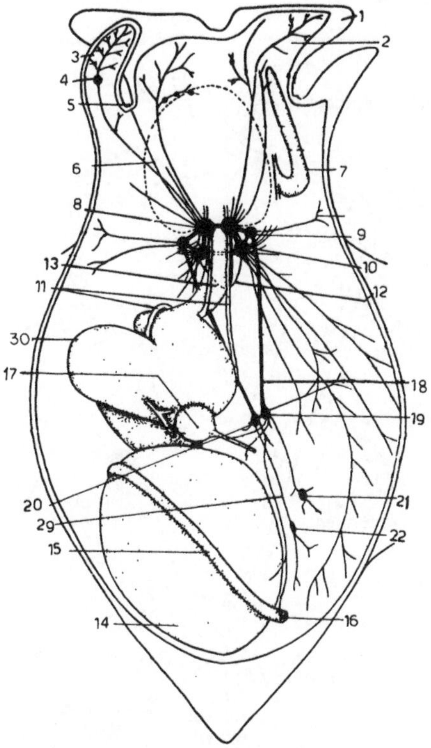

FIG. 11.2. Diagram of the nervous system of the *Aplysia:* (1) anterior tentacles, (2) rhinophore, (3) rhinophoral nerve, (4) rhinophoral ganglion, (5) eye, (6) optic nerve, (7) penis, (8) cerebral ganglion, (9) pleural ganglion, (10) abdominal ganglion, (11) foregut glands, (12) parapedal commissure, (13) esophagus, (14) midgut, (15) intestine, (16) anus, (17) copulatory bursa, (18) pleurovisceral loop, (19) supraintestinal ganglion, (20) subintestinal ganglion, (21) osphradial ganglion, (22) genital ganglion, (29) and nerve, (30) foregut. (From Hyman, 1940; after Eales.)

The extraordinary feature that has made these ganglia so interesting is that they are composed of relatively few cells (the abdominal ganglion contains about 2000 neurons), and some of them are very large; a few are actually macroscopic—almost a millimeter in diameter. Furthermore, the connections and positions of the larger neurons seem to be relatively constant from preparation to preparation. Almost as a bonus, the neurons are sufficiently pigmented to be seen without special staining. Furthermore, many of the cell bodies of particular interest lie on the surface, rather than in the interior of the ganglion. Because of these anatomic oddities, it has been possible to individually identify and number the larger neurons, providing a chain of research continuity from animal to animal that is quite unlike anything available in vertebrate neuron systems.

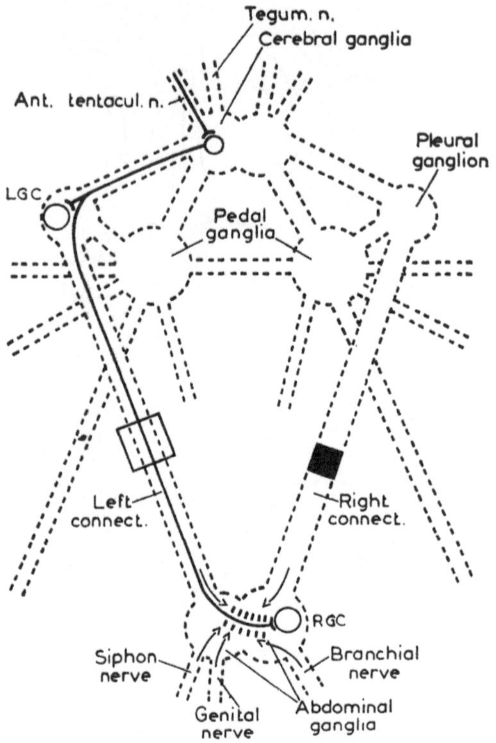

FIG. 11.3. Enlarged drawing of the ring of ganglia known as the central nervous system of the *Aplysia,* showing the interconnecting run of a single neuron from the cerebral to the abdominal ganglion. LGC and RGC are the left and right giant cells respectively. (From Tauc & Epstein, 1967.)

Although the general neuroanatomy of the *Aplysia* has been known for almost 100 years, the animal was not introduced as a model preparation in neurophysiology until the early 1950's when Chalazonitis and Arvanitaki (Chalazonitis & Arvanitaki, 1951; Arvanitaki & Chalazonitis, 1955) did so.

Figure 11.4 is a dual schematic drawing of the *Aplysia*'s abdominal ganglion, but here the larger, identified cells have been labelled. The two figures show the abdominal ganglion from both the dorsal and the ventral surface. In addition to the cell bodies in the ganglion itself, some accessory cells located on the connective descending from the esophageal ganglia also can be seen. These clusters of cells are referred to as *bag cells.* Though bag cells are clearly neurons, their major function seems to be to act as neurosecretory organs; substances extracted from them are effective in inducing egglaying (Kupfermann, 1967). Several other neurons within the ganglion proper are also neurosecretory cells that emit substances of unknown function: These are known as *white cells* because their coloration

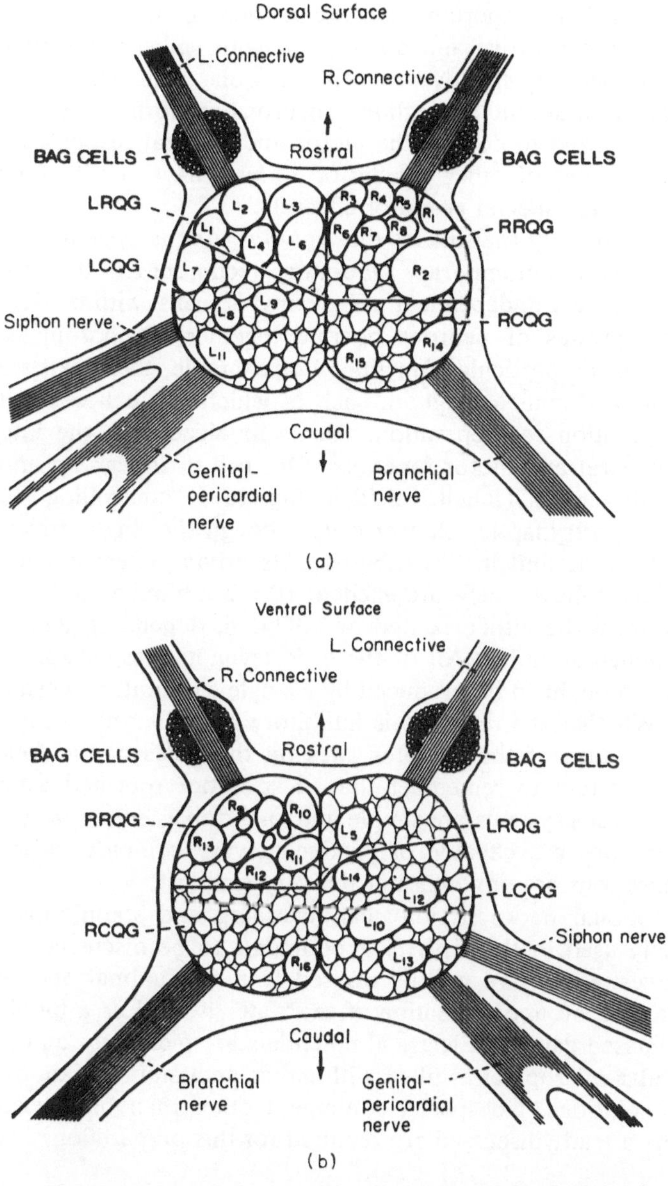

FIG. 11.4. Further enlarged dorsal and ventral views of the abdominal ganglia, showing the identified and numbered cells that are repeatedly present from one preparation to another. RQG = rostral quarter-ganglion; CQG = caudal quarter-ganglion. (From Frazier, Kandel, Kupfermann, Waziri, & Coggeshall, 1967.)

is distinguishable from the typical orange cast of most other *Aplysia* ganglionic neurons. The important fact about these neurosecretory white cells is that they are neurons and do respond with spike action potentials but seem to interact only slightly with other neurons of the ganglia.

The other abdominal ganglionic neurons perform various integrative functions required to control the motor and visceral responses of the animal. The neurons of other ganglia of *Aplysia* have similar functions, but each serves some special group of effectors.

Another property that makes the *Aplysia* nervous system especially easy to analyze with contemporary laboratory techniques is that there appear to exist relatively independent subgroups of cells within each ganglion. The latter groups of neurons operate together as coordinated subunits to achieve some particular function. For example, there exists a group of cells in the abdominal ganglion, each of which is labelled and localizable from preparation to preparation, that is involved with the integration of various visceral and motor functions. One cell of this group appears to be a master driver cell (labelled *L10* in Fig. 11.4) controlling the action of a group of postsynaptic follower cells. Though the single driver cell is the only presynaptic unit in this subgroup, its driving effect is quite complex. Three of the follower cells are excited, 10 are inhibited, and the remaining cell appears to be either excited or inhibited, depending upon the nature of the applied stimulus. All of these postsynaptic responses, it should be noted, are thought to be produced by a single transmitter substance, acetylcholine. Whether the response is inhibitory or excitatory seems to depend upon the nature of the receptor sites on the postsynaptic follower cells.

It is important to remember that, despite our repeated assertions that *Aplysia* nervous systems are simple and analyzable, even a very small number of neurons are capable of producing very complex responses if the interconnections are anything but the very simplest.

Conventional neurophysiological recording and stimulating equipment is typically used in the *Aplysia* experiments to be discussed. The ganglia under investigation are usually excised from the animal and pinned to a block of wax; an approximation of sea water is used as a bathing solution for the excised ganglia. Electrical potentials are picked up quite easily with intracellular micropipettes filled with either potassium sulfate or potassium chloride solutions. No special equipment other than basic electrical preamplifiers already discussed are required for this preparation.

B. THE BIOCHEMISTRY OF SYNAPTIC ACTION

Although our discussion has already covered the general biochemistry of synaptic action, the *Aplysia* has been so especially useful as a model system in acquiring new knowledge about ionic mechanisms that it is well worth considering some new developments before turning to the network aspects of integration in this animal.

To review briefly chemical synaptic action, the reader should remember that it is now believed that presynaptic activation of a neuron stimulates it to release graded amounts of various transmitter substances that migrate across the very small synaptic cleft to the postsynaptic membrane, where they are picked up by specialized receptor sites. These transmitter substances are capable of modifying the postsynaptic resting potential by varying the permeability of the receptor sites in a way that leads directly to the generation of spike action potentials at other neuronal sites specialized for this conversion.

In this section we shall concentrate on the nature of the transmitter substances and of the receptor sites, and briefly mention possible chemical mechanisms that might account for the alteration of the membrane potential on postsynaptic neurons of *Aplysia,* in particular. First, however, we should reexamine, briefly, a general notion that now appears to be in need of considerable modification.

It has been assumed for many years, since the earliest synaptic transmitter substances were identified, that there were separate excitatory and inhibitory chemicals. The idea was that a given substance could operate on a generalized postsynaptic receptor site to produce either an inhibitory or an excitatory response but not both. The polarity of the response was supposed to be determined by the nature of the transmitter substance rather than that of the receptor site. Thus, some substances such as acetylcholine were assumed to be exclusively excitatory; others such as GABA were assumed to be exclusively inhibitory. As we saw in Chapter 9, recent work, particularly in invertebrate preparations (Wachtel & Kandel, 1967), indicates that this notion is probably incorrect. It now appears that a single transmitter substance is capable of producing either inhibitory or excitatory effects, depending upon the frequency of presynaptic stimulation and, presumably, the nature of the receptor site upon which the substance impinges.

This question of multiple effects of each transmitter substance, of course, is intimately related to the question of whether or not a single neuron is capable of producing more than one kind of transmitter substance. A generally accepted rule of synaptic organization has been that a neuron is capable of producing only one transmitter substance. This rule has been referred to as "Dale's principle." It appears that Dale's principle must still be considered to be valid—a single neuron is a chemical factory for only one type of transmitter. The fact that a single neuron may both excite and inhibit seems to be solely due to the differential response of the postsynaptic receptor sites to the single chemical.

A much more detailed understanding of the varied nature of *Aplysia* postsynaptic receptor sites has been forthcoming in recent years. In 1972 a very important series of papers was published by Kehoe (1972a, 1972b, 1972c) and Ascher (1972). In brief, from their findings, it appears that there are as many as five different kinds of receptor sites in *Aplysia,* three of which are sensitive to acetylcholine and two of which are sensitive to another substance, dopamine, known also to have considerable synaptic efficacy. Usually, more than one of the five types of receptors, often including both excitatory and inhibitory sites, appear to be present on the same cell. Cells in certain locations display predominately excitatory sensitivity, while others elsewhere in a ganglion appear to have a finer inhibitory sensitivity. Indeed, different parts of the same cells may exhibit specific sensitivities to different transmitter substances.

Kehoe (1972b) has shown that a single transmitter substance, acetylcholine, iontophoretically injected into an *Aplysia* pleural ganglion neuron, can evoke three kinds of responses. This suggests the existence of three kinds of receptor sites. In cells located near the middle of the ganglion, a two-component inhibitory response is produced, consisting first of a rapid, spike-like postsynaptic hyperpolarizing potential, and next a much longer-lasting (as long as 20 seconds) response. This longer response is a hyperpolarization when the resting membrane potential is more positive than —60 mV. However, when the membrane potential is more negative than —60 mV, even though the rapid component remains much the same as it was at more depolarized levels, the long component is reduced in amplitude. At even greater clamped membrane hyperpolarizations it is converted to a depolarization. Figure 11.5a shows a combined rapid and slow inhibitory response, while Figs. 11.5b, c, d, and e show the effect on the slow potential of increasing the hyperpolarization with a voltage clamp.

Kehoe reports that in anterior portions of the ganglion some cells seem to respond only with a depolarizing excitatory potential when acetylcholine is injected. Hyperpolarizing the membrane potential in these cells leads only to an increase in the size of this quick, depolarizing excitatory response.

In another paper in this series, Kehoe (1972c) shows that both inhibitory and excitatory postsynaptic potentials can be produced in the same cell in the abdominal ganglion and that the electrical response observed must be a mixture of interacting excitatory and inhibitory effects. Indeed, Kehoe reports that all seven possible combinations of the three types of responses to acetylcholine can be produced, with the single exception of a composite response consisting solely of the slow inhibitory response. He believes this to be evidence of a scattered combination of three types of receptor sites, each of which is sensitive in a particular way to acetylcholine. One site responds with a rapid change in permeability of the membrane, which produces the fast inhibitory depolarization. Another alters membrane permeability in a different way to produce the slow inhibitory response. Finally, a third type produces excitatory depolarization. The fact

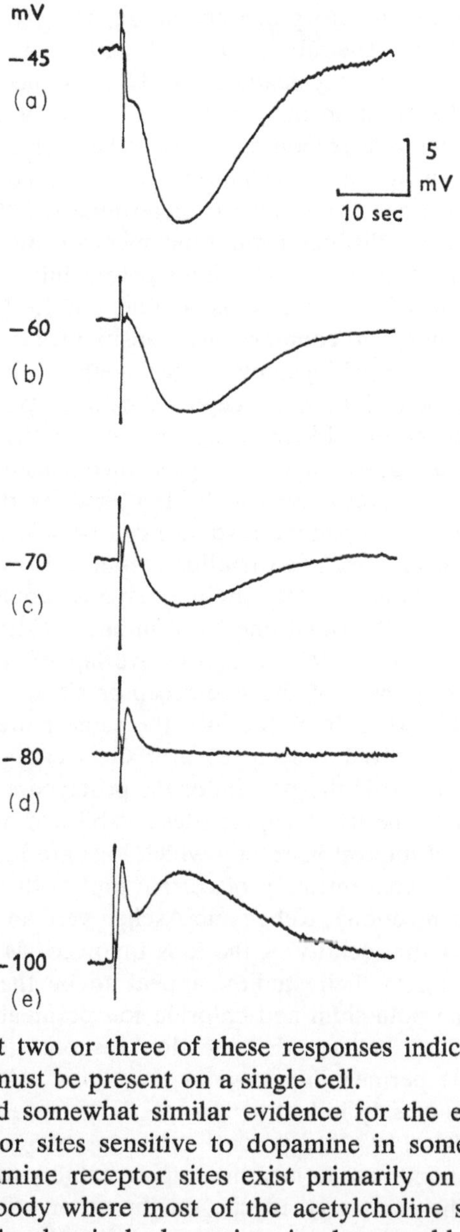

FIG. 11.5. Records of postsynaptic potentials produced by the application of acetylcholine to a postsynaptic region in *Aplysia.* A two-component response is observed, consisting of a rapid spike-like response of short latency and a long slow wave. Depending upon the clamped resting membrane potential, the polarity of the slow wave may be reduced or even reversed, but the spike remains unchanged. (From Kehoe, 1972b.)

that a single neuron can exhibit two or three of these responses indicates that two or three receptor sites must be present on a single cell.

Ascher (1972) has produced somewhat similar evidence for the existence of two specialized receptor sites sensitive to dopamine in some of these same *Aplysia* cells. Dopamine receptor sites exist primarily on the axons, rather than on the cell body where most of the acetylcholine sites seem to be localized. The single chemical, dopamine, is also capable of producing either excitatory or inhibitory responses, or a biphasic combination of the two, again apparently depending upon the presence of the two types of receptor sites. In the pleural ganglia, once again, the responses seemed to be segregated on the basis of the anatomic location of the particular neuron under study. Anterior cells generally responded to dopamine with an excitatory depolarization, while cells located in the middle of the ganglion generally responded with a slow inhibitory hyperpolarization. Posterior cells typically showed a combined biphasic wave.

The question that now arises concerns the origin of these five types of postsynaptic response. It would be highly illuminating if one could determine what ionic mechanisms are involved in each response. It would also be of great significance to determine whether the "receptor sites" actually represent distinct membrane mechanisms or if they are simply different manifestations of some single permeability changing process.

One of the best ways to find out if the receptor sites are distinct or the same is to compare their sensitivities to various pharmacological agents. In a series of ingenious experiments, Kehoe and Ascher have determined that each of the five receptor sites does exhibit a unique pattern of sensitivity to certain blocking agents—drugs that can selectively "turn off" the various postsynaptic receptor mechanisms. For example, the excitatory dopamine response can be inhibited by tubocurarine and strychnine; the inhibitory dopamine response can be selectively turned off by ergometrine. The excitatory acetylcholine response is inhibited by hexamethonium. The slow inhibitory acetylcholine response could be blocked by methyloxylcholine, and the rapid one by a number of drugs that also affected the excitatory response. Whatever the overlap of the effects of blocking agents on the responses of the five receptor sites, in no single situation did two of the blocking drugs produce the same pattern of effects. Thus, the mechanism involved at each receptor site seems to be an individual specialization of that particular portion of the postsynaptic neuronal membrane.

With the five receptor sites established as distinct mechanisms, the question of interest is exactly which ions are involved in each process. By altering the concentration of certain ions both outside and inside (by iontophoretic injection), Kehoe and Ascher were able to come to fairly firm decisions about the identity of the ions involved. Most of the inhibitory effects, according to their studies, appear to be the result of the changes induced in the potassium and chloride ion permeabilities of the membrane. Kehoe (1972a) believes that the slow response is produced by changes in membrane permeability to potassium ions, whereas the rapid responses seem to be closely linked to changes in permeability to chloride ions. Similarly, Ascher (1972) reports that the inhibitory effect of dopamine is mediated by an increase in membrane permeability to potassium ions.

The excitatory effects, on the other hand, appear to be mainly due to increases in membrane permeability to sodium ions (Chiarandini, Stefani, & Gerschenfeld, 1967; Blankenship, Wachtel, & Kandel, 1971). One exception to this generalization has been shown: The membranes of the so-called *D* cells of some snails seem selectively to increase their permeability to chloride ions upon excitation. However, this is an anomaly. The notion of sodium permeability increases upon excitation is more consistent with the Hodgkin-Huxley theory of membrane action.

This concludes a brief introduction to the nature and variety of at least some of the receptor sites, both excitatory and inhibitory, that exist in *Aplysia* neurons. It should be clear that the effects of even a simple, monosynaptic transmission can drastically alter a conducted signal pattern. Depending upon the frequency of the presynaptic spike action potentials and other inputs from collateral neurons, the ability of even a very simple nervous system to modify an input information pattern is considerable. In the normal condition, with feedback and multiple inputs, the potential complexities of networks of even small numbers of neurons, then, is seen to be enormous.

C. PATTERNS OF EXCITATION AND INHIBITION

The fact that the action of a driver cell may be inhibitory or excitatory is only one suggestion that the analysis of even simple nerve networks is not a routine straightforward task. Indeed, an analogy may be drawn between the many-body problem in classical physics and the problem of the multiple neuron interaction. Just as the mathematical solution of the many-body problem becomes increasingly intractable as the number of interactions increases, so does the problem of interacting neurons increase in complexity as the number of neurons and the kinds of interactions increase. There are many ways in which nervous interactions are more complicated than gravitational ones. In this section we consider exemplary empirical observations that illustrate some of these complexities and shall try to clarify just how much actual progress has been made so far in understanding them in *Aplysia*.

Perhaps the most thorough analysis of any nervous net in any organism is that done (e.g., Kandel, Frazier, Waziri, & Coggeshall, 1967) on a small system of cells in the abdominal ganglion of the *Aplysia*, which centers about the action of driver cell *L10*. These investigators have made sufficient progress to allow them actually to draw a tentative wiring diagram of the group of involved neurons. This diagram, which would have been impossible ten years ago, illustrates, perhaps for one of the first times in human history, a situation in which the action of specific, identified neurons is used to account for the molar aspects of behavior.

The group of neurons, consisting of the driver neuron *L10* and its 14 follower cells already mentioned, is itself part of a larger system, according to Kandel and his colleagues. Although *L10* is a driver in the subgroup, it too receives inputs from other interneurons; Table 11.2 indicates that it is actually a follower for at least four other neurons (identified only as numbered interneurons *II, III, IV,* and *V,* since their specific anatomic

TABLE 11.2
Postulated Interneurons and the Follower Cells Associated with Them in the Abdominal Ganglion of the *Aplysia*[b]

Interneuron no.	Identification	Follower cell	Action [a]
I	L10 (LCQG)	L1, L2, L3, L4, L5 L6, L8, L11, L12, L13, LB group	I
		L9, R15, R16, RB group, LC group	E
		L7	E–I
II	?	L2, L3, L4, L5, L6 L7, L8, L9, L10 LB group, LC group	I
		L11, R2, R16	E
		R15	E–I
III	?	L10	I
		R16	E
IV	?	L10	I
		R16	E
V	?	L10	I(L)
		L2, L3, L4, L5, L6	E
VI	L15 (LCQG)	L5, L3	E(S)
VII	?	L2, L3, L4, L6	I(S)
VIII	?	L12, L13	E
IX	?	R1, R2	E(L)
X	?	L1, R1, R2	E(L)
XI	?	RB group	E

[a]I = inhibitory connection; E = connection; ? = unidentified (inferred) interneuron. (L) = light-sensitive interneurons; these are set into activity by light offset. (S) = silent interneurons; these fire rarely even following repeated stimulation of connectives and nerves. (LCQG) = left caudal quarter ganglion.

[b]From Kandel, Frazier, Waziri, and Coggeshall (1967).

locus is not yet known). Their existence is inferred on the basis of their effects on other neurons.

In addition to the individual neurons mentioned, there are some clusters of cells, all of which seem to respond synchronously, that also are driven by one or another known or inferred driver cell. For example, *L10* also influences the action of a group of cells labelled *RB* in Table 11.2, so that they all respond in unison. Note also that Table 11.2 indicates that cells that are followers of *L10* are also affected by other drivers. Obviously the total system is quite a bit more complex than was suggested in the original study of the *L10* subgroup.

A further complication is that some neurons feed back upon neurons that excite them, or even feed forward to others skipping over an interspersed neuron in a chain. Figure 11.6 demonstrates how feed forward and feed backward occur, and how they can be combined even within a single system.

In such a network it is also possible that disinhibitory mechanisms that imitate excitatory ones might be encountered. In fact Frazier, Waziri, and Kandel (1965) have observed what they believe to be two different kinds of disinhibition mechanisms in the *Aplysia* abdominal ganglion. In one case, the disinhibition is presumed to be presynaptic; that is, inhibition of an inhibitory driver cell lowers its response rate and thereby reduces

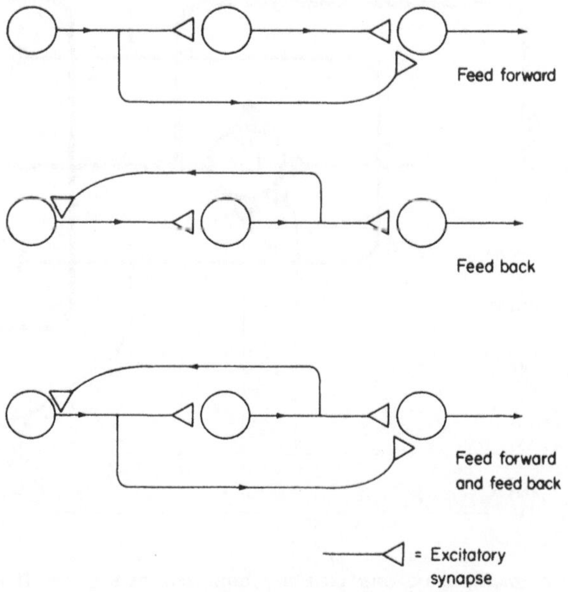

FIG. 11.6. Circuit diagram showing the possible forms of feed back, feed forward, and combinations of the two that can complicate system analysis in even simple nervous systems like that of *Aplysia*. (From Kandel & Wachtel, 1968.)

its inhibitory effect on a follower. The follower then responds more actively. In the other case, the disinhibitory mechanism is presumed to be postsynaptic. Stimulation through one of the ganglionic nerves reduces the inhibitory effect of the driver, even though its response rate remains constant.

The patterns of inhibition and excitation that have been so briefly cited here produce situations that are extremely complicated, and there is always the possibility that, when the total system is reduced to a component subgroup, the essentials of the network's function may have been lost. Nevertheless, an attempt to put all of this material together into some sort of summary diagram is appropriate. Figure 11.7 summarizes various processes that have been discussed in this section. The figure is only partial (too many lines would obscure the details), but even at this level of abstraction it may still be too complicated to allow systematic predictions. Nevertheless, this is an extraordinary illustration of an integrative neuronal network; few similar ones for other networks are currently available.

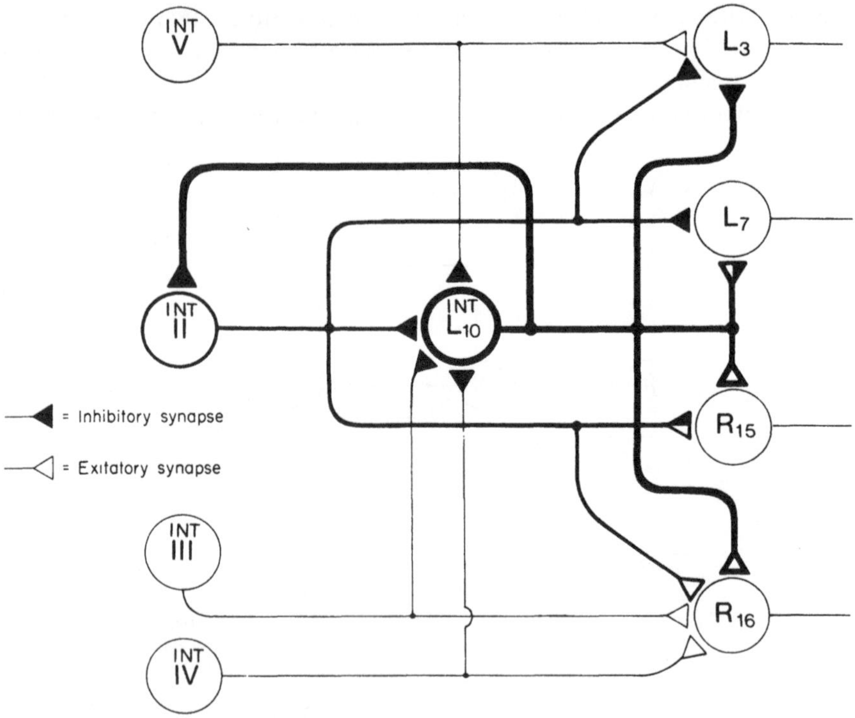

FIG. 11.7. An example of one partially analyzed neural net from the *Aplysia* abdominal ganglion, showing how even a few cells can interact in exceedingly complicated manners. The letters in cells are identifying codes. (From Kandel & Wachtel, 1968.)

D. MOTOR CONTROL AND REFLEXES

Now that there is some familiarity with the details of at least part of the neural network in this extraordinary preparation, it is timely to consider how these networks might act to control the behavior of the *Aplysia.* Unfortunately, promising though such an analytic approach is, only a very few reports are available in which the specific details of the neuronal network anatomy have been compared to the behavior of the total organism. Although this type of research is at the current frontier of psychobiological interest, specific studies, in which the behavior of an organism is linked to the specific properties of neurons, are, as yet, few and far between. Nevertheless, several reports have been produced in which the anatomic simplicity of *Aplysia* (and a closely related animal, the *Triton*) have been used to make this comparison.

Before discussing the details of these important experimental studies it should be noted that no analogous studies have ever been done on the vertebrate nervous system. As yet, too little is known about the identity of specific vertebrate neurons and their interconnections to allow similar studies to be performed. Vertebrate neurons are so small that they can be detected only blindly, and it is highly unlikely that any single cell has ever been impaled, even fortuitously, more than once, much less repeatedly targeted in a premediated manner.

The difference in size and number between vertebrate and invertebrate neuronal systems may mean that the two possess entirely different modes of organization and operation. The small number of neurons involved and the almost automatic control of behavior in the central nervous system of invertebrates may reflect an organization more comparable to the reflexes of the spinal cord than to the action of the cerebral hemispheres of a primate. *Aplysia* and other invertebrates may be said to lack real "brains," the term "brain" being used here only as an analogy for the central, largest ganglion.

The reader is cautioned not to assume that all the behavioral processes observed in the *Aplysia* can be directly extended to analogous human cognition. The reflexive operation in the *Aplysia* may, in fact, be only slightly relevant to the way humans think. Self-awareness may be an emergent property that arises only when the number of neurons in the nervous system is much closer to 10^{10} than the 10^4 of *Aplysia*'s rudimentary nervous system. Nevertheless, only in a model preparation such as *Aplysia* has it been possible even to begin making comparisons between nerve nets and behavior; and thus it is to these animals that we now turn for information about the interaction of known nets of neurons.

Feeding into the abdominal ganglion of the *Aplysia* is a nerve known as the *ctenidio-genital* nerve. This nerve conveys signals to the gills, the gonads, and other structures of the reproductive tract. Peretz (1969) has carried out two kinds of experiments to determine the involvement of the abdominal ganglion neurons in the motor control of the gill. His first experiment consisted of a conventional electrophysiological recording situation, in which spike action potentials were picked up from specifically identified neurons during spontaneous gill movements. In this way, Peretz was able to determine which cells were involved with gill motion as sensors. Armed with this information, he was able to perform a more complex experiment, using electrical stimulation with intracellular microelectrodes to define which of the cells were specifically involved in controlling motor responses.

Comparisons of electrophysiological responses to stimulation and direct behavioral observation of gill movements led Peretz to the conclusion that four different types of neurons are involved in this motor system. The first type is capable of eliciting gill responses, as well as responding during spontaneous gill movements. The second type was able to elicit gill movements when stimulated, but when intracellular recordings were made it was seen to respond with a diminution in its rate of activity during spontaneous movements. The third type was inhibited in the same way during spontaneous gill movements, but could not evoke any gill movement when stimulated; the fourth type increased its frequency of spike firing during gill movements, but could not evoke gill movement.

At about the same time, Kupfermann and Kandel (1969), working with a similar *Aplysia* preparation, reported the involvement of other neurons in controlling movements, not only of the gill, but also of siphon and mantle shelf. In its normal environment, the *Aplysia* exhibits reflex withdrawal of this group of organs whenever they are touched—a useful defensive response. With the information known about the abdominal ganglion, Kupfermann and Kandel were able not only to demonstrate the details of the neural network underlying this reflexive withdrawal but also to show certain details of the operation of the net that reflect a new important mechanism: This simple net can operate on the basis either of a simple peripheral reflex or on the basis of signals controlled by and initiated from the abdominal ganglion. The alteration between these two modes of operation also appears to be centrally controlled. Figure 11.8 is an up-to-date schematic summary of their work from Kandel (1973). It will be used as a vehicle for the discussion to follow.

According to Kupfermann and Kandel, this sensory motor system may be considered to be composed of the siphon, the mantle shelf, and the gill, along with their respective sensors, the afferent neurons and a group of 5 motor neurons that are driven by the sensors either directly or through one or two interposed interneurons. The important discovery made by Kupfermann and Kandel was that there are, indeed, two possible routes for afferent information to take on its way to the motor cells. These routes mediate two behavioral modes. If the animal is stimulated externally in what may be a threatening manner, it is desirable for it to respond rapidly and reflexively by immediate withdrawal. In this case, the connection to all five of the motor neurons must be a simple excitatory one, so that the muscular actions necessary for withdrawal can be elicited instantaneously.

In other instances, these same muscular actions are required not as parts of a massive withdrawal reflex, but as a more specific, coordinated response. If the five motor neurons were activated by purely excitatory connections under all conditions, self-stimulation of the receptors might be expected to lead to a continued (tonic) contraction of all of the muscles once the animal was touched. For these reasons, some sort of switching mechanism is useful so that under certain conditions some of the neurons can be inhibited, rather than all being excited by afferent signals. Under conditions of central control and initiation (as opposed to reflexive control), the interconnections of the afferent signal to the motor neuron must, therefore, be different than the connection to the same neuron during reflexive responses. This biological switching action is symbolized by the three drawings of Fig. 11.8. In Figs. 11.8b and c the afferent signals are routed to the motor neurons directly to produce an excitatory reflex withdrawal response. Figure 11.8a shows the case in which the sensory signal is routed through an interneuron. The output of this interneuron on the system is bipolar; that is, it excites some cells and inhibits others.

The exact nature of the hypothetical mechanism that allows the nervous system to switch between the two modes of action is still not known, of course, but it is clear that such a mechanism adds a great deal of flexibility to the response repertoire of *Aplysia,* even as it is constrained by the limited number of neurons available. The reflexive withdrawal can be rapid and massive as needed for a defensive reaction; yet at another time, the integration of inhibitory and excitatory responses of individual neurons can allow the animal to perform coordinated movements. This should not be misinterpreted as a sort of "free will" on the part of the animal but some process does allow the alternative modes of response. Kupfermann and Kandel suggest that the adaptive value of the internally generated coordinated action may be to produce the pattern of movements required for locomotion, respiration, and clearing the mantle cavity of foreign objects.

Having discussed how more or less simple movements are produced by known neural nets, it would be appropriate to attempt to relate some more complex behavior in the *Aplysia* to their neural correlates. Unfortunately, however, no such studies are yet available on this animal; and, therefore, an exception is made here to the general rule of this chapter that we would discuss only findings from this genus. Fortunately, the *Triton,* another Opisthobranchian mollusk very similar in general anatomy to *Aplysia,* has been used as a model for studies of a more complex response, namely, swimming. The research, carried out by Willows (1967, 1968, 1969), represents a remarkable technical *tour de force*.

To compare a complex motor response with its neural correlates requires that the animal be sufficiently free to perform the motions; yet the nervous system must be sufficiently restrained to allow precision in the placement and maintenance of stimulating and recording electrodes. The two require-

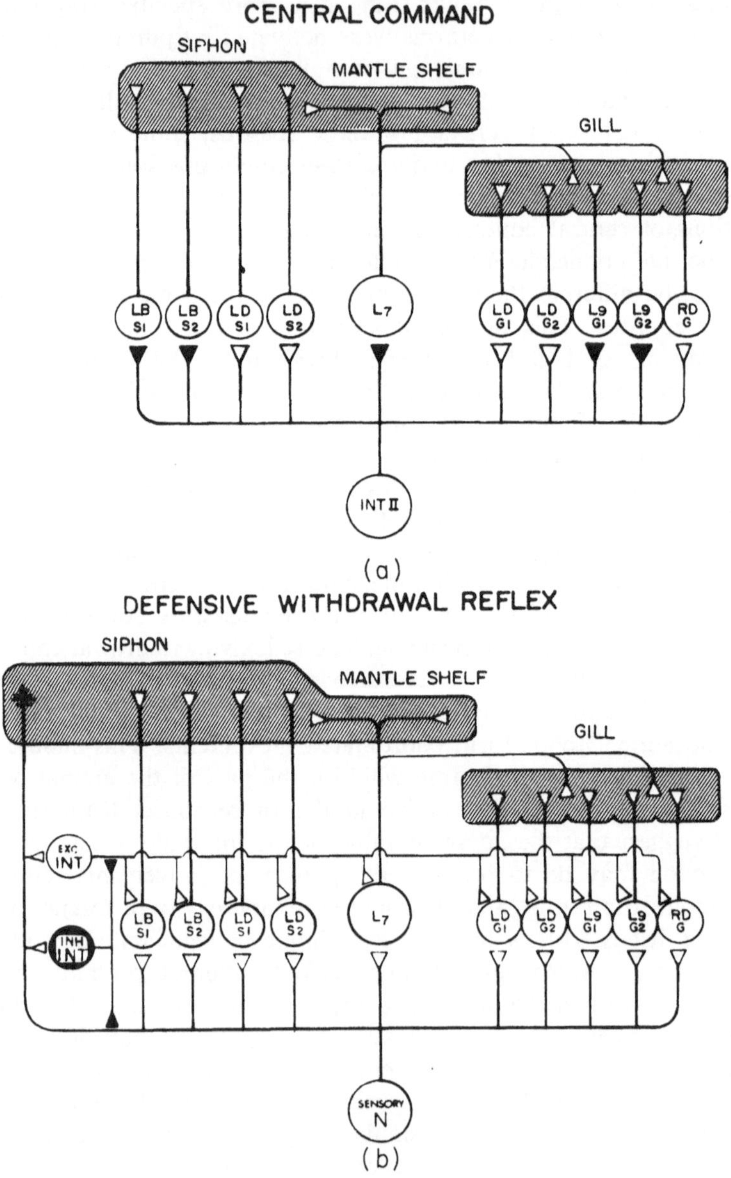

FIG. 11.8a and b

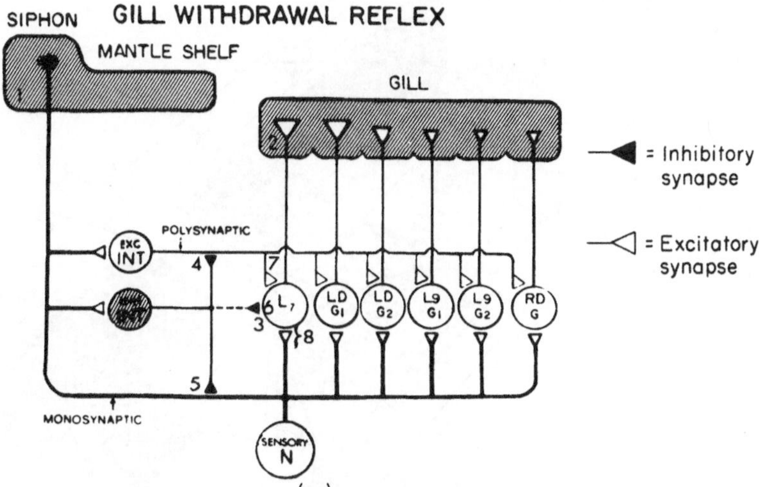

(c)

FIG. 11.8c

FIG. 11.8. Proposed circuit diagrams of a dual control mechanism for control of the *Aplysia*'s siphon, mantle, and gill. Central commands allow coordinated movements of the structures for sand removal (for example) using one circuit arrangement (a), but the system may be switched into a massive reflex mode to allow rapid withdrawal of either the entire mass of soft tissue (b) or only the gill (c). (From Kandel, 1973.)

ments seem contradictory, but Willows' solution to the dilemma was ingenious. The animal was literally "hung" by its nervous system. A ring of hooks was inserted into the mantle surrounding the cluster of cerebral, pleural, and pedal ganglia, as shown in Figs. 11.9a and b. The structure of this cluster of ganglia is sufficiently different from that of *Aplysia* that we have reproduced Willows original drawing in Fig. 11.9c. This diagrammatic sketch also indicates by number the identified cells of these ganglia. A pair of wires secured at each end of the ring was then passed under the ganglia and lifted to immobilize the nervous system without immobilizing the animal itself. Microelectrodes were then inserted into various cells to determine their specific motor effect.

Willows observed a number of different responses to various cells when they were electrically stimulated. Some cells appeared to be neither sensory nor motor, so he concluded that they were neurosecretory units, like the white bag cells of the *Aplysia*. Some cells (particularly numbers 3, 4, 33, and 34) produced muscular contractions in localized regions of the animal's integument. This seemed to be the result of a direct motor output from these units, but in a totally uncoordinated manner. The size of the areas that responded was thought to reflect simply the size of the motor fields of these units. A group of other cells (1, 2, 29, 30, 15, and 24) produced particular patterns of motor contraction of the gill plumes, whereas cells 3 and 37 produced turning responses. The nature of all of

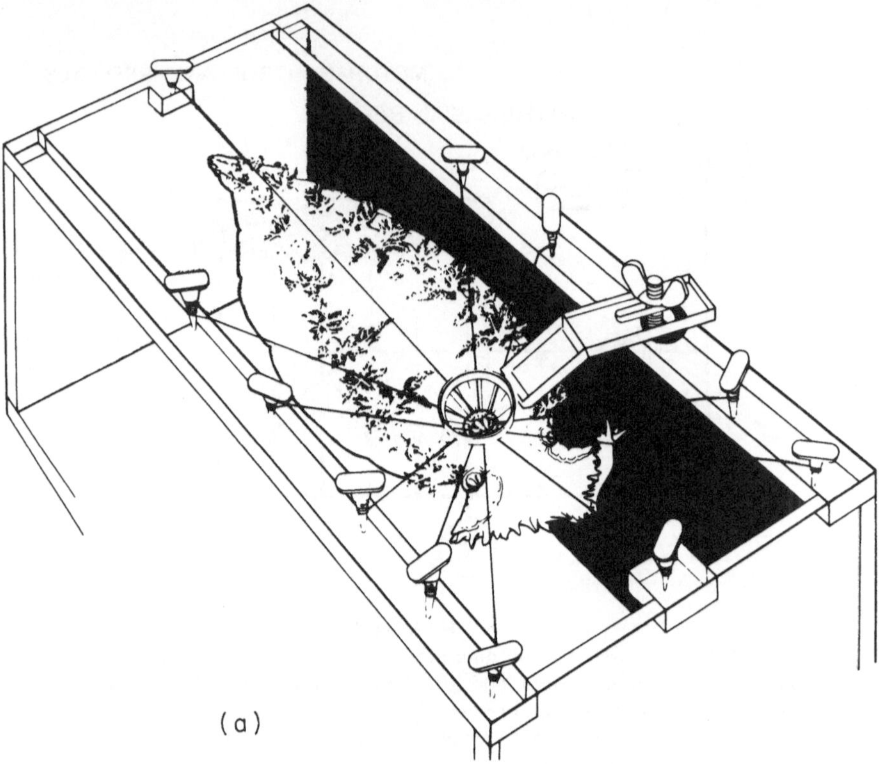
(a)

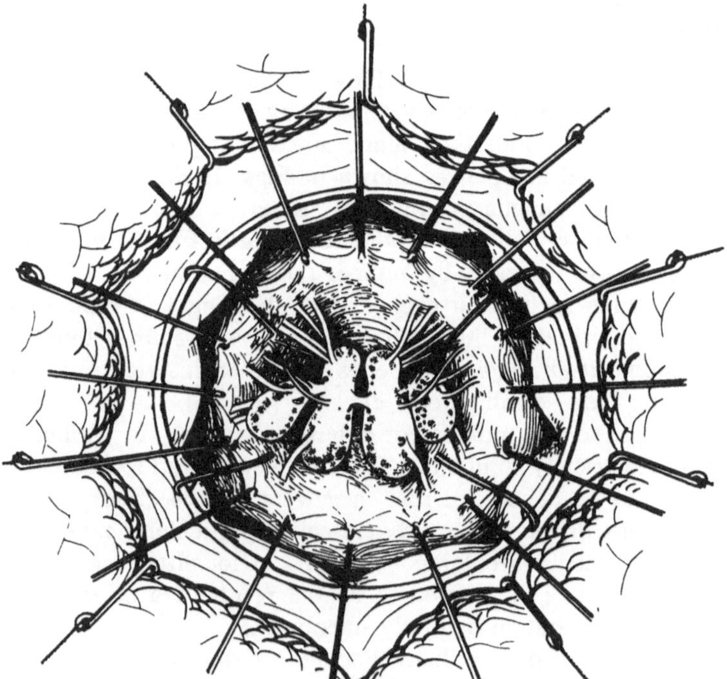
(b)

FIG. 11.9a and b

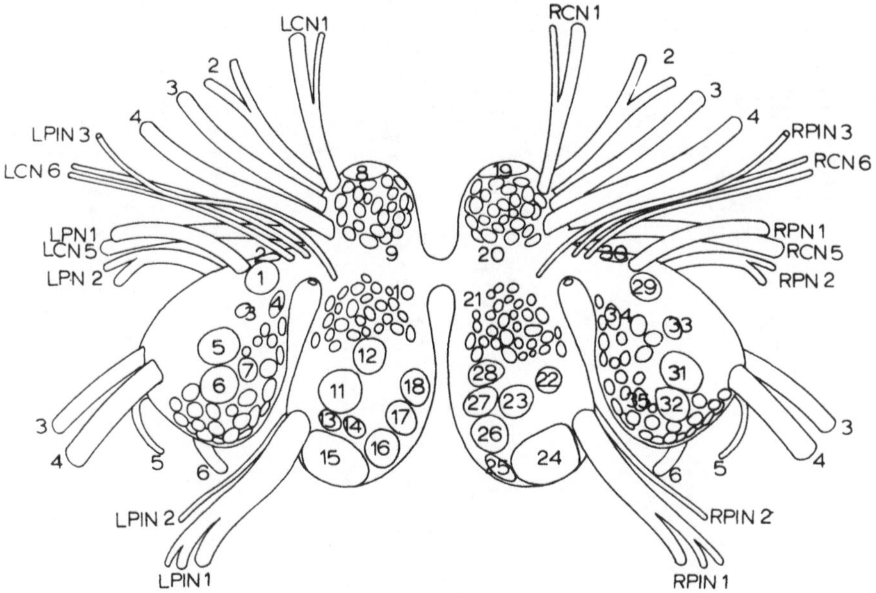

(c)

FIG. 11.9c

FIG. 11.9. Drawing of Willows' *Triton* preparation, showing both the general arrangement of the animal so that its body is free to respond (a) and the ring that simultaneously immobilizes the central nervous system (b); (c) is a drawing of the identified cells of the *Triton's* central ganglia. The other symbols indicate various nerves. (From Willows, 1968.)

these responses is shown in Fig. 11.10. A number of other cells (10, 21, 15, 16, 17, 18, 23, 24, 25, 26, 27, 28) produced a generalized tensing of the entire body of the animal—a sort of protective reaction that could also be elicited by chemicals secreted from certain starfish known to be predators of the *Triton.*

The most complicated responses that could be evoked by single stimuli, however, were extraordinarily complex, considering the simplicity of the triggering stimulus and of the neural mechanism. Whereas the responses already described exhibited a sort of mechanical, deterministic constancy, it was possible to produce an integrated series of motions that looked very much like the *Triton*'s natural swimming motions, by simultaneously stimulating several members of two groups of smaller neurons (numbered 10 and 21 in Fig. 11.9c). The sequence consisted of contractions alternating between one direction and another, as shown in Fig. 11.11. The sequence of swimming motions occasionally lasted for as long as 30 seconds.

Curiously, the 60 or so cells involved in this complex behavior appeared to be interconnected in a relatively nonspecific manner, by what Willows

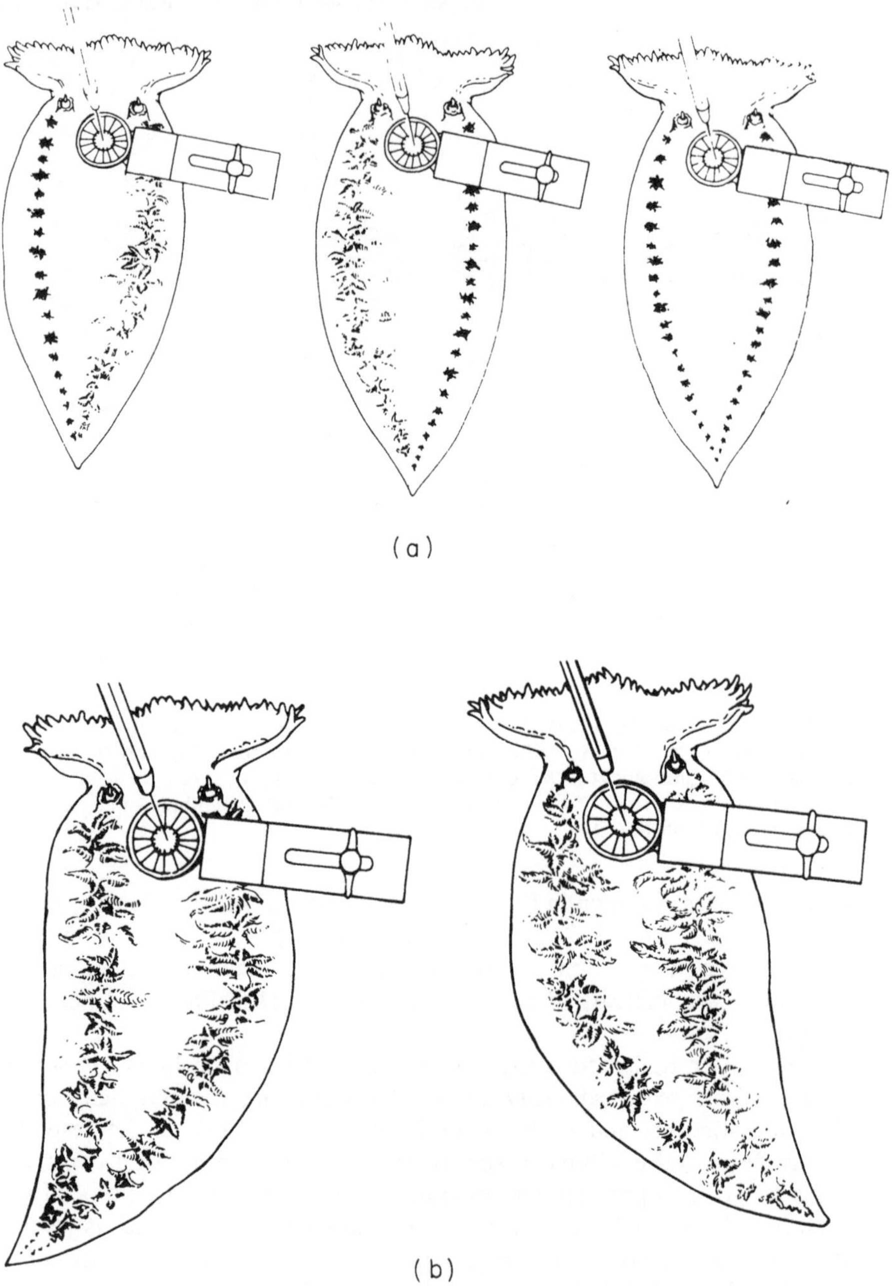

FIG. 11.10. Major responses elicited by stimulation of single cells in the *Triton*. (a) Single stimuli to cell 15 or 24 produced total bronchial tuft withdrawal; single stimulus to cell 1 causes tufts on left side of animal to be withdrawn; single stimulus to cell 19 causes tufts on right side of animal to be withdrawn. (b) Single stimulus to cell 3 stimulates the animal to flex body to the left while a single stimulus to cell 33 stimulates the animal to flex body to the right. (From Willows, 1968.)

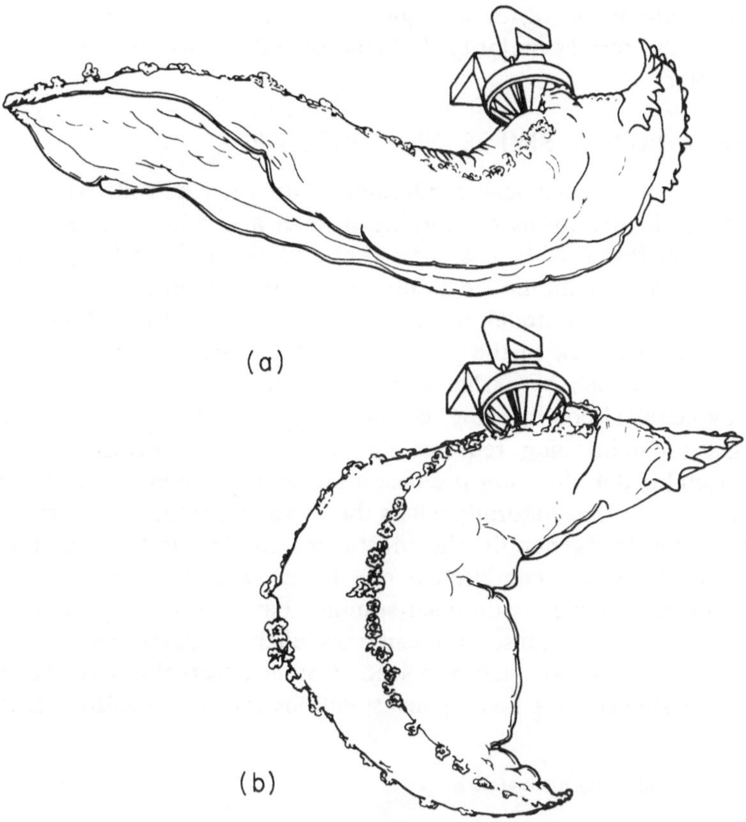

FIG. 11.11. Two still drawings of a complex swimming motion that occasionally can be produced in the *Triton* by a single stimulus to a single neuron. (From Willows, 1968.)

thought might be electrical synapses. The stimulation of one cell produced a depolarization in that cell that spread very rapidly through the entire group with the diminishing voltage pattern typical of electrotonic spread. Surprisingly, however, the two groups of related cells in either of the two pleural ganglia appeared to be as closely electrically coupled to each other as related neurons in the same ganglion were.

Willows notes further that the elicitation of this complex swimming response pattern required the simultaneous stimulation of several cells and that even a large depolarization of a single cell was usually ineffective. He speculates that the occasional appearance of the swimming behavior pattern without external stimulation may be due to the accidental aggregation of spontaneous activity within the neuron clusters. Thus, the clusters operate as an integrative and decision-making network by distinguishing between the synchronous firing of many neurons of the cluster and the occasional random firing of a single one. This may be a true biological example of the sort of majority decision logic discussed earlier in the previous chapter.

E. "LEARNING" IN SIMPLE NEURONAL NETS

For many years, classical conditioning had been considered to be the prototype of higher forms of learning in man and other animals. Though it is no longer fashionable to assume that all learning is nothing more than a concatenation of conditioned responses, classical conditioning is still used by many workers as a model to elaborate some of the temporal and systematic variables involved in the processes through which the behavior of an organism is modified by earlier experience. As a brief review of what is meant by classical conditioning, consider Fig. 11.12a. The basic paradigm of classical conditioning requires the synchronous pairing of a conditioned stimulus that does not produce a particular response with an unconditioned stimulus that naturally elicits the same response. If the timing conditions are correct (generally, the effect is maximal when the unconditioned stimulus precedes the conditioned one by .5 second or so), it is subsequently found that the conditioned stimulus has become "associated" with the unconditioned response and can now elicit it. Although there are a very large number of artifacts and second-order effects that have been considered and studied in a host of investigations into the manifold problems

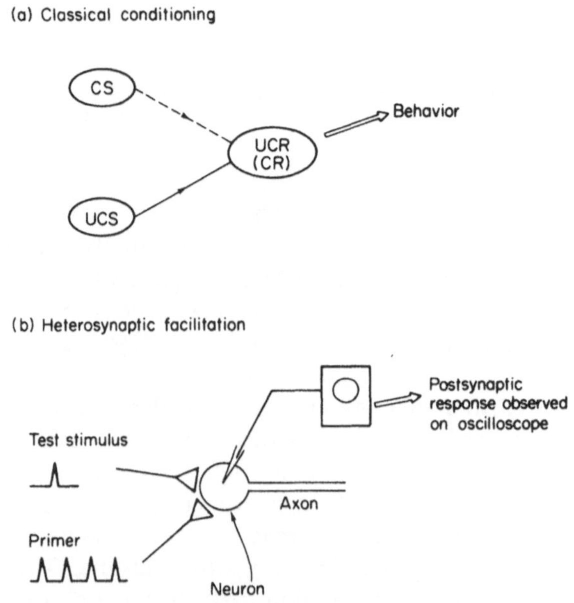

FIG. 11.12. Drawings of the paradigms of (a) classical conditioning and of (b) heterosynaptic facilitation.

of classical conditioning, the essentials of the classical conditioning paradigm are relatively simple.

The *Aplysia's* nervous system exhibits an analogous neurophysiological process that has been designated *heterosynaptic facilitation.* It is important to emphasize that classical conditioning and heterosynaptic facilitation are only process analogs of each other, and few researchers believe that they are homologous. That is, while the two exhibit many common features in their operation, it is probably not yet appropriate to link the two in a way that identifies heterosynaptic facilitation as the underlying neuronal mechanism for whatever classical behavioral conditioning might be observed in *Aplysia* or any other animal.

Heterosynaptic facilitation is a phenomenon observed in neural nets in which the effect of a weak electrical stimulus, analogous to a conditioned stimulus, applied to a presynaptic neuron can be enhanced in its ability to evoke a postsynaptic potential. This enhancement is achieved by pairing the weak stimulus with a vigorous stimulus applied to another presynaptic neuron. Heterosynaptic facilitation, therefore, is quite comparable to post-tetanic potentiation in terms of its results. The latter process is quite distinct, however, for it involves a change in the synaptic effectiveness solely as a result of exercise of that single synapse whereas heterosynaptic facilitation is a network process involving several neurons and synapses. Nevertheless, heterosynaptic facilitation and post-tetanic potentiation are alike in that they both produce similar effects—a transitory increase in the response magnitude of the postsynaptic neuron to a stimulus that previously had been less effective. For this reason, both processes have been repeatedly suggested as possible cellular mechanisms mediating some sort of adaptive or learned behavior.

The distinction between the two was not always so clear. Indeed, during the early work on the phenomenon of heterosynaptic facilitation, there was an assumption that it might simply be another version of post-tetanic facilitation. However, Tauc and Epstein (1967) and Epstein and Tauc (1970) clearly showed that the two were separate processes and that heterosynaptic facilitation occurred under conditions, such as low temperature or substitution of lithium for sodium in the bathing solutions, in which it was impossible to obtain post-tetanic potentiation. Even more important, work by these investigators showed that the heterosynaptic facilitory effect was not due to a mere change in the postsynaptic membrane, but required the action of some sort of intermediary neuron.

Heterosynaptic *inhibition* is also known to occur in some cases. In such instances the response is decreased rather than increased. However, all the other properties of such an inhibition are sufficiently comparable to facilitation to warrant no further discussion.

Since the terminology and methodology of the typical heterosynaptic experiment is somewhat specialized, Fig. 11.12b is presented to clarify the technology and define some of the relevant terms.

To demonstrate heterosynaptic facilitation, a stimulus is applied to one of the nerves feeding into the selected ganglion. The amplitude of this test stimulus is then adjusted to a relatively low value so that it produces only a small postsynaptic potential in a cell impaled with a recording microelectrode. This relatively weak test stimulus is then paired with another stronger stimulus, applied to another of the nerves feeding into the ganglion. This second stimulus, usually referred to as the *priming* stimulus, in the terminology originally suggested by Tauc and his co-workers, is adjusted so that it is capable of producing a vigorous response in the responding cell. If the test and the priming stimulus are paired repeatedly over approximately a three-minute period, a subsequent presentation of the test stimulus alone will produce a greatly enhanced postsynaptic potential. The typical time course of the facilitation is shown in Fig. 11.13. As can be seen in this figure, the facilitation can last for over 15 minutes—an extraordinarily long time—and also, at its maximum, closely approximates the response level of the original priming stimulus.

According to Kandel and Tauc (1964, 1965), only certain cells found within the abdominal ganglion of *Aplysia* display heterosynaptic facilitation. Most cells of the ganglion exhibit no change in response, but in no case did any cells exhibit a facilitation without the paired stimuli. The fact

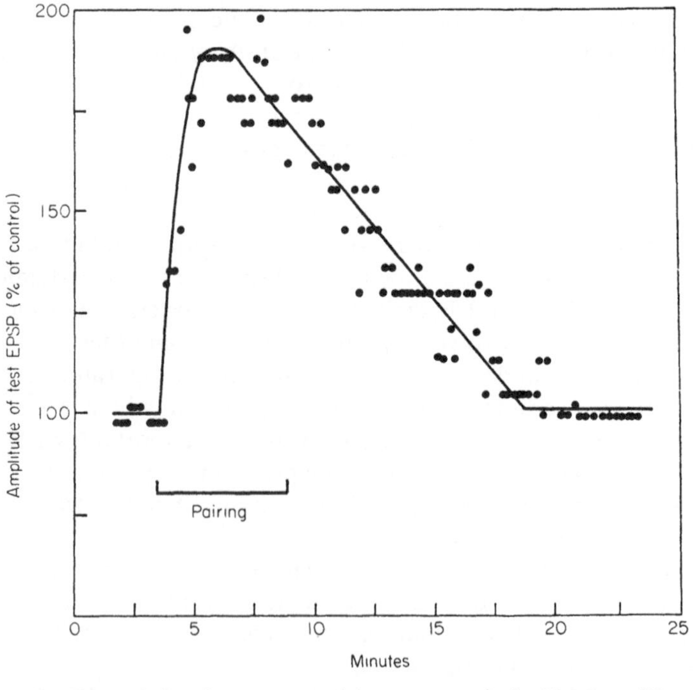

FIG. 11.13. Plot of the time course of heterosynaptic facilitation. (From Kandel & Tauc, 1965a.)

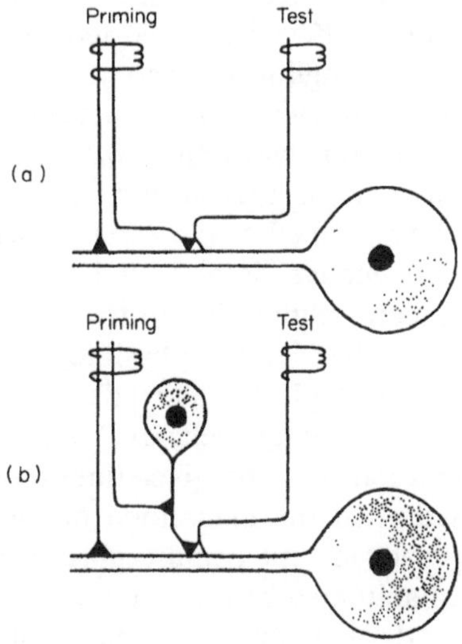

FIG. 11.14. Drawings of two possible circuits that might account for heterosynaptic facilitation: (a) does not but (b) does involve an intermediary neuron. (From Kandel & Tauc, 1965b.)

that the facilitation could not be produced by the test stimulus alone, even greatly enlarged and under repeated activation, suggested that some kind of intermediary synaptic mechanism was involved in those instances in which pairing was effective. This led Kandel and Tauc to believe that the phenomenon of heterosynaptic facilitation is most likely mediated by a neural circuit similar to one shown in Figs. 11.14a and b. Two possible forms of presynaptic interaction are diagrammed in these figures.

In Fig. 11.14a, the presynaptic interaction is in the form of a direct facilitation. The neuron conveying the priming stimulus impinges directly on the presynaptic region of the junction between the neuron conducting the test stimulus (the test neuron) and the neuron in which the response occurs (the response neuron). By repeated activation with the priming stimulus, the synaptic junction between the test neuron and the response neuron acquires a heightened ability to produce postsynaptic potentials in the response cell when stimulated by the test stimulus alone.

Figure 11.14b diagrams a similar sort of neural circuit, except that here an intermediary neuron is presumed to exist between the priming stimulus input and the presynaptic junction on the test neuron. Such an additional neuron in the circuit, Kandel and Tauc note, might have the advantage of maintaining the facilitory response over longer periods of time, due to repetitive firing in this additional interneuron.

Clearly this sort of plasticity in the nervous system is but one of many possible means of altering synaptic effectiveness and thus possibly mediating learning and memory behavior. Moreover, while presynaptic facilitory interconnections resulting from pairing can noticeably affect the response pattern of a single cell, that single cell's increased propensity to fire may have still other effects in a heavily interconnected tree of neurons. One such additional effect was reported by Kristan (1971), who showed that in the pleural ganglion of *Aplysia* a marked increase in the synchrony of the signals from two impaled neurons is produced by a single stimulus when that stimulus was paired with a priming stimulus as described above and shown in Figs. 11.14a and b. This increase in synchrony is thought by Kristan to be mediated by the presence of an interneuron that, though originally ineffective, is activated to provide an additional indirect route of simultaneous stimulation to the two neurons.

There are other known physiological processes in the *Aplysia* that also appear to reflect plastic and adaptive alterations in this animal's simple nervous system. Many "reflexive" responses have been shown to *habituate* under conditions of repetitive stimulation. Habituation refers to a progressive reduction in the magnitude of the response as stimuli are repeated. A closely related, but equally plastic, response is *dishabituation*—the revival of a habituated response upon the presentation of a different or novel stimulus pattern. However, neither habituation nor dishabituation require stimulus pairing as does heterosynaptic facilitation nor even, in the case of dishabituation, repeated presentation of the novel stimulus. Instead, the key factor seems to be some aspect of the "sameness" of the sequence of stimuli.

Were it not for the known existence of dishabituation, habituation might have been viewed as nothing more than a form of fatigue. Exactly what are habituation and dishabituation, in terms of the plastic neural processes that most certainly underlie this sort of adaptive behavior, is not yet definable. Peretz (1970) has shown that habituation can occur for the *Aplysia* gill withdrawal response even when it is severed from the central nervous system, but Lukowiak and Jacklet (1972) have, nothwithstanding Peretz's findings, shown that the *Aplysia* siphon withdrawal response is controlled by interactions between the central ganglion and the peripheral nerves. Clearly, much additional work must be done to explicate the details of this important mechanism.

F. CODING AND DECODING

All of the advantages described that make the *Aplysia* so useful as a model nervous system also make it particularly valuable for studying specific questions concerning neural coding—the representation of information patterns by patterns of neural signals. One of the most fruitful applications of the *Aplysia* preparation has been in helping to determine whether or not synaptic junctions exhibit any special sensitivity to the statistical characteristics of a train of presynaptic spike action potentials. Parameters such as the frequency of the spike train are, of course, known to have substantial

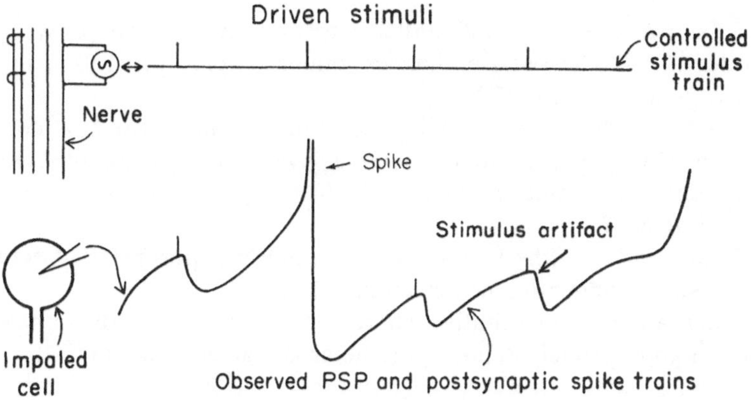

FIG. 11.15. Drawing of the experimental paradigm for the experiments in which the effect of pulse pattern on synaptic conductivity was examined. (From Segundo & Perkel, 1969.)

differential effects, but what about the standard deviation of the intervals between spikes in the train? If the mean frequency is held constant, is the postsynaptic response sensitive to the degree of irregularity of the presynaptic spike train? The answer to this question could partially supply the answer to some very important questions concerning neural coding.

Segundo, Moore, Stensaas, and Bullock (1963), and Segundo and Perkel (1969) have used the familiar abdominal ganglion preparation in what has now become a classic inquiry into the ability of synaptic junctions to decode such higher-order statistical information. The general procedure used in all of their experiments involved a simple monosynaptic, two-neuron link. The presynaptic neuron was used as an input; electrical pulse stimuli were applied through rather large electrodes to one of the nerves leading into the ganglion. Electrical stimulation procedures allowed Segundo and his colleagues to define precisely the sequence and timing of the spike action potentials in the neuron—an achievement not possible had the stimuli been applied via the usual receptor transduction processes. Recordings of postsynaptic and spike action potentials were picked up from the second, postsynaptic neuron by means of the usual microelectrode preamplifier recording system. Both types of electropotential were used as measures of the response in their various experiments. Figure 11.15 shows the stimulating and recording arrangement, along with a schematic representation of the type of electrical responses that were detected.

Segundo and Perkel set out to answer several related but independent questions concerning the effectiveness of various dimensions of the stimulus-driven afferent nerve action potential pattern on the postsynaptic responses. Their list (Segundo & Perkel, 1969, p. 350) of experimental goals included:

a. What are the ". . . influential statistics; that is, the presynaptic interval statistics whose changes influence spike generation in the postsynaptic neurons"?
b. What are ". . . the defining conditions under which a certain statistic is strongly influential at a given junction, as opposed to the defining conditions under which it is poorly influential or not read at all"?
c. What are ". . . the bounds within which an appropriately defined statistic is influential at a given junction"?
d. What are the quantitative characteristics of ". . . the influence of an appropriately defined parameter at a given junction"?

Segundo and Perkel and their colleagues tested a number of different statistical patterns (or, as we have called them elsewhere [Uttal, 1973] dimensions) to determine if and how such patterns might influence synaptic conductivity. Over years of research, they have found that the mean frequency, the standard deviation, and the auto- and cross-correlations between and among two or more afferent neurons are among the dimensions that have some independent and selective ability differentially to effect the postsynaptic response. To illustrate their general approach, let us consider a situation in which the mean frequency and the standard deviation of a stimulus train are dealt with separately. Figure 11.16 shows six stimulus trains, each of which consists of the same number of pulses and lasts for the same duration, but each of which displays a different pattern of pulse spacing. These six stimulus trains, of course, all possess the same mean frequency, but the standard deviation of each pulse train is quite different from the others.

Segundo's determination of the effects of these six patterns indicates that only the patterns C, E, and F are capable of eliciting spike action potentials. The remaining patterns are simply ineffective and, though they produce some postsynaptic potentials, they do not evoke a spike. As a simple comparison, the effect on the postsynaptic intracellular potential of two of the pulse trains is shown in Fig. 11.17. In the first train, pulses are evenly spaced. The second train has the same average frequency but the pulses, instead of being evenly spaced, are paired. Though the average frequency is the same, the paired pulse train is seen to be considerably more effective in eliciting spikes. In the same time period (and with the same number of pulses), the paired stimulus train twice produces a postsynaptic potential that exceeds the horizontal line indicating the threshold for the elicitation of a spike. On the other hand, the postsynaptic potential produced by the evenly spaced train never exceeded this threshold. Therefore, in this case, despite the fact that the energy content of the two trains is the same, the temporal pattern is seen to be a powerful determinant of the response.

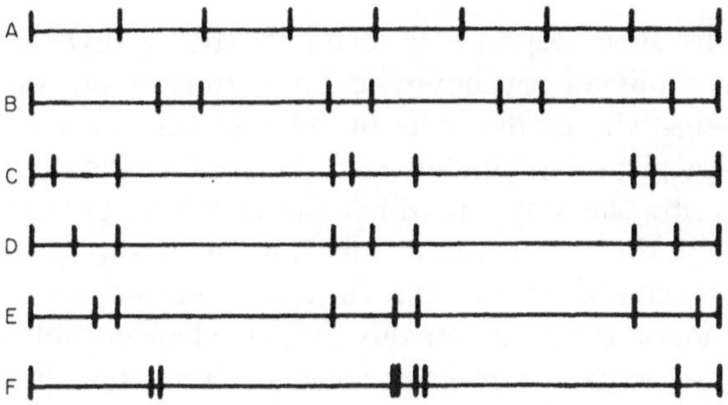

FIG. 11.16. Drawing of a group of pulse stimulus train patterns all of which possess the same mean frequency but differ in the degree of internal irregularity. (From Segundo & Perkel, 1969.)

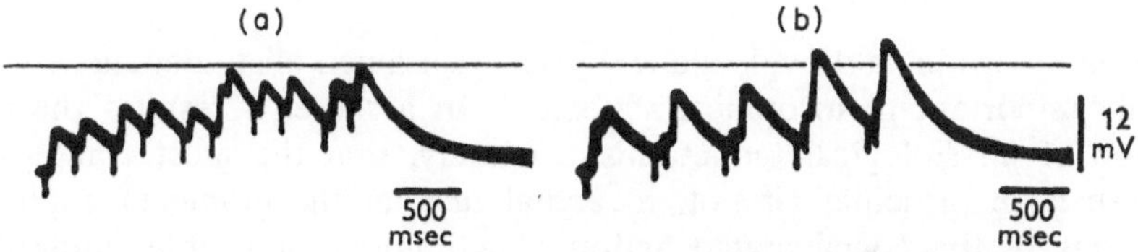

FIG. 11.17. Oscilloscopic display of the results of stimulating with two of the patterns shown in Fig. 11.16: (a) evenly spaced patterns do not exceed the threshold for a spike (indicated by the horizontal line); (b) however, when the stimulus pulses are paired, the spike threshold for the spike action potential is exceeded twice, even though the mean frequency is the same in both cases. (From Segundo & Perkel, 1969.)

Segundo and Perkel have also shown that, while other characteristics of the interval histogram of the presynaptic stimulus pulse train are equally strong determinants of the response of a postsynaptic cell, some neural decoders are idiosyncratic and the nature of the effect will also vary, depending upon the particular type of postsynaptic cell studied. For example, if the postsynaptic cell is a pacemaker, spontaneously beating out some biological rhythm, the effect of the stimulus pattern may be only to modulate that intrinsic rhythm. On the other hand, if the cell is not spontaneously active, the stimulus pulse train may simply be replicated (or demultiplied) by a regular input stimulus.

In another important discovery, working with triplets of stimuli, these investigators were able to show that the temporal range within which these effects occurred was not unlimited. Rather, the effects of the stimulus pattern were observable only if the intervals between the stimulus pulses were within certain sharply defined limits. For example, the *Aplysia* nervous system seems to be maximally sensitive to these sorts of timing patterns when the intervals between pulses are less than .5 second but greater than

.25 second. This slow response is characteristic of invertebrate preparations. Analogous human psychophysical experiments put the limits on interval pattern sensitivity on the order of milliseconds (Uttal & Smith, 1967).

Electrophysiological experiments such as those carried out in Segundo's laboratory describe the ways in which the decoding processes occur, but they do not speak directly to the underlying membrane properties that account for the specific course of the various responses. However, Segundo makes a very important point on this matter when he notes that most of the processes, no matter how complex and subtle the decoding appears to be, may be accounted for in terms of very simple temporal summation mechanisms. No new mechanisms, he says, other than those already well-known to be characteristic of neural membranes, are required to explain the exquisite decoding discriminability exhibited by *Aplysia* postsynaptic cells.

This assertion, although made by an experimentalist, nonetheless is a highly important philosophical statement. In a sense, it restates the basic notion of physiological reductionism, namely, that the most complex response is in principle (if not, in actual fact, at the moment) explicable in terms of the concatenated action of a number of simpler units. This concept of reductionism can also be extended upwards, and these cellular decoding mechanisms, so well modeled by this simple mollusc, may also serve as the building blocks of all that we know as "behavior," including man's intrapersonal self-awareness.

Perhaps, that is not a proper note upon which to conclude this chapter. Certainly, the reader will have realized that, while enormous strides have been made towards understanding the cellular and integrative physiology of the nervous system, we have hardly begun to attack more important questions dealing with the relationship between cellular neurophysiological data and the behavior of higher vertebrates, man among them. The work reviewed here that has accumulated from studies of the *Aplysia* in particular is illuminating and serves as a guidepost for the future. But new technological and conceptual advances are required before similar analyses can be applied to higher organisms. Some of the very basic premises of the work on *Aplysia* and other model preparations may ultimately turn out to be not even first approximations of what we shall find in vertebrate preparations. It has yet to be demonstrated, for example, that the relationship between specifically identified cells and specific functions, such as found in this simple invertebrate preparation, is also characteristic of the vertebrate nervous system. It is quite conceivable that completely different organizing principles may apply in more complex brains, and that models based on our elementary friend, the *Aplysia,* may be too simplistic to answer the questions man asks of himself. The process analogies that seem so promising now may prove not to be the structural homologies we desire.

12
A SUMMARY OF GENERAL PRINCIPLES

We have now completed an interpretive and introductory survey of the science of cellular neurophysiology and integration. The material has been presented in a way that, hopefully, emphasizes the meaning of the various experiments while still making clear the procedures and results. The book has been organized in such a way that a conceptual thread is maintained, progressing from the simple to the more complex and from the periphery to the central nervous systems. Necessarily, in presenting a body of knowledge of such complexity, the material must be spread over many pages and, probably, many separated hours of the reader's time. Very often the effect of this dispersion is to dissipate what, perhaps, may be the most important content of a text. That essential content is, of course, the general perspective that emerges as one is exposed to the many details of the experimental material. In other words, the "trees" of the experiments obscure the "forest" of a general theoretical overview. It may also be the case that each individual experiment is not of great importance; any one of their number might make the same point—as, indeed, it must if a body of knowledge is to become a structured science rather than remain only a disordered collection of isolated and trivial facts.

For these reasons, it would be useful to summarize what are, in the author's opinion, the most important general concepts of cellular neurophysiology. These will be stated in a list of brief recapitulations. No attempt will be made to justify these principles anew (if that was not done adequately to this point, the book has failed its intent). The principles will simply be presented *ex cathedra*. This particular list of general conclusions, of course, reflects the personal biases, interests, and experiences of the author as well as the empirical findings surveyed in this text. Nevertheless, we hope that our perspective is relatively consistent with that of the rest of the neurophysiological community, and that this list will be a fair statement of what most neurophysiologists would agree are the most important principles of the science today.

1. Cellular neurophysiology is the study of the function, structure, and biochemistry of neurons—the cells of an organism that are specialized for the transduction, communication, and integration of stimulus information.

2. Though the science of cellular neurophysiology is an interesting and exciting field of inquiry in its own right, recent enthusiasm has been directed at its implications for establishing a reductionistic basis for molar behavior. In G. A. Horridge's words, "The aim of neurobiology is the explanation of behavior."

3. This aspect of the goals of the science of neurophysiology is founded upon a specific philosophical statement about the nature of the mind. That statement is the assertion of structural monism: All aspects of mind, consciousness, behavior, and performance are nothing more (nor less) than the function of the tissues that make up the nervous system. Though many neurophysiologists compartmentalize their personal beliefs and their scientific work, this monistic premise is implicit in the work of all contemporary researchers in the field.

4. Modern neuron theory assumes a system of discrete neurons that are not protoplasmically interconnected but that interact through specific points of contact known as synapses.

5. Neurons are specialized cells characterized by a high degree of electrochemical reactivity. Their function in the nervous system is primarily communication and integration of information patterns, but it is becoming increasingly evident that many neurons also play a secretory role.

6. Neurons carry out their communication and integration functions by means of electrochemical reactions. The flow and concentration of a few ions (sodium, potassium, and chloride) mainly define the direction and magnitude of the voltages measured during any neural response.

7. Neurons respond in a number of different ways. Both regenerative spike action potentials and graded potentials are to be found at different points on the cell. Some of these responses are activated by mechanisms intrinsic to the cell (e.g., spontaneous or pacemaker action potentials), while others are produced by outside stimulation (e.g., receptor or postsynaptic action potentials).

8. The category of "action potential" (contrary to popular usage) includes responses other than the spike action potential. In fact, any deviation from the potential level of the resting membrane that is not related

to the metabolic state of the membrane may be considered to be an information-bearing action potential. Specifically included in the category of action potentials are nonregenerative graded potentials.

9. The electrical signals picked up from neurons are correlates of the dynamic and steady-state distributions of the involved ions. However, the electrical signal should not be considered to be the primary aspect of neural activity. We are now learning that the electrochemical action of the neuron is more chemical than it is electrical. Our fascination with the electrical concomitants is due to the fact that our recording technology is more sensitive to volts than to ions.

10. Although the electrical correlates of nervous activity have been guessed at for several hundred years, it was not until the introduction of modern electronic recording instruments that a specific understanding of the electrochemistry of neurons was achieved. Both the anatomical oddity represented by the giant squid axon and the invention of the intracellular microelectrode contributed greatly to the attainment of this understanding.

11. Electrophysiological instruments have had a dual influence on the development of both the theory and data of neurophysiology. In many cases a particular device has opened the door to detailed understanding of one or another process while at the same time inhibiting the development of a broad conceptual understanding of the action of neurons. This limitation has been especially unfortunate with regard to multicellular processes, since so many of our instruments are sensitive only to the temporal response of single points in space.

12. The electrochemical events underlying nervous action are currently assumed to be most directly mediated by the neuronal or plasma membrane. This assumption is known as the membrane theory. In particular, variations in the impedance of the membrane to the flow of sodium, potassium, and chloride ions seem to explain much of what we know of neural processes.

13. Neurons exhibit a wide variety of anatomic forms, depending upon their particular function. Transmission neurons typically possess elongated axons, whereas integrating neurons more often display extensive arborization of the dendritic trees. Receptor neurons may or may not have elongated axons. Generalizations about anatomic form, however, usually turn out to have many exceptions.

14. Receptors are tuned to respond to vanishingly small amounts of a given type of physical energy—their adequate stimulus—and in some cases are as sensitive as, theoretically, they can be (e.g., a rod of the vertebrate retina seems to produce a response to a single quantum of light).

15. Many receptors exhibit ultramicroscopic structures at their synaptic junctions that are suggestive of both afferent and efferent synaptic

interconnections. Information, therefore, may travel both into and out of a receptor, and neuronal integration probably occurs even in these most peripheral units.

16. Many receptors have elongated or ciliated structures. This characteristic feature produces two advantages: the protrusion of the receptive tissue into an area in which it is likely to encounter stimuli, and an increase in the area of the sensitive surfaces.

17. Some receptors (e.g., the rod of the vertebrate eye and the receptor cells of the taste bud) are continually in a dynamic process of decay and regeneration.

18. Transmission neurons can conduct in both the orthodromic and the antidromic directions. However, synapses are not, in general, bidirectional. Synaptic rectification, therefore, is the main basis of a directional nervous system. The refractory period following a spike response also helps to make the nervous system directional by preventing retrofiring.

19. Transmission neurons often are covered with a myelin sheath composed of multiple layers of Schwann cell membranes. This myelin sheath acts to produce a rapid form of neural transmission by allowing the ionic currents to skip between nodes of Ranvier—the discontinuities in the sheath.

20. The plasma or cell membrane of a neuron is a region of demarcation between the inside and the outside of the cell. Since the membrane can just be resolved at the limits of electronmicroscopy, there is no consensus concerning its structure. A number of competing theories of membrane structure are current. They all involve a combination of lipid and protein molecules, perhaps arranged as a bilayer of lipid molecules penetrated by protein molecules.

21. An ionic concentration equilibrium is established across the membrane by a metabolically driven pumping action operating in a direction opposite to the passive forces of diffusion, osmosis, and potential driving. The establishment of ion concentration difference by the "pump" produces a resting potential characteristic of the unactivated neuron.

22. The exact nature of the "pump" itself is not yet fully understood, but it does appear to pump out of the neuron at least two ions of sodium (creating a sodium deficiency inside the cell) for each ion of potassium pumped in (creating a potassium ion surplus inside the cell). Chloride ion concentrations seem to follow more or less passively and are determined by the final resting potential, although the existence of a separate chloride pump has also been suggested.

23. The transmembrane channels, through which sodium and potassium ions flow, appear to be different, since it is possible to inhibit selectively one or the other with various pharmacological agents.

24. All deviations from the resting potential are the direct result of ion concentration shifts caused by changes in the permeability of the membrane.

25. In general, all receptor processes also involve an alteration in the receptor membrane permeability, so that the cell's resting potential is changed. The resulting receptor or generator graded potential is a necessary precursor of subsequent propagated spike activity.

26. Each receptor possesses some sort of specific trigger action, or primary sensory action, which leads directly to a change in the membrane permeability and thence to the generator or receptor potential.

27. Neural transmission may occur by either of two processes. The first is a passive, decremental spread of electrical signals by ionic currents. This process is suitable only for transmission over very short distances. The second is the propagation of regenerative spike action potentials. This is a nondecremental process, suitable for communication over very long distances.

28. Spike action potentials are produced by a sequential, but independent, breakdown in the membrane's permeability, first to sodium and then, a millisecond or so later, to potassium. The increase in sodium flow, according to the Hodgkin-Huxley theory, accounts for the leading edge of the spike, while the increase in potassium flow accounts for the rapid recovery of the spike back to almost the resting potential level.

29. Long-term recovery of ion concentrations to resting potential levels, however, appears to be dependent upon an extended period of sodium and potassium pumping.

30. The triggering of the breakdown in membrane permeability that leads to a spike is a threshold phenomenon. Once a local depolarization has achieved that threshold, the remainder of the process is spontaneous and requires no further introduction of extracellular energy. Since the process produces larger electropotentials than those needed to trigger it, the spike is able to propagate to distant locations without decrement.

31. The amplitude of a spike is independent of the stimulus; it is a function only of the state of the neuron. This fact is known as the all-or-none law.

32. The speed of spike propagation is a function of the size and degree of myelinization of the neuron as well as of its general state.

33. Compound and pooled potentials reflect the sum of the responses of many neurons. If the number of responses is few, then the individual spikes can be seen in a pooled record. If not, the continuous waveform, characteristic of a compound action potential, will be produced. The duration of the compound response depends upon the degree of synchronization and the nature of the contributing action potentials.

34. A mathematical proof by Cox and Smith has shown that pooled spike action potentials cannot be separated solely on the basis of the intervals between successive spikes. Any attempt to use a computer analyzer to achieve this goal would be fruitless. Some success is possible in special cases, if amplitude information is included in the analytic criteria.

35. Compound responses often can be detected by computer averaging techniques, even though such responses are considerably smaller than the electrical noise in which they are embedded.

36. Compound brain potentials can be shown to be related to the spike responses of individual neurons. Whether this is due to the fact that the spikes electrically summate with each other to form the compound potential, or whether both are derived functions of some other, more fundamental process, is not yet known.

37. Hypersynchronization of peripheral nerve spike action potentials into a compound action potential by means of "nonphysiological" electrical pulse stimulation introduces no artifacts that could distort comprehension of the coding mechanisms of the individual axons.

38. Both electrical and chemical means of synaptic transmission are known. The two types of synapse differ in the spacing between the involved neurons, the rapidity of synaptic transmission, the degree of electrical coupling, and the relative sizes of the pre- and postsynaptic regions.

39. Chemically mediated synapses work by diffusion of transmitter molecules secreted by the presynaptic region to receptor sites on the postsynaptic neuron.

40. A single neuron does not produce more than a single kind of transmitter substance. This fact is known as Dale's principle.

41. A single transmitter substance may produce either inhibitory or excitatory responses, depending upon the nature of the postsynaptic receptor sites.

42. The transmitter substance appears to be bundled into packets of several hundred molecules each and produces quantal miniature postsynaptic responses associated with the arrival of the contents of individual packets at the receptor sites. These miniature quantal responses, however, may sum with each other to produce postsynaptic potentials of virtually continuously graded amplitude.

43. Excitatory postsynaptic responses appear to be always depolarizations, but inhibitory potentials may be either hyper- or depolarizing, depending on the level of the resting potential. In the vertebrate retina, however, some excitatory potentials are hyperpolarizing. Therefore, no simple relationship exists between excitatory and inhibitory responses, on the one hand, and hyperpolarizing or depolarizing membrane potentials, on the other.

44. Inhibitory postsynaptic responses generally seem to involve an increase in the membrane's permeability to potassium and chloride ions. Excitatory postsynaptic responses seem to involve changes in the permeability of the membrane to sodium and potassium ions, changes which are similar to those occurring in other membrane actions. However, there are notable differences between action responses in synapses and axons.

45. Neural integration, occurring primarily at synaptic junctions, represents what is perhaps the most significant neurobiological process. It is the basis of all adaptive behavior and, thus, of all cognitive functions. Neural integration is defined as the cumulative effect of multiple inputs such that the conducted information pattern is modified in an adaptive manner.

46. Integration is a property of most neurons, including receptors and motor units that receive multiple inputs, as well as interneurons, the latter being defined as "neurons that connect neurons to neurons."

47. A number of processes that tend to modulate the flow of information occur within the neuron. The major integrative factors modulating information flow, however, occur at the synapses. Some of these synaptic processes mimic some attributes of learning at the molar level.

48. Several physical alterations that occur at the synapse have been suggested to account for the plastic behavior of neurons and, perhaps, of molar learning. These changes include swelling of dendritic boutons, growth of new synaptic boutons, decreases in synaptic resistance, increases in synaptic conductivity as a result of the availability of additional transmitter substance, and selection of a few synapses from an original pool of many. Post-tetanic potentiation and heterosynaptic facilitation are examples of the synaptic efficiency changes that might result from these physical alterations.

49. Because of the complexity of mammalian nerve nets and the small size of the constituent neurons, model preparations (usually invertebrate) have been used to explicate the integrative action of nervous tissue. Among the many model preparations, the *Aplysia,* an Opisthobranchian mollusk, has been especially useful because of the small number and large size of the neurons comprising its simple ganglia.

50. A number of cellular processes that are similar to the various aspects of molar behavior have been observed in such model preparations. Whether these processes are structural homologues or mere process analogues of the behavior is not certain. Nevertheless, these processes do serve, at least, as conceptual models of a nervous function that might underlie the organism's behavior. At best, they represent insights into the cellular basis of human experience that are of utmost importance.

This last statement in the summary is fraught with import. The frontier of contemporary psychobiological research lies exactly in this area. Have we finally made the breakthrough so long awaited, and are we now really beginning to unlock the cellular basis of human experience? Or, on the other hand, have we merely set up another situation in which an analogous process is assumed to explain more than it really can?

If there is a final comment on which it is most appropriate to end this book, it would be the caveat that this great philosophical issue—the relation between the microscopic responses of the neuron and the macroscopic responses of the organism—is yet unresolved. Though a considerable amount of technical information has been gained in regard to the details of neural function, we still must face the grander issue of what it all means. The search for a solution to this problem may well represent the major intellectual challenge of the next century.

BIBLIOGRAPHY

Aidley, D. J. *The physiology of excitable cells.* New York: Cambridge University Press, 1971.

Albers, R. W., Siegel, G. T., Katzman, R., & Agranoff, B. (Eds.), *Basic neurochemistry.* Boston: Little, Brown, 1972.

Amassian, V. E., Waller, H. J., & Macy, J., Jr. Neural mechanism of the primary somatosensory evoked potential. *Annals of the New York Academy of Sciences,* 1964, **112,** 5–32.

Amatniek, E. Measurement of bioelectric potentials with micorelectrodes and neutralized input capacity amplifiers. *I.R.E. Transaction Medical Electronics,* 1958, **PGME10,** 3–14.

Amoore, J. E. The stereochemical specificities of human olfactory receptors. *Perfumery and Essential Oil Record,* 1952, **43,** 321–323.

Amoore, J. E. Elucidation of the stereochemical properties of the olfactory receptor sites. *Proceedings of the Scientific Section of the Toilet Goods Association,* 1962 (October), Special Supplement to #37, 13–23.

Amoore, J. E. Current status of the steric theory of odor. *Annals of the New York Academy of Sciences,* 1964, **116,** 457–476.

Amoore, J. E. Psychophysics of odor. *Cold Spring Harbor Symposium on Quantitative Biology,* 1965, **30,** 623–636.

Amoore, J. E. *Molecular basis of odor.* Springfield, Ill.: Charles C Thomas, 1970.

Amoore, J. E., Johnston, J. W., Jr., & Rubin, M. The stereochemical theory of odor. *Scientific American,* 1964, **210,** 42–49.

Amoore, J. E., & Venstrom, D. Correlations between stereochemical assessments and organoleptic analysis of odorous compounds. In T. Hayashi (Ed.), *Olfaction and taste II.* Oxford: Pergamon Press, 1967. Pp. 3–17.

Andres, K. H. Discussion. In G. E. W. Wolstenholme & J. Knight (Eds.), *Taste and smell in vertebrates: A Ciba foundation symposium.* London: J. & A. Churchill, 1970. Pp. 25–30.

Armstrong, C. M., & Bezanilla, F. Currents related to movement of the gating particles of the sodium channels. *Nature,* 1973, **242,** 459–461.

Arvanitaki, A., & Chalazonitis, N. Potentiels d'activité du soma neuronique géant (*Aplysia*). *Archives des Sciences Physiologiques,* 1955, **9,** 115–144.

Arvanitaki, A., Takeuchi, H., & Chalazonitis, N. Specific unitary osmereceptor potentials and spiking patterns from giant nerve cells. In T. Hayashi (Ed.), *Olfaction and taste II.* Oxford: Pergamon Press, 1967. Pp. 573–598.

Ascher, P. Inhibitory and excitatory effects of dopamine on *Aplysia* neurones. *Journal of Physiology,* 1972, **225,** 173–200.

Baker, P. F., Hodgkin, A. L., & Shaw, T. I. Replacement of the axoplasm of giant nerve fibres with artificial solutions. *Journal of Physiology,* 1962, **164,** 330–354. (a)

Baker, P. F., Hodgkin, A. L., & Shaw, T. I. The effects of changes in internal ionic concentrations on the electrical properties of perfused giant axons. *Journal of Physiology,* 1962, **164,** 355–374. (b)

Bard, P. Emotion. In C. Murchison (Ed.), *Foundations of experimental psychology.* Worcester, Mass.: Clark University Press, 1929.

Barlow, H. B., & Levick, W. R. The mechanism of directionally selective units in rabbit's retina. *Journal of Physiology,* 1965, **178,** 477–504.

Barlow, H. B., Narasimhan, R., & Rosenfeld, A. Visual pattern recognition in machines and animals. *Science,* 1972, **177,** 567–575.

Beidler, L. M. Anion influences on taste receptor response. In T. Hayashi (Ed.), *Olfaction and taste II.* Oxford: Pergamon Press, 1967. Pp. 509–534.

Bennett, M. V. L., Aljure, E., Nakajima, Y., & Pappas, G. D. Electrotonic junctions between teleost spinal neurons: Electrophysiology and ultrastructure. *Science,* 1963, **141,** 262–264.

Bernard, M. C. *Lessons on the physiology and pathology of the nervous system.* London: Baillière Tindall Ltd., 1858.

Bernstein, J. Untersuchungen zur Thermodynamik der bioelektrischen Strome. Erster Theil. *Pflueger's Archives of the European Journal of Physiology,* 1902, **92,** 521.

Bezanilla, F., & Armstrong, C. M. Gating currents of the sodium channels: Three ways to block them. *Science,* 1974, **183,** 753–754.

Birks, R., Huxley, H. E., & Katz, B. The fine structure of the neuromuscular junction of the frog. *Journal of Physiology,* 1960, **150,** 134–144.

Blankenship, J. E., Wachtel, H., & Kandel, E. R. Ionic mechanisms of excitatory, inhibitory, and dual synaptic actions mediated by an identified interneuron in abdominal ganglion of *Aplysia. Journal of Neurophysiology,* 1971, **34,** 76–92.

Bodian, D. Introductory survey of neurons. *Cold Spring Harbor Symposium on Quantitative Biology,* 1952, **17,** 1–13.

Bodian, D. The generalized vertebrate neuron. *Science,* 1962, **137,** 323–326.

Bodian, D. Neurons, circuits, and neuroglia. In G. C. Quarton, T. Melnechuk, & F. O. Schmitt (Eds.), *The neurosciences: A study program.* New York: Rockefeller University Press, 1967. Pp. 6–24.

Boeckh, J. Inhibition and excitation of single insect olfactory receptors, and their role as a primary sensory code. In T. Hayashi (Ed.), *Olfaction and taste II.* Oxford: Pergamon Press, 1967. Pp. 721–735.

Bradbury, S. *The evolution of the microscope.* Oxford: Pergamon Press, 1967.

Brazier, M. A. B. *The electrical activity of the nervous system: A textbook for students.* London: Pitman & Sons, 1951.

Brazier, M. A. B. The historical development of neurophysiology. In J. Field (Ed.), *Handbook of physiology. Section I: Neurophysiology.* Vol. I. Washington, D.C.: American Physiological Society, 1959. Pp. 1–58.

Bretscher, M. S. Membrane structure: Some general principles. *Science,* 1973, **181,** 622–629.

Brown, J. E. Rat retinal ganglion cells: Receptive field organization and maintained activity. *Journal of Neurophysiology,* 1965, **28,** 1091–1100.

Brown, K. T., & Murakami, M. A new receptor potential of the monkey retina with no detectable latency. *Nature,* 1964, **201,** 626–628.

Brown, P. K., & Wald, G. Visual pigments in single rods and cones of the human retina. *Science,* 1964, **144,** 145–151.

Bullock, T. H. Neuron doctrine and electrophysiology. *Science,* 1959, **129,** 997–1002.

Bullock, T. H., & Horridge, G. A. (Eds.) *Structure and function in the nervous system of invertebrates.* San Francisco: Freeman, 1965.

Bunge, R. P. Glial cells and the central myelin sheath. *Physiological Review,* 1968, **48,** 197–251.

Butcher, L. L., & Fox, S. S. Motor effects of copper in the caudate nucleus: Reversible lesions with ion exchange resin beads. *Science,* 1968, **160,** 1237–1239.

Caldwell, P. C., Hodgkin, A. L., Keynes, R. D., & Shaw, T. I. The effects of injecting 'energy-rich' phosphate compounds on the active transport of ions in the giant axons of *Loligo. Journal of Physiology,* 1960, **152,** 561.

Cannon, W. B. *Bodily changes in pain, hunger, fear and rage.* New York: D. Appleton & Co., 1915.

Cannon, W. B., & Bacq, Z. M. Studies on the conditions of activity in endocrine organs. XXVI. A hormone produced by sympathetic action on smooth muscle. *American Journal of Physiology,* 1931, **96,** 392–412.

Carpenter, D. O., & Alving, B. O. A contribution of an electrogenic Na^+ pump to membrane potential in *Aplysia* neurons. *Journal of General Physiology,* 1968, **52,** 1–21.

Casby, J. U., Siminoff, R., & Houseknecht, T. R. An analogue cross-correlator to study naturally induced activity in intact nerve trunks. *Journal of Neurophysiology,* 1963, **26,** 432–448.

Cauna, N. The free penicillate endings of the human hairy skin. *Journal of Anatomy,* 1973, **115,** 277–288.

Chalazonitis, N., & Arvanitaki, A. Identification et localisation de quelques catalyseurs respiratoires dans le neurone d'*Aplysia.* Institute Océanographique. *Bulletin,* 1951, **996,** 1–20.

Chiarandini, D. J., Stefani, E., & Gerschenfeld, H. M. Ionic mechanisms for cholinergic excitation in molluscan neurons. *Science,* 1967, **156,** 1597–1599.

Cole, K. S. Dynamic electrical characteristics of the squid axon membrane. *Archives des Sciences Physiologiques,* 1949, **3,** 253–258.

Cole, K. S. *Membranes, ions, and impulses.* Berkeley: University of California Press, 1968.

Cone, R. A. Early receptor potential: Photoreversible charge displacement in rhodopsin. *Science,* 1967, **156,** 536 (abstract).

Cone, R. A., & Brown, P. K. Dependence of the early receptor potential on the orientation of rhodopsin. *Science,* 1967, **156,** 536 (abstract).

Cone, R. A., & Cobbs, W. H., III. Rhodopsin cycle in the living eye of the rat. *Nature,* 1969, **221,** 820–822.

Conger, A. C., & Wells, M. A. Radiation and aging effect on taste, structure and function. *Radiation Research,* 1969, **37,** 31–49.

Coombs, J. S., Eccles, J. C., & Fatt, P. The specific ionic conductances and the ionic movements across the motoneuronal membrane that produce the inhibitory post-synaptic potential. *Journal of Physiology,* 1955, **130,** 326–373.

Couteaux, R. Morphological and cytochemical observations on the post-synaptic membrane at motor end-plates and ganglionic synapses. *Experimental Cellular Research,* 1958, Suppl. 5, 294–322.

Cox, D. R., & Smith, W. L. On the superposition of renewal processes. *Biometrika,* 1954, **41,** 91–99.

Curtis, H. J., & Cole, K. S. Membrane resting and action potentials from the squid giant axon. *Journal of Cellular and Comparative Physiology,* 1942, **19,** 135–144.

Dale, H. H., & Feldberg, W. Chemical transmission at motor nerve endings in voluntary muscle? *Journal of Physiology,* 1934, **81,** 39P.

Dale, H. H., Feldberg, W., & Vogt, M. Release of acetylcholine at voluntary motor nerve endings. *Journal of Physiology,* 1936, **86,** 353–380.

Dallos, P., Billone, M. C., Durrant, J. D., Wang, C.-y., & Raynor, S. Cochlear inner and outer hair cells: Functional differences. *Science,* 1972, **177,** 356–358.

Dastoli, F. R. On comparison of interactions of a "bitter-sensitive protein" from porcine tongues with human taste thresholds. *Nature,* 1969, **223,** 524–525.

Dastoli, F. R., Lopiekes, D. V., & Price, S. A sweet-sensitive protein from bovine taste buds. Purification and partial characterization. *Biochemistry,* 1968, **7,** 1160–1164.

Dastoli, F. R., & Price, S. Sweet-sensitive protein from bovine taste buds: Isolation and assay. *Science,* 1966, **154,** 905–907.

Davies, J. T. A theory of the quality of odours. *Journal of Theoretical Biology,* 1965, **8,** 1–7.

Davies, J. T. Recent developments in the "penetration and puncturing" theory of odour. In G. E. W. Wolstenholme & J. Knight (Eds.), *Taste and smell in vertebrates.* London: J. & A. Churchill, 1970. Pp. 265–291.

Davis, H. A model for transducer action in the cochlea. *Cold Spring Harbor Symposium on Quantitative Biology,* 1965, **30,** 181–190.

Davis, H., & Silverman, S. R. *Hearing and deafness.* (3rd ed.) New York: Holt, Rinehart & Winston, 1970.

Davis, L., Jr., & Lorente de No, R. Contribution to the mathematical theory of the electrotonus. In R. Lorente de No, *A study of nerve physiology.* (Studies from the Rockefeller Institute for Medical Research, Vol. 131.) New York: Rockefeller University Press, 1947.

Davson, H., & Danielli, J. F. *The permeability of natural membranes.* New York: Cambridge University Press, 1952.

Dawson, G. D. Cerebral responses to nerve stimulation in man. *British Medical Bulletin,* 1950, **6,** 326–329.

Dawson, G. D. A summation technique for the detection of small evoked potentials. *Electroencephalography and Clinical Neurophysiology,* 1954, **1,** 65–84.

Dawson, G. D., & Scott, J. W. The recording of nerve action potentials through skin in man. *Journal of Neurology, Neurosurgery, and Psychiatry,* 1949, **12,** 259–273.

de Lorenzo, A. J. Electron microscopy of the olfactory and gustatory pathways. *Annals of Otology, Rhinology, and Laryngology,* 1960, **69,** 410–420. (a)

de Lorenzo, A. J. The fine structure of synapses in the ciliary ganglion of the chick. *Journal of Biophysics, Biochemistry, and Cytology,* 1960, **7,** 31–36. (b)

de Lorenzo, A. J. Studies on the ultrastructure and histophysiology of cell membranes, nerve fibers and synaptic junctions in chemoreceptors. In Y. Zotterman (Ed.), *Olfaction and taste.* Oxford: Pergamon Press, 1963. Pp. 5–17.

De Robertis, E. Some observations on the ultrastructure and morphogenesis of photoreceptors. *Journal of General Physiology,* 1960, **43** (suppl.), 1.

De Robertis, E. Molecular biology of synaptic receptors. *Science,* 1971, **171,** 963–971.

De Robertis, E., & Bennett, H. S. Submicroscopic vesicular component in the synapse. *Federation Proceedings,* 1954, **13,** 35 (abstract).

Del Castillo, J., & Katz, B. Quantal components of the end-plate potential. *Journal of Physiology,* 1954, **124,** 560–573.

Donaldson, P. E. K. *Electronic apparatus for biological research.* New York: Academic Press, 1958.

Dzendolet, E. A structure common to sweet-evoking compounds. *Perception and Psychophysics,* 1968, **3,** 65–68.

Eccles, J. C. *The neurophysiological basis of mind.* Oxford: The Clarendon Press, 1956.

Eccles, J. C. *The physiology of synapses.* Berlin: Springer-Verlag, 1964.

Eccles, J. C., Eccles, R. M., Iggo, A., & Lundberg, A. Electron-physiological investigations of Renshaw cells. *Journal of Physiology,* 1961, **159,** 461–478.

Engström, H., Ades, H. W., & Hawkins, J. E., Jr. Cellular pattern, nerve structures, and fluid spaces of the organ of corti. In W. D. Neff (Ed.), *Contributions to sensory physiology.* Vol. I. New York: Academic Press, 1965.

Engström, H., Bergström, B., & Ades, H. W. Macula utriculi and macula sacculi in the squirrel monkey. *Acta Oto-Larynogologica,* 1972, Supplement 301.

Epstein, R., & Tauc, L. Heterosynaptic facilitation and post-tetanic potentiation in *Aplysia* nervous system. *Journal of Physiology,* 1970, **209,** 1–23.

Erlanger, J., & Gasser, H. S. Action currents of individual fibers. In J. F. Fulton & L. G. Wilson (Eds.), *Selected readings in the history of physiology.* Springfield, Ill.: Charles C Thomas, 1924. Pp. 242–246.

Erlanger, J., & Gasser, H. *Electrical signs of nervous activity.* Philadelphia: University & L. G. Wilson (Eds.), *Selected readings in the history of physiology.* Springfield, of Pennsylvania Press, 1937.

Fatt, P., & Katz, B. Spontaneous subthreshold activity at motor nerve endings. *Journal of Physiology,* 1952, **117,** 109–128.

Fox, S. S., & Norman, R. J. Functional congruence: An index of neural homogeneity and a new measure of brain activity. *Science,* 1968, **159,** 1257–1259.

Fox, S. S., & O'Brien, J. H. Duplication of evoked potential waveform by curve of probability of firing of a single cell. *Science,* 1965, **147,** 888–890.

Frank, K., & Fuortes, M. C. G. Presynaptic and postsynaptic inhibition of monosynaptic reflexes. *Federation Proceedings,* 1957, **16,** 39–40 (abstract).

Frazier, W. T., Kandel, E. R., Kupfermann, I., Waziri, R., & Coggeshall, R. E. Morphological and functional properties of identified neurons in the abdominal ganglion of *Aplysia californica. Journal of Neurophysiology,* 1967, **30,** 1288–1351.

Frazier, W., Waziri, R., & Kandel, E. R. *Two mechanisms for disinhibition in identifiable neurons in Aplysia.* Boston: Harvard Medical School, 1965.

Friedman, L., & Miller, J. G. Odor incongruity and chirality. *Science,* 1971, **172,** 1044–1046.

Fullerton, P. M., & Gilliatt, R. W. Axon reflexes in human motor nerve fibres. *Journal of Neurology, Neurosurgery, and Psychiatry,* 1965, **28,** 1–11.

Fulton, J. F., & Wilson, L. G. *Selected readings in the history of physiology.* Springfield, Ill.: Charles C Thomas, 1966.

Galvani, L. *Commentary on the effect of electricity on muscular motion* (as translated by R. M. Green). New Haven, Conn.: Elizabeth Licht, 1953.

Gasser, H. S. Recruitment of nerve fibers. *American Journal of Physiology,* 1938, **121,** 193–202.

Gasser, H. S., & Grundfest, H. Axon diameter in relation to the spike dimensions and the conduction velocity in mammalian A fibers. *American Journal of Physiology,* 1939, **127,** 393–414.

Gourlay, K., Uttal, W. R., & Powers, M. VRS—A programming system for visual neuron receptive field research. *Behavior Research Methods and Instrumentation,* 1974, **6,** 281–287.

Graham, J., & Gerard, R. W. Membrane potentials and excitation of impaled single muscle fibers. *Journal of Cellular and Comparative Physiology,* 1946, **28,** 99–117.

Grundfest, H. Electrical inexcitability of synapses and some consequences in the central nervous system. *Physiological Reviews,* 1957, **37,** 337–361.

Hagins, W. A., Penn, R. D., & Yoshikami, S. Dark current and photocurrent in retinal rods. *Biophysical Journal,* 1970, **10,** 380–410.

Hámori, J., & Szentágothai, J. The Purkinje cell baskets: Ultrastructure of an inhibitory synapse. *Acta-Biologica* (Academiae Scientiarum Hungaricae), 1965, **15,** 465–479.

Hartline, H. K., Wagner, H. G., & Ratliff, F. Inhibition in the eye of *Limulus. Journal of General Physiology,* 1956, **39,** 651–673.

Hensel, H., Ström, L., & Zotterman, Y. Electrophysiological measurements of depth of thermoreceptors. *Journal of Neurophysiology,* 1951, **14,** 423–429.

Hernández-Peón, R. Neurophysiological correlates of habituation and other manifestations of plastic inhibition. *Electroencephalography and Clinical Neurophysiology,* 1960, Suppl. 13, 101–114.

Hille, B. Ionic channels in nerve membranes. In J. A. V. Butler & D. Noble (Eds.), *Progress in biophysics and molecular biology.* Vol. 21. Oxford: Pergamon Press, 1970. Pp. 3–32.

Hillman, D. E., & Lewis, E. R. Morphological basis for a mechanical linkage in otolithic receptor transduction in the frog. *Science,* 1971, **174,** 416–419.

Höber, R. Eine Methode, die elektrische Leitfähigkeit im Innern von Zellen zu messen. *Archive fuer die gesamte Physiologie,* 1910, **133,** 237–259.

Höber, R. Ein zweites Verfahren, die Leitfähigkeit im Innern von Zellen zu messen. *Archiv fuer die gesamte Physiologie,* 1912, **148,** 189–221.

Hodgkin, A. L. Ionic movements and electrical activity in giant nerve fibres. *Proceedings of the Royal Society* (London), 1958, Series B, **148,** 1–37.

Hodgkin, A. L. *The conduction of the nervous impulse.* Springfield, Ill.: Charles C Thomas, 1967.

Hodgkin, A. L., & Huxley, A. F. Action potentials recorded from inside a nerve fibre. *Nature,* 1939, **144,** 710.

Hodgkin, A. L., & Huxley, A. F. The components of membrane conductance in the giant axon of *Loligo. Journal of Physiology,* 1952, **116,** 473. (a)

Hodgkin, A. L., & Huxley, A. F. A quantitative description of membrane current and its application to conduction and excitation in nerve. *Journal of Physiology,* 1952, **117,** 500–544. (b)

Hodgkin, A. L., Huxley, A. F., & Katz, B. Measurement of current-voltage relations in the membrane of the giant axon of *Loligo. Journal of Physiology,* 1952, **116,** 424.

Hodgkin, A. L., & Keynes, R. D. The potassium permeability of a giant nerve fibre. *Journal of Physiology,* 1955, **128,** 61–88.

Horn, G., Rose, S. P. R., & Bateson, P. P. G. Experience and plasticity in the central nervous system. *Science,* 1973, **181,** 506–514.

Horridge, G. A. *Interneurons: Their origin, action, specificity, growth, and plasticity.* San Francisco: Freeman, 1968.

Horridge, G. A. The interpretation of behavior in terms of interneurons. In M. A. B. Brazier (Ed.), *The interneuron.* Berkeley: University of California Press, 1969. Pp. 1–20.

Horvath, W. J., Halick, P., Peretz, B., & Miller, J. G. Precision measurements of latency in single nerve fibers. Preprint #115, Mental Health Research Institute, The University of Michigan, August, 1963.

Hubel, D. H. Tungsten microelectrode for recording from single units. *Science,* 1957, **125,** 549–550.

Hubel, D. H., & Wiesel, T. N. Receptive fields of single neurons in the cat's striate cortex. *Journal of Physiology,* 1959, **148,** 574–591.

Hubel, D. H., & Wiesel, T. N. Integrative action in the cat's lateral geniculate body. *Journal of Physiology,* 1961, **155,** 385–398.

Hubel, D. H., & Wiesel, T. N. Receptive fields, binocular interaction and functional architecture in the cat's visual cortex. *Journal of Physiology,* 1962, **160,** 106–154.

Hubel, D. H., & Wiesel, T. N. Receptive fields and functional architecture in two nonstriate visual areas (18 and 19) of the cat. *Journal of Neurophysiology,* 1965, **28,** 229–289.

Hubel, D. H., & Wiesel, T. N. Receptive fields and functional architecture of monkey striate cortex. *Journal of Physiology,* 1968, **195,** 215–243.

Hursh, J. B. Conduction velocity and diameter of nerve fibers. *American Journal of Physiology,* 1939, **127,** 131–139.

Huxley, A. F., & Stämpfli, R. Evidence for saltatory conduction in peripheral myelinated nerve fibres. *Journal of Physiology,* 1949, **108,** 315–339.

Hyden, H., & Eghazi, E. Changes in the base composition of nuclear ribonucleic acid of neurons during a short period of enhanced protein production. *Journal of Cell Biology,* 1962, **15,** 37–44.

Hyman, L. H. *The invertebrates: Mullusca I.* Vol. VI. New York: McGraw-Hill, 1940.

Iggo, A., & Muir, A. R. The structure and function of a slowly adapting touch corpuscle in hairy skin. *Journal of Physiology,* 1969, **200,** 763–796.

Ito, M., & Olds, J. Unit activity during self-stimulation behavior. *Journal of Neurophysiology,* 1971, **34,** 263–273.

Jasper, H. H., & Ajmone-Marsan, C. *A stereotaxic atlas of the diencephalon of the cat.* Ottawa: The National Research Council of Canada, 1960.

John, E. R. *Mechanisms of memory.* New York: Academic Press, 1967.

Kandel, E. R. An invertebrate system for the cellular analysis of simple behaviors and their modifications. In F. O. Schmitt & F. G. Worden (Eds.), *The neurosciences: Third study program.* Cambridge: M.I.T. Press, 1973. Chapter 31.

Kandel, E. R., Frazier, W. T., Waziri, R., & Coggeshall, R. E. Direct and common connections among the identified neurons in the abdominal ganglion of *Aplysia. Journal of Neurophysiology,* 1967, **30,** 1352–1376.

Kandel, E. R., & Tauc, L. Mechanism of prolonged heterosynaptic facilitation. *Nature,* 1964, **202,** 145–147.

Kandel, E. R., & Tauc, L. Heterosynaptic facilitation in neurons of the abdominal ganglion of *Aplysia depilans. Journal of Physiology,* 1965, **181,** 1–27. (a)

Kandel, E. R., & Tauc, L. Mechanism of heterosynaptic facilitation in the giant cell of the abdominal ganglion of *Aplysia depilans. Journal of Physiology,* 1965, **181,** 28–47. (b)

Kandel, E. R., & Wachtel, H. The functional organization of neural aggregates in *Aplysia.* In F. D. Carlson (Ed.), *Physiological and biochemical aspects of nervous integration.* Englewood Cliffs, N.J.: Prentice-Hall, 1968. Pp. 17–65.

Karten, H. J., & Hodos, W. *A stereotaxic atlas of the brain of the pigeon (Columba livia)*. Baltimore: The Johns Hopkins Press, 1967.

Katz, B. *Nerve, muscle, and synapse.* New York: McGraw-Hill, 1966.

Keehn, D. G. An iterative spike separation technique. *IEEE Transactions on Bio-Medical Engineering,* 1966, **BME-13,** 19–28.

Kehoe, J. Ionic mechanisms of a two-component cholinergic inhibition in *Aplysia* neurones. *Journal of Physiology,* 1972, **225,** 85–114. (a)

Kehoe, J. Three acetylcholine receptors in *Aplysia* neurones. *Journal of Physiology,* 1972, **225,** 115–146. (b)

Kehoe, J. The physiological role of three acetylcholine receptors in synaptic transmission in *Aplysia. Journal of Physiology,* 1972, **225,** 147–172. (c)

Kennedy, D. Crayfish interneurons. *The Physiologist,* 1971, **14,** 5–30.

Keynes, R. D. The ionic movements during nervous activity. *Journal of Physiology,* 1951, **114,** 119.

Keynes, R. D. Chloride in the squid giant axon. *Journal of Physiology,* 1963, **169,** 690–705.

Keynes, R. D. Excitable membranes. *Nature,* 1972, **239,** 29–32.

Keynes, R. D., & Rojas, E. Characteristics of the sodium gating current in the squid giant axon. *Journal of Physiology,* 1973, **233,** 28.

Kiang, N. Y.-S., *Discharge patterns of single fibers in cat's auditory nerve.* (M.I.T. Research Monograph #35) Cambridge, Mass.: M.I.T. Press, 1965.

Kristan, W. B., Jr. Plasticity of firing patterns in neurons of *Aplysia* pleural ganglion. *Journal of Neurophysiology,* 1971 (May), 321–336.

Kuffler, S. W. Discharge patterns and functional organization of mammalian retina. *Journal of Neurophysiology,* 1953, **16,** 37–68.

Kuffler, S. W. Mechanisms of activation and motor control of stretch receptors in lobster and crayfish. *Journal of Neurophysiology,* 1954, **17,** 558–574.

Kuffler, S. W., & Nicholls, J. G. The physiology of neuroglial cells. *Reviews of Physiology, Biochemistry and Experimental Pharmacology,* 1966, **57,** 1–90.

Kuhn, T. S. *The structure of scientific revolutions.* Chicago: University of Chicago Press, 1962.

Kupfermann, I. Stimulation of egg laying: Possible neuroendocrine function of bag cells of abdominal ganglion of *Aplysia californica. Nature,* 1967, **216,** 814–815.

Kupfermann, I., & Kandel, E. R. Neuronal controls of a behavioral response mediated by the abdominal ganglion of *Aplysia. Science,* 1969, **164,** 847–850.

Landowne, D. Movement of sodium ions associated with the nerve impulse. *Nature,* 1973, **242,** 457–459.

Larrabee, M. G., & Bronk, D. W. Prolonged facilitation of synaptic excitation in sympathetic ganglia. *Journal of Neurophysiology,* 1947, **10,** 139–154.

Leão, A. A. P. Further observations on the spreading depression of activity in the cerebral cortex. *Journal of Neurophysiology,* 1947, **10,** 409–414.

Lettvin, J. Y., Maturana, H. R., McCulloch, W. S., & Pitts, W. H. What the frog's eye tells the frog's brain. Institute of Radio Engineers. *Proceedings,* 1959, **47,** 1940–1951.

Lillie, R. S. Factors affecting transmission and recovery in the passive iron nerve model. *Journal of General Physiology,* 1925, **7,** 473–507.

Ling, G., & Gerard, R. W. The normal membrane potential of frog sartorius fibers. *Journal of Cellular and Comparative Physiology,* 1949, **34,** 383–385.

Livingston, R. B. Central control of receptors and sensory transmission systems. In J. Field (Ed.), *Handbook of physiology. Section I: Neurophysiology.* Vol. I. Washington, D.C.: American Physiological Society, 1959. Pp. 741–760.

Llinás, R., Nicholson, C., Freeman, J. A., & Hillman, D. E. Dendritic spikes and their inhibition in alligator Purkinje cells. *Science,* 1968, **160,** 1132–1135. (a)

Llinás, R. Nicholson, C., & Precht, W. Preferred centripetal conduction of dendritic spikes in alligator Purkinje cells. *Science,* 1968, **163,** 184–187. (b)

Loewenstein, W. R., & Mendelson, M. Components of receptor adaptation in a Pacinian corpuscle. *Journal of Physiology,* 1965, **177,** 337–397.

Loewenstein, W. R., & Rathkamp, R. The sites for mechano-electric conversion in a Pacinian corpuscle. *Journal of General Physiology,* 1958, **41,** 1245–1265.

Loewi, O. Über humorale Übertragbarkeit der Herznervenwirkung. I. Mitteilung. *Pflueger's Archives of the European Journal of Physiology,* 1921, **189,** 239–242.

Lukowiak, K., & Jacklet, J. W. Habituation and dishabituation: Interactions between peripheral and central nervous systems in *Aplysia. Science,* 1972, **178,** 1306–1308.

Luparello, T. J. *Stereotaxic atlas of the forebrain of the guinea pig.* Baltimore: Williams & Wilkins, 1967.

Maffei, L., Fiorentini, A., & Bisti, S. Neural correlate of perceptual adaptation to gratings. *Science,* 1973, **182,** 1036–1038.

Majorossy, K., Réthelyi, M., & Szentágothai, J. The large glomerular synapse of the pulvinar. *Journal für Hirn Forschung,* 1965, **7,** 415–432.

Marks, W. B., Dobelle, W. H., & MacNichol, E. F. Visual pigments of single primate cones. *Science,* 1964, **143,** 1181–1183.

Marx, J. L. Cyclic AMP in brain: Role in synaptic transmission. *Science,* 1972, **178,** 1188–1190.

Maynard, D. M. The occurrence and functional characteristics of heteromorph antennules in an experimental population of spiny lobsters, *Panulirus argus. Journal of Experimental Biology,* 1965, **43,** 79–106.

McConnell, J. V. Memory transfer through cannibalism in planarians. *Journal of Neuropsychiatry,* 1962, **3** (Suppl. 1), 542–548.

McLennan, H. *Synaptic transmission.* (2nd ed.) Philadelphia: Saunders, 1970.

Moncrieff, R. W. What is odor? A new theory. *American Perfumery and Essential Oil Review,* 1949, **54,** 453–454.

Mountcastle, V. B. (Ed.) *Medical Physiology.* (12th ed.) St. Louis: C. V. Mosby Co., 1968.

Mulloney, B., & Selverston, A. Antidromic action potentials fail to demonstrate known interactions between neurons. *Science,* 1972, **177,** 69–72.

Munger, B. L. Patterns of organization of peripheral sensory receptors. In W. R. Loewenstein (Ed.), *Principles of receptor physiology.* New York: Springer-Verlag, 1971.

Nastuk, W. L. Some ionic factors that influence the action of acetylcholine at the muscle end-plate membrane. *Annals of the New York Academy of Sciences,* 1959, **81,** 317–327.

Nathan, P. W., & Sears, T. A. Differential nerve block by sodium-free and sodium-deficient solutions. *Journal of Physiology,* 1962, **164,** 375–394.

Neild, T. O., & Thomas, R. C. New design for a chloride-sensitive micro-electrode. *Journal of Physiology,* 1973, **231**(1), 7-8P.

Nicholls, J. G., & Kuffler, S. W. Na and K content of glial cells and neurons determined by flame photometry in the central nervous system of the leech. *Journal of Neurophysiology,* 1965, **28,** 519–525.

Nishi, S., & Koketsu, K. Electrical properties and activities of single sympathetic neurons in frogs. *Journal of Cellular and Comparative Physiology,* 1960, **55,** 15–30.

Norton, W. T. Myelin. In R. W. Albers, G. J. Siegel, R. Katzman, & B. W. Agranoff (Eds.), *Basic neurochemistry*. Boston: Little, Brown, 1972. Pp. 365–386.

Orkand, R. K., Nicholls, J. G., & Kuffler, S. W. Effect of nerve impulses on the membrane potential of glial cells in the central nervous system of amphibia. *Journal of Neurophysiology,* 1966, **29,** 788–806.

Ottoson, D. Analysis of the electrical activity of the olfactory epithelium. *Acta Physiologica Scandinavica,* 1956, **35** (Suppl. 122), 1–83.

Ozeki, M. Hetero-electrogenesis of the gustatory cell membrane in rat. *Nature,* 1970, **228,** 868–869.

Pappas, G. D. Electron microscopy of neuronal junctions involved in transmission in the central nervous system. In K. Rodahl & B. Issekutz (Eds.), *Nerve as a tissue*. New York: Harper & Row, 1966. Pp. 49–87.

Patton, H. D. Special properties of nerve trunks and tracts. In T. C. Ruch & H. D. Patton (Eds.), *Physiology and biophysics.* (19th ed.) Philadelphia: Saunders, 1965. Pp. 73–94.

Peretz, B. Central neuron initiation of periodic gill movements. *Science,* 1969, **166,** 1167–1172.

Peretz, B. Habituation and dishabituation in the absence of a central nervous system. *Science,* 1970, **169,** 379–381.

Peters, A., Palay, S. L., & Webster, H. de F. *The fine structure of the nervous system, the cells and their processes.* New York: Harper & Row, 1970.

Pfaffmann, C. The sense of taste. In J. Field (Ed.), *Handbook of physiology. Section I: Neurophysiology.* Vol. I. Washington, D.C.: American Physiological Society, 1959. Pp. 507–534.

Pinching, A. J. Myelinated dendritic segments in the monkey olfactory bulb. *Brain Research,* 1971, **29,** 133–138.

Plonsey, R. *Bioelectric phenomena.* New York: McGraw-Hill, 1969.

Porter, K. R., & Bonneville, M. A. *Fine structure of cells and tissues.* (4th ed.) Philadelphia: Lea & Febiger, 1968.

Purpura, D. P. Nature of electrocortical potentials and synaptic organizations in cerebral and cerebellar cortex. *International Review of Neurobiology,* 1959, **1,** 47–163.

Purpura, D. P. Comparative physiology of dendrites. In G. C. Quarton, T. Melnechuk, & F. O. Schmitt (Eds.), *The neurosciences: A study program.* New York: Rockefeller University Press, 1967. Pp. 372–393.

Purpura, D. P. Interneuronal mechanisms in thalamically induced synchronizing and desynchronizing activities. In M. A. B. Brazier (Ed.), *The interneuron.* Berkeley: University of California Press, 1969. Pp. 467–496.

Quarton, G. C., Melnechuk, T., & Schmitt, F. O. (Eds.) *The neurosciences: A study program.* New York: Rockefeller University Press, 1967.

Ramón y Cajal, S. *Histologie du système nerveux de l'homme et des vertébrés.* Madrid: Instituto Ramón y Cajal del Consejo Superior de Investigaciones Cientificas, 1955.

Ranson, S. W. *The anatomy of the nervous system: Its development and function,* as revised by S. L. Clark. (10th ed.) Philadelphia: Saunders, 1959.

Ratliff, F. *Mach bands: Quatitative studies on neural networks in the retina.* San Francisco: Holden-Day, 1965.

Rose, S. P. R. Effect on incorporation of tritiated lysine into protein. *Nature,* 1967, **215,** 253–255.

Rushton, W. A. H. Visual pigments in man. *Scientific American,* 1962, **207,** 120–132.

Russell, G. F., & Hills, J. I. Odor differences between enantiomeric isomers. *Science,* 1971, **172,** 1043–1044.

Scheibel, M. E., & Scheibel, A. B. Some structuro-functional correlates of development in young cats. *Electroencephalography and Clinical Neurophysiology,* 1963, Suppl. 24, 235–246.

Scheibel, M. E., & Scheibel, A. B. Some structural and functional substrates of development in young cats. *Progress in Brain Research,* 1964, **9,** 6–25.

Schmidt, E. M., & Stromberg, M. W. Computer dissection of peripheral nerve bundle activity. *Computers and Biomedical Research,* 1969, **2,** 446–455.

Schmitt, F. O., & Samson, F. E., Jr. Brain cell microenvironment: A report based on an NRP work session. Neurosciences Research Program. *Bulletin,* 1969, **7**(4).

Schmitt, O. H., & Dubbert, D. R. Tissue stimulators utilizing radio frequency coupling. *Review of Scientific Instruments,* 1949, **20,** 170–173.

Schwanzara, S. A. Visual pigments of tropical freshwater fishes. *Life Sciences,* 1967, **6,** 157–162.

Segundo, J. P., Moore, G. P., Stensaas, L. J., & Bullock, T. H. Sensitivity of neurones in *Aplysia* to temporal pattern of arriving impulses. *Journal of Experimental Biology,* 1963, **40,** 643–667.

Segundo, J. P., & Perkel, D. H. The nerve cell as an analyzer of spike trains. In M. A. B. Brazier (Ed.), *The interneuron.* Berkeley: University of California Press, 1969. Pp. 349–390.

Shibuya, T. Dissociation of olfactory neural response and mucosal potential. *Science,* 1964, **143,** 1338–1340.

Singer, S. J., & Nicholson, G. L. The fluid mosaic model of the structure of cell membranes. *Science,* 1972, **175,** 720–731.

Smith, D. S., Järlfors, U., & Beránek, R. The organization synaptic axoplasm in the lamprey (*Petromyzon marinus*) central nervous system. *Journal of Cell Biology,* 1970, **46,** 199–219.

Snider, R. S., & Lee, J. C. *A stereotaxic atlas of the monkey brain.* Chicago: The University of Chicago Press, 1961.

Sotelo, C., & Palay, S. L. The fine structure of the lateral vestibular nucleus in the rat. II. Synaptic organization. *Brain Research,* 1970, **18,** 93–115.

Sperling, G. The description and luminous calibration of cathode ray oscilloscope visual displays. *Behavior Research Methods and Instrumentation,* 1971, **3,** 148–151. (a)

Sperling, G. Flicker in computer-generated visual displays: Selecting a CRO phosphor and other problems. *Behavior Research Methods and Instrumentation,* 1971, **3,** 151–153. (b)

Sperling, G. Stereoscopic visual displays: Principles, viewing devices, alignment procedures. *Behavior Research Methods and Instrumentation,* 1971, **3,** 154–158. (c)

Spinelli, D. N. Receptive field organization of ganglion cells in the cat's retina. *Experimental Neurology,* 1967, **19,** 291–315.

Spinelli, D. N., Bridgeman, B., & Owens, S. Technical note: A simple single-unit microelectrode recording system. *Medical and Biological Engineering,* 1970, **8,** 599–602.

Stevens, C. F. *Neurophysiology: A primer.* New York: Wiley, 1966.

Swammerdam, J. Experiment to demonstrate contraction of frog muscle. In J. F. Fulton & L. G. Wilson (Eds.), *Selected readings in the history of physiology.* Springfield, Ill.: Charles C Thomas, 1966. Pp. 211–213.

Takagi, S. F. Are EOG's generator potentials? In T. Hayashi (Ed.), *Olfaction and taste II.* Oxford: Pergamon Press, 1967. Pp. 167–179.

Takagi, S. F., & Shibuya, T. "On"- and "off"-responses of the olfactory epithelium. *Nature,* 1959, **184,** 60.

Takagi, S. F., Shibuya, T., Higashino, S., & Arai, T. The stimulative and anaesthetic actions of ether on the olfactory epithelium of the frog and the toad. *Japanese Journal of Physiology,* 1960, **10,** 571–584.

Tasaki, I. *Nervous transmission.* Springfield, Ill.: Charles C Thomas, 1953.

Tauc, L., & Epstein, R. Heterosynaptic facilitation as a distinct mechanism in *Aplysia. Nature,* 1967, **214,** 724–725.

Thomas, R. C. Measurement of current produced by the sodium pump in a snail neurone. *Journal of Physiology,* 1968, **195,** 23–24P.

Thomas, R. C. New design for sodium-sensitive glass micro-electrode. *Journal of Physiology,* 1970, **210**(2), 82–83P.

Thompson, R. F. *Foundations of physiological psychology.* New York: Harper & Row, 1967.

Tilney, L. G., & Porter, K. R. Studies on the microtubules in heliozoa. II: The effect of low temperature on these structures in the formation and maintenance of the axopodia. *Journal of Cell Biology,* 1967, **34,** 327–343.

Toennies, J. F. Reflex discharge from the spinal cord over the dorsal roots. *Journal of Neurophysiology,* 1938, **1,** 378–390.

Tosteson, D. C. (Ed.) *The molecular basis of membrane function: Papers.* Englewood Cliffs, N.J.: Prentice-Hall, 1969.

Truex, R. C., & Carpenter, M. B. *Strong and Elwyn's human neuroanatomy.* (5th ed.) Baltimore: Williams & Wilkins, 1964.

Uttal, W. R. A comparison of neural and psychophysical responses in the somesthetic system. *Journal of Comparative and Physiological Psychology,* 1959, **52,** 485–490.

Uttal, W. R. Oscillations in the amplitude of human peripheral nerve action potentials during repetitive stimulation. *Kybernetik,* 1966, **3,** 24–27.

Uttal, W. R. *Real-time computers: Technique and applications in the psychological sciences.* New York: Harper & Row, 1968.

Uttal, W. R. Masking of alphabetic character recognition by dynamic visual noise (DVN). *Perception & Psychophysics,* 1969, **6**(2), 121–128.

Uttal, W. R. The psychobiological silly season—or—what happens when neurophysiological data become psychological theories. *Journal of General Psychology,* 1971, **84,** 151–166.

Uttal, W. R. *The psychobiology of sensory coding.* New York: Harper & Row, 1973.

Uttal, W. R., & Cook, L. Systematics of the evoked somatosensory cortical potential: A psychophysical-electrophysiological comparison. In R. Katzman (Ed.), *Sensory evoked response in man.* New York: New York Academy of Sciences, 1964. Pp. 60–80.

Uttal, W. R., & Kasprzak, H. The caudal photoreceptor of the crayfish: A quantitative study of responses to intensity, temporal and wavelength variables. *AFIPS Conference Proceedings,* 1962, **21,** 159–169.

Uttal, W. R., & Smith, P. On the psychophysical discriminability of somatosensory nerve action potential patterns with irregular intervals. *Perception & Psychophysics,* 1967, **2,** 341–348.

von Békésy, G. Microphonics produced by touching the cochlear partition with a vibrating electrode. *Journal of the Acoustical Society of America,* 1951, **23,** 29–35.

von Békésy, G. *Experiments in hearing.* New York: McGraw-Hill, 1960.

Wachtel, H., & Kandel, E. R. A direct synaptic connection mediating both excitation and inhibition. *Science,* 1967, **158,** 1206–1208.

Wald, G. Vitamin A in the retina. *Nature,* 1933, **132,** 316–317.

Wald, G. The photoreceptor process in vision. In J. Field (Ed.), *Handbook of physiology. Section I: Neurophysiology.* Vol. I. Washington, D.C.: American Physiological Society, 1959. Chapter 28.

Wald, G. Molecular basis of visual excitation. *Science,* 1968, **162,** 230–239.

Weisstein, N. What the frog's eye tells the human brain: Single cell analyzers in the human visual system. *Psychological Bulletin,* 1969, **72,** 157–176.

Wever, E. G., & Bray, C. W. Action currents in the auditory nerve in response to acoustical stimulation. National Academy of Sciences. *Proceedings,* 1930, **16,** 344–350.

Wever, E. G., Bray, C. W., & Lawrence, M. The nature of cochlear activity after death. *Annals of Otology, Rhinology, and Laryngology,* 1941, **50,** 317–329.

Wiersma, C. A. G. On the functional connections of single units in the central nervous system of the crayfish, *Procambarus clarkii* (Girard). *Journal of Comparative Neurology,* 1958, **110,** 421–471.

Willows, A. O. D. Behavioral acts elicited by stimulation of single, identifiable brain cells. *Science,* 1967, **157,** 570–574.

Willows, A. O. D. Behavioral acts elicited by stimulation of single identifiable nerve cells. In F. D. Carlson (Ed.), *Physiological and biochemical aspects of nervous integration.* Englewood Cliffs, N.J.: Prentice-Hall, 1968.

Willows, A. O. D. Neuronal network triggering a fixed action pattern. *Science,* 1969, **166,** 1549–1551.

Wilson, D. M. An approach to the problem of control of rhythmic behavior. In C. A. G. Wiersma (Ed.), *Invertebrate nervous systems: Their significance for mammalian neurophysiology.* Chicago: University of Chicago Press, 1967. Pp. 219–229.

Woodbury, J. W. The cell membrane: Ionic and potential gradients and active transport. In T. C. Ruch & H. D. Patton (Eds.), *Physiology and biophysics.* (19th ed.) Philadelphia: Saunders, 1965. Chapter 1.

Worden, F. G. Attention and auditory electrophysiology. In E. Stellar & J. M. Sprague (Eds.), *Progress in physiological psychology.* Vol. I. New York: Academic Press, 1966. Pp. 45–116.

Wright, R. H. *The science of smell.* New York: Basic Books, 1964.

Young, J. A. The giant nerve fibres and epistellar body of cephalopods. *Quarterly Journal of Microscopical Science,* 1936, **78,** 367.

Young, R. W. Visual cells. *Scientific American,* 1970, **223,** 80–91.

[illegible]

[illegible]

[illegible]

[illegible]

[illegible]

[illegible]

[illegible]

[illegible]

[illegible]

[illegible]

[illegible]

[illegible]

[illegible]

AUTHOR INDEX

The numbers in *italics* refer to the pages on which the complete reference is cited.

SUBJECT INDEX

B

C

D

N

O

P

S

T